METAL IONS IN BIOLOGICAL SYSTEMS

VOLUME 29

Biological Properties of
Metal Alkyl Derivatives

METAL IONS IN BIOLOGICAL SYSTEMS

Edited by

Helmut Sigel
and **Astrid Sigel**
Institute of Inorganic Chemistry
University of Basel
CH-4056 Basel, Switzerland

VOLUME 29

Biological Properties of
Metal Alkyl Derivatives

MARCEL DEKKER, INC.   New York • Basel • Hong Kong

ISBN 0-8247-9022-7

This book is printed on acid-free paper.

MARCEL DEKKER, INC.
270 Madison Avenue, New York, New York 10016

Current printing (last digit):
10 9 8 7 6 5 4 3 2 1

PRINTED IN THE UNITED STATES OF AMERICA

Preface to the Series

Recently, the importance of metal ions to the vital functions
of living organisms, hence their health and well-being, has become
increasingly apparent. As a result, the long-neglected field of
"bioinorganic chemistry" is now developing at a rapid pace. The
research centers on the synthesis, stability, formation, structure,
and reactivity of biological metal ion-containing compounds of low
and high molecular weight. The metabolism and transport of metal
ions and their complexes is being studied, and new models for complicated natural structures and processes are being devised and
tested. The focal point of our attention is the connection between
the chemistry of metal ions and their role for life.

No doubt, we are only at the brink of this process. Thus,
it is with the intention of linking coordination chemistry and
biochemistry in their widest sense that the series METAL IONS IN
BIOLOGICAL SYSTEMS reflects the growing field of "bioinorganic
chemistry." We hope, also, that this series will help to break
down the barriers between the historically separate spheres of
chemistry, biochemistry, biology, medicine, and physics, with the
expectation that a good deal of the future outstanding discoveries
will be made in the interdisciplinary areas of science.

Should this series prove a stimulus for new activities in
this fascinating "field," it would well serve its purpose and would
be a satisfactory result for the efforts spent by the authors.

Fall 1973

Helmut Sigel
Institute of Inorganic Chemistry
University of Basel
CH-4056 Basel, Switzerland

Preface to Volume 29

This book is entirely devoted to the biological properties of metal alkyl derivatives. The occurrence of these organometallic species in the environment is due to two processes: anthropogenic input and biogeochemical cycling involving the methylation (or alkylation) of inorganic or organometallic precursors. Biomethylation occurs almost exclusively by microorganisms, most commonly by anaerobic bacteria, predominantly in sediments at the bottom of bodies of waters (estuaries, harbors, rivers, lakes, and oceans). Depending on the metal or metalloid, either the biological or the anthropogenic production may dominate; for example, the related elements germanium, tin, and lead from the same main group of the periodic table differ considerably in this respect as well as in their properties. Field observations and laboratory experiments suggest that methylgermanium is produced by microbial methylation of inorganic germanium in anoxic aquatic environments, while organotin and organolead compounds are overwhelmingly (yet not exclusively) man-made, the most well-known derivatives being tetramethyllead and tetraethyllead, which are used (although to a lesser extent than formerly) as antiknock agents for gasoline.

Next to the three mentioned elements, the alkyl derivatives of arsenic and antimony, selenium and tellurium, cobalt (vitamin B_{12} derivatives), nickel, and mercury are covered in individual chapters, with, in the case of Ni, the emphasis being on the natural formation of methane via the coenzyme F_{430}. Derivatives of indium, thallium, bismuth, and various transition metals appear in sections of the more general chapters, where the corresponding derivatives of the nonmetals silicon, phosphorus, and sulfur are also briefly mentioned.

The volume closes with a somewhat different aspect, i.e., the biosynthesis of halomethanes by fungi and plants, which includes the

so-called haloperoxidase route, and these peroxidases contain metals
(Fe, V) at their active site. The realization that halogenated gases
(such as the man-made chlorofluorocarbons) have increased the rate
of ozone destruction in the stratosphere has stimulated a growing
interest in identifying and quantifying the natural sources of volatile halogenated compounds, and in terms of atmospheric abundance,
the gaseous monohalomethanes, CH_3Cl, CH_3Br, and CH_3I, are undoubtedly
the most important. In fact, the predominant volatile halohydrocarbon in the atmosphere is CH_3Cl: between 2.5 and 5 million tonnes
per year must originate from natural sources; the contribution from
industrial sources of about 30 thousand tonnes per year is insignificant in comparison.

The last-mentioned example clearly demonstrates the importance
of this kind of research and it indicates also how rudimentary our
present knowledge on the biosynthesis of all sorts of alkyl derivatives still is. It is compulsory that we learn to understand the
biological formation of metal alkyl derivatives—as well as their
properties, including the physiological and toxic characteristics—
and that we reveal the role that metal ions are playing in the biosynthesis of alkyl derivatives of other elements. It is the hope of
the authors and editors of this volume that it will serve as a stimulus for further research on these fascinating topics.

Helmut Sigel
Astrid Sigel

Contents

Contributors

Numbers in parentheses indicate the pages on which the authors' contributions begin.

Fumio Arai Department of Public Health, St. Marianna University, School of Medicine, 2-16-1 Sugao, Miyamae-ku, Kawasaki 216, Japan (137)

Yasuaki Arakawa Department of Hygiene and Preventive Medicine, Faculty of Health Sciences, University of Shizuoka, 52-1 Yada, Shizuoka-shi, Shizuoka 422, Japan (101)

Peter J. Craig Department of Chemistry, Leicester Polytechnic, P.O. Box 143, Leicester LE1 9BH, England (37)

William T. Frankenberger, Jr. Department of Soil and Environmental Sciences, University of California, Riverside, CA 92521, USA (185)

David B. Harper Department of Food and Agricultural Chemistry, The Queen's University of Belfast, Newforge Lane, Belfast BT9 5PX, Northern Ireland (345)

Bernhard Jaun Organic Chemistry Laboratory, ETH-Zürich, Universität-strasse 16, CH-8029 Zürich, Switzerland (287)

Ulrich Karlson Department of Marine Ecology and Microbiology, National Environmental Research Institute, P.O. Box 358, Frederiksborgvej 399, DK-4000 Roskilde, Denmark (185)

Brent L. Lewis College of Marine Studies, University of Delaware, Lewes, DE 19958, USA (79)

Erminio Marafante Commission of the European Communities Joint Research Centre, Environment Institute, I-210 20 Ispra, Italy (161)

H. Peter Mayer Max-Planck Institut für Chemie, Saarstrasse 23, D-6500 Mainz, Germany (79)

Darren Mennie Department of Chemistry, Leicester Polytechnic, P.O. Box 143, Leicester LE1 9BH, England (37)

John M. Pratt Department of Chemistry, University of Surrey, Guildford GU2 5XH, England (229)

John S. Thayer Department of Chemistry, University of Cincinnati, Mail Location 172, Cincinnati, OH 45221-0172, USA (1)

Marie Vahter Institute of Environmental Medicine, Karolinska Insti-
        tutet, P.O. Box 60208, S-104 01 Stockholm, Sweden (161)

Osamu Wada Department of Hygiene and Preventive Medicine, Faculty
        of Medicine, University of Tokyo, 7-3-1 Hongo, Bunkyo-ku,
        Tokyo 113, Japan (101)

Yukio Yamamura Department of Public Health, St. Marianna University,
        School of Medicine, 2-16-1 Sugao, Miyamae-ku, Kawasaki 216,
        Japan (137)

Contents of Other Volumes

_____

*Out of print

Other volumes are in preparation.

Comments and suggestions with regard to contents,
topics, and the like for future volumes of the series
are welcome.

The following Marcel Dekker, Inc. books are also of
interest for any reader dealing with metals or other
inorganic compounds:

HANDBOOK ON TOXICITY OF INORGANIC COMPOUNDS edited by
Hans G. Seiler and Helmut Sigel, with Astrid Sigel

In 74 chapters, written by 84 international authorities,
this book covers the physiology, toxicity, and levels of
tolerance, including prescriptions for detoxification,
for all elements of the Periodic Table (up to atomic
number 103). The book also contains short summary
sections for each element, dealing with the distribution

of the elements, their chemistry, technological uses,
and ecotoxicity as well as their analytical chemistry.

In preparation:

HANDBOOK ON METALS IN CLINICAL CHEMISTRY edited by Hans
G. Seiler, Astrid Sigel, and Helmut Sigel

METAL IONS IN
BIOLOGICAL SYSTEMS

VOLUME 29

Biological Properties of
Metal Alkyl Derivatives

1

Global Bioalkylation of the Heavy Elements

John S. Thayer

Department of Chemistry
University of Cincinnati
Mail Location 172
Cincinnati, Ohio 45221-0172 USA

1. INTRODUCTION

1.1. Scope

In its broadest sense, the term *biological alkylation* (herein con-
tracted to *bioalkylation*) would refer to processes involving living
cells that cause an alkyl group to become directly bonded through a
carbon atom to some "heavy element" (herein defined as any element
with an atomic number larger than 10). Since this definition includes
metals, metalloids, and nonmetals, *metal(loid)* will be used as a generic term.

The methyl group is the simplest alkyl group, and *biological
methylation* (*biomethylation*) occurs extensively in nature. Because
biomethylation has been investigated in great detail, it will be

considered separately, and the term *bioalkylation* will hereafter
refer specifically to those processes involving alkyl groups *other*
than methyl. *Biometh/alkylation* will be used as an inclusive term.
No examples of *bioarylation* of heavy elements except the halogens
have been reported.

All biometh/alkylation reactions reported to date involve the
transfer in vivo of a preformed alkyl group from some precursor
moiety to a metal(loid) atom. From a chemist's perspective, such
transfers are simple exchange reactions that don't necessarily
require biological systems in order to take place; abiotic alkyl
exchange reactions are known to occur under environmental conditions
[1]. However, the fact that these reactions do occur within *biological* systems makes a difference, a big difference, in the products
formed, the rate of their formation, and their subsequent fate.

As Fig. 1 indicates, a considerable number of elements will
undergo biomethylation; a smaller number will also undergo bio-

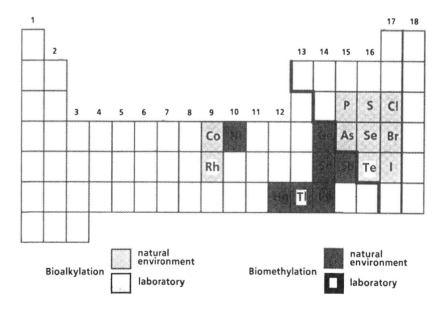

FIG. 1. Elements undergoing biological meth/alkylation.

alkylation [1]. At present, except for cobalt, rhodium, and (possibly) antimony, only metalloids are known to accept alkyl groups in addition to methyl under biological conditions.

Occasionally an organic group already attached to a metal(loid) atom might itself undergo biotransformation, as in the enzymatic hydroxylation of phenyltrimethylsilane [2a]:

$$C_6H_5Si(CH_3)_3 \longrightarrow p\text{-}HOC_6H_4Si(CH_3)_3 \tag{1}$$

Such transformations have been employed for biochemically controlled organometal(loid) synthesis [3] and have led to the use of ferrocene in various biological assays [3].

Various examples of biological cleavage of metal(loid)-carbon bonds (*biodealkylation*) have been reported [2b,3-5]; these are discussed in Sec. 4.

1.2. Effects of Metal(loid)-Carbon Bond Formation on Physical Properties

Methylation of a metal(loid) atom tends to make the resulting species more volatile [6]. Permethyl derivatives of heavy elements exist as gases or volatile liquids at ambient temperatures (Table 1). They are stable toward, and insoluble in, water, although decomposition might occur in aqueous systems through acid cleavage or through exchange:

$$(CH_3)_4Pb + H^+ \longrightarrow CH_4 + (CH_3)_3Pb^+ \tag{2}$$

$$(CH_3)_4Sn + Hg^{2+} \longrightarrow CH_3Hg^+ + (CH_3)_3Sn^+ \tag{3}$$

In the atmosphere, methyl metal(loid) compounds will undergo photolysis [1a]. However, the rate at which this occurs varies widely. Dimethylsulfide and the methyl halides can exist for long periods of time in the atmosphere; methylation represents a major route for the environmental distribution of these elements [7]. Even

TABLE 1

Melting and Boiling Points (°C) of Some
Permethyl Metal(loid)s

Compound	Melting point	Boiling point
$(CH_3)_2Hg$	ℓ^a	92
$(CH_3)_3Tl$	38.5	d95
$(CH_3)_4Si$	-102.2	26.5
$(CH_3)_4Ge$	-88	43.4
$(CH_3)_4Sn$	-55	78
$(CH_3)_4Pb$	-27.5	110
$(CH_3)_3P$	-85	37.8
$(CH_3)_3As$	-87	52
$(CH_3)_3Sb$	-62	82
$(CH_3)_3Bi$	-86	110
$(CH_3)_2S$	-98.3	37.3
$(CH_3)_2Se$	ℓ	55/753 torr
$(CH_3)_2Te$	ℓ	93.5/749 torr
CH_3Cl	-97.1	-24.2
CH_3Br	-93.6	3.6
CH_3I	-66.4	42.4

$^a\ell$ = liquid at 20°C.

the permethyl metals, with their weaker bond energies, can resist both oxygen and photons long enough to travel considerable distances.

The presence of a lipophilic methyl group on a metal(loid) atom increases its solubility in lipids. One measure of this is the distribution coefficient between water and 1-octanol [8]; the higher the value, the more lipophilic the material. For the series

$HgCl_2$, CH_3HgCl, $(CH_3)_2Hg$, the coefficients are 0.61, 2.54, and 191.3, respectively [6]. Such changes can alter, often drastically, distribution pathways of the element in question within biological systems, and can contribute to the marked difference in toxicity between inorganic derivatives and methylated derivatives of most elements [1b,2c].

Such arguments can readily be extended to other alkyl metal(loid) derivatives, although less information is available for these species. The longer the alkyl chain, the more readily the molecule will dissolve in hydrocarbons, and the less it will dissolve in water. While boiling points will be higher than for corresponding methyl species, alkyl metal(loid)s will still be volatile species.

1.3. Effect of Metal(loid)-Carbon Bond(s) on Chemical Reactivity

An alkyl group bonded to a metal(loid) atom usually lowers the tendency of that atom to undergo further reaction. Metal(loid)-carbon bonds, once formed, are kinetically more stable than many metal-ligand linkages. For both steric and electronic reasons, they alter the tendency of the central metal(loid) atom to form donor-acceptor bonds. Metalloids act as electron pair donors. Alkyl groups promote this activity; thousands of complexes between transition metals and alkyl derivatives of P, As, S, Se, and I are known. Metals act as electron pair acceptors. Biometh/alkylation often enhances toxicity by altering the metal's ability to bind to active sites on enzymes [1b,2c] and/or to be sequestered by defensive biological chelating agents. This has been observed in greatest detail for alkyltin species [1b,2c,9], where the toxicity changes not only from one alkyl group to another, but from one organism to another! Alkyllead compounds show similar relationships [10].

Many main group elements have stable oxidation states two units apart. Formation of element-carbon bonds tends to favor the more positive state and to change the ability of the element to interconvert between them. This is especially important for the heaviest elements. Inorganic lead compounds found in nature are invariably Pb(II) species, whereas alkyllead compounds in nature are invariably Pb(IV) derivatives. A similar relationship holds for Sn, As, and Sb: inorganic derivatives for two oxidation states are found for these elements, whereas the alkyl derivatives occur predominantly in the higher state [trimethylarsine, $(CH_3)_3As$, is an exception]. Thus biometh/alkylation plays an important part in the environmental interconversions of various elements between different oxidation states, the full extent of which remains to be determined.

1.4. Occurrence of Biometh/alkylation in the Environment

Much has already been written about the extent of biometh/alkylation in nature [1,2,11]. For metals, biomethylation occurs predominantly in sediments at the bottom of bodies of waters—estuaries, harbors, rivers, lakes, and oceans. Soils can also be a source of such activity [12]. Methylating biota are almost exclusively microorganisms, most commonly anaerobic bacteria. Few, if any, higher organisms appear to be capable of methylating true metals, with the significant exception of cobalt in vitamin B_{12} [4].

For those elements capable of undergoing both biomethylation and bioalkylation, the situation is quite different. Methylarsenicals, for example, can be formed by most living creatures, including humans [11,13,14]. The same appears to be true for selenium and tellurium. Frederick Challenger, the pioneer worker in this area, used fungi in his initial studies [15], but many organisms, both prokaryotes and eukaryotes, can methylate these elements [1,2,16]. Methylation of phosphorus has been reported thus far only for soil microorganisms [17]. Likewise, the halogens appear to undergo methylation primarily by fungi, plants, and marine kelps [7].

Alkylarsenicals occur in a wide variety of marine organisms but are less common elsewhere [13,14]. Phosphorus can undergo bioalkylation in marine environments and in soils [17]. Selenium, because of its similarity to sulfur, probably can undergo bioalkylation by more

organisms than any element in this category, while tellurium is more
restricted [16]. Numerous biogenic (as opposed to anthropogenic)
organohalogen compounds have been reported, largely in marine
organisms [18].

Some of the apparent limitations mentioned above may be arte-
facts of circumstances. Most reports on biometh/alkylation are
quite recent; also, possible environmental sites actually investigated are relatively few. As discussed in the article by Mennie and
Craig [19], quantitative measurement and structural determination of
environmentally occurring organometal(loid) compounds has been a
challenging and complicated matter. Advances have been slow and
nonuniform. Undoubtedly much more remains to be discovered.

2. BIOMETHYLATION

2.1. Methylating Agents and Mechanisms

Most investigations involving biological methylation have resulted
from concerns over public health issues. The role of biomethylation
in the functioning of DNA [20-23] relates to its application in
cancer therapy. Challenger's pioneering work in the biomethylation
of arsenic [15] grew out of earlier reports on poisonings by "Gosio-
gas"* [1d]. The tragic poisonings at Minamata Bay and elsewhere
spurred research into the formation and biological effects of methylmercuric compounds [1d]. Similar practical considerations have generated more recent work on biomethylation of tin, selenium, and lead.

Under ordinary biological conditions, the great majority of
transmethylation reactions involve carbon, nitrogen, oxygen, or
sulfur atoms as methyl acceptors [24]. Transmethylation of heavier
elements is less common. Two compounds serve as the primary methylating agents for such elements; a handful of other compounds can
serve as biomethylating agents under certain conditions.

---

*An early name for biologically generated trimethylarsine.

The most important methylating agent is S-adenosylmethionine (SAM) (Fig. 2). This molecule has an extensive chemistry and has been the subject of many investigations [23,24]. Chemists would view SAM as a sulfonium salt, and the methyl group would be transferred as a carbocationic intermediate. Any recipient atom must therefore be nucleophilic, which usually requires an available pair of electrons in the valence shell:

$$(HO)_n E: + CH_3^{\oplus} \longrightarrow [(HO)_n ECH_3]^{\oplus} \tag{4a}$$

$$[(HO)_n ECH_3]^{\oplus} \longrightarrow O(HO)_{n-1} ECH_3 + H^{\oplus} \tag{4b}$$

This mechanism is often termed the Challenger mechanism, in honor of the late Frederick Challenger who first formulated it [15]. It has been firmly established for arsenic and selenium, where biomethylation apparently acts as a detoxifying process. The most poisonous forms of As and Se are arsenious and selenious acids (or their salts, esters, etc.), which occur commonly in nature and possess electron pairs in their valence shells. Methylation of these elements lowers toxicity drastically and facilitates removal, either by excretion [e.g., $(CH_3)_2 AsO_2^-$, cacodylate anion; $(CH_3)_3 Se^+$, selenonium ion] or by volatilization [e.g., trimethylarsine; dimethylselenide, $(CH_3)_2 Se$]. Details of this are discussed elsewhere [1,2,11,13-16]. Treatment of hamsters with SAM greatly enhanced the methylation of orally administered arsenious oxide [25]. The Challenger mechanism also almost certainly applies to the methylation of tellurium, probably to the methylation of phosphorus [17], and quite possibly to metallic elements having an available lone pair, such as Sn(II) or Sb(III).

The second major methylating agent is methylcobalamin, also shown in Fig. 2. This derivative of vitamin B_{12} has an extensive biochemistry [26] and has been widely studied in many laboratories as a methyl transfer agent [1,2,11]. Available evidence indicates that the methyl group is transferred as a carbanionic species and that acceptor atoms must be electrophilic. Methylcobalamin has been established as the methylating agent for mercury and may well be involved in the biomethylation of higher valent Ge, Sn, Pb, and Sb.

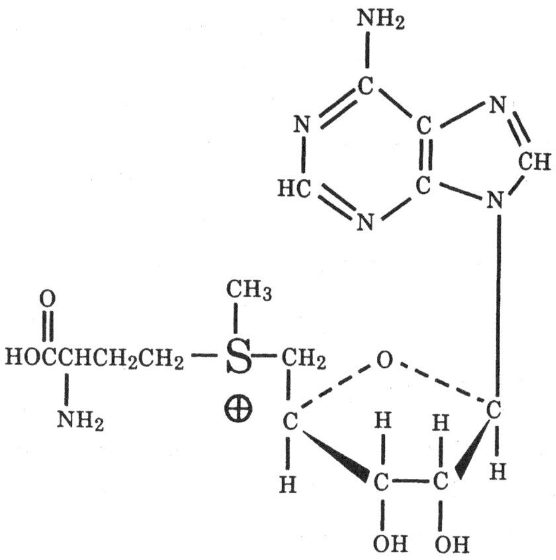

S-ADENOSYLMETHIONINE

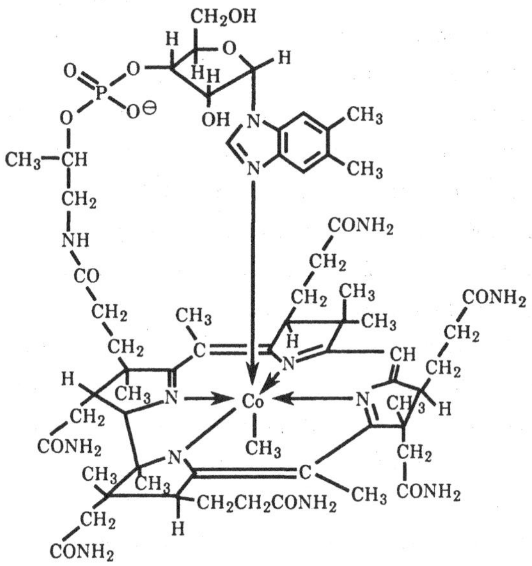

METHYLCOBALAMIN

FIG. 2. Two biological methylating agents.

Other methylating agents have been reported in special situations. Dimethyl-β-propiothetin will generate methyl iodide from iodide ion [27]. Coenzyme F-430 forms a methylnickel species that generates methane [28]. Selenoadenosylmethionine (the selenium analog of SAM) is reported to be a methylating agent in cells [29]. Regardless of the methyl donor used, only one methyl group is transferred initially. Further methylation of the same acceptor atom often occurs, but the rate and extent will depend on the specific systems involved.

The possibility of other transmethylation reactions occurring within, or upon, the boundaries of cell walls has not received the attention it deserves. A methyl metal(loid) formed by one of the "normal" processes mentioned above may well encounter another metal(loid) atom within a cell and transfer its methyl group. Alternatively, a biologically generated methyl metal(loid) might encounter an acceptor species on or near the cell wall and undergo methyl transfer. Such exchange reactions are most likely for true metals, whose linkages to carbon are kinetically labile. Much research effort remains to be done before we can claim anything approaching complete knowledge about conditions and mechanisms of biological methylation.

2.2. Arsenic

The biological methylation of arsenic has been extensively reviewed [12,13,30]. Biomethylation appears to occur in virtually every organism investigated; however, the final products formed and their proportions vary considerably. For terrestrial organisms, biomethylation leads to removal either by volatilization [as methylarsine, CH_3AsH_2, dimethylarsine, $(CH_3)_2AsH$, or trimethylarsine] or by excretion [as water-soluble methylarsonic acid, $CH_3AsO(OH)_2$, or cacodylic acid, $(CH_3)_2AsOOH$]. A strain of *Penicillium* isolated from evaporation pond waters generated trimethylarsine, suggesting a possible use in bioremediation [31]. Aquatic organisms, especially marine organisms, may likewise excrete water-soluble species, or they may sequester arsenic by incorporating methylarsenic groups into sugar or lipid molecules (cf. Sec. 3.2). Arsenic is the only element shown to bind four methyl groups through biomethylation [in tetramethylarsonium ion, $(CH_3)_4As^+$] [14].

2.3. Mercury

More effort has been devoted to studying the biomethylation of mercury than of any other element, primarily because of extensive poisonings by methylmercuric compounds [1,2,11,32]. Methylation of Hg(II) proceeds in two steps; kinetic investigations with methylcobalamin showed that the first methyl group is taken up much more rapidly than the second [33,34]. Methylmercuric compounds are encountered in the environment far more frequently than dimethylmercury.

Much current research concentrates on determining the effects of local conditions on rates of mercury biomethylation. The primary generators of methylmercuric species are microorganisms in sediments; recent reviews report that both lowered pH and higher sulfate concentrations enhance biomethylation [35,36]. This is also supported by reports [37-39] that sulfate-reducing bacteria are the primary biomethylating organisms in anoxic sediments. Biomethylation can occur under both aerobic and anaerobic conditions, although rates are faster in the latter state [40]. The nature of the sediments involved, especially the relative abundances of clays and humic materials, will also determine how rapidly and extensively methylmercuric derivatives are formed and released [41]. A recent report proposes that the methyl group of methylmercuric compounds formed by *Desulfovibrio desulfuricans* originates from C-3 of serine and transfers via either methylcobalamin or a closely related cobalt porphyrin [42].

2.4. Selenium

Biomethylation of selenium has been receiving increased attention in recent years. Although its chemistry parallels that of arsenic [16], selenium biomethylation shows different dependencies on pH and redox conditions than arsenic biomethylation [43]. Volatile dimethyl-selenide and the water-soluble trimethylselenonium ion are the most important methylselenium species formed by biomethylation; $(CH_3)_3Se^+$ has been found in natural waters [44]. Much current research on selenium biomethylation concentrates on its potential role in the removal of toxic selenium compounds from soils or wastewater [16,45-47]. In this context it should be noted that while selenium is a required trace element, only minor enhancement of Se concentration above required levels will induce toxic effects. In an experiment with Wistar rats, dimethylselenide exhalation depended on SAM and was enhanced by the presence of methylmercuric compounds [48].

2.5. Tin

Research on the biomethylation of tin has grown out of the use of tri-n-butyltin compounds as active agents in antifouling coatings on ships and boats [49]. These compounds are leached out into the surrounding waters and eventually enter the sediments, where they are debutylated and/or methylated. The alkyltin compounds are taken up by resident organisms (especially shellfish), thereby entering aquatic food chains. The grim examples of methylmercury poisonings have generated great concern that a similar tragedy might happen with tin [50-52].

Biomethylation of tin compounds occurs under a variety of conditions and with a variety of substrates. Sulfate-reducing bacteria in estuarine sediments generated methyltin compounds under anaerobic conditions [53], but methylation is known to occur under aerobic conditions as well [54]. Two different Bacillus spp. combined to convert aqueous CH_3SnCl_3 to $(CH_3)_3SnX$ [54]. Neither species methylated $CH_3Sn(IV)$ by itself, nor did the combination methylate either aqueous $SnCl_4$ or $(CH_3)_2SnCl_2$. Some mixed methylbutyltin compounds have been detected [55], but the scarcity of such reports suggests that butyltin compounds per se rarely undergo biomethylation. They appear instead to undergo stepwise loss of butyl groups, and the inorganic tin-containing end product then might undergo biomethylation. But what is the nature of this end product? Both Sn(II) and Sn(IV) derivatives can be found in nature. The yeast *Saccharomyces cerevisiae* generated small quantities of methyltin compounds from Sn(II) precursors but not from Sn(IV) precursors [56]. Both Sn(II) [57] and Sn(IV) [58] compounds can undergo transmethylation in the presence of methylcobalamin. Both Sn(II) [59] and Sn(IV) [60] can undergo methylation under simulated environmental conditions. One study reported that addition of inorganic Sn(IV) to a system containing the macroalga *Enteromorpha* spp. markedly enhanced CH_3Sn production, whereas addition of metallic Sn enhanced $(CH_3)_3Sn$ production [61]. At present, there is no conclusive evidence as to the mechanistic route (routes?) by which tin compounds undergo biomethylation, although available data tend to favor Sn(II) and SAM as methyl acceptor and donor, respectively. Despite the substantial body of reported research on tin biomethylation, much remains to be discovered.

2.6. Lead

While methyllead compounds have been detected in the environment, much controversy has arisen as to whether lead derivatives undergo biomethylation under environmental conditions [1,2,11]. Two factors have complicated investigations into this question:

1. *The instability of monomethyllead(IV) species.* Unlike Tl(III) or Sn(IV), no compound of formula CH_3PbX_3 has ever been isolated. If such species were formed in a

reaction, they would either rearrange rapidly or undergo
reductive elimination, abetted by the vast excess of
chloride ion present in living cells and natural waters:

$$CH_3Pb^{3+} + 3Cl^- \longrightarrow CH_3PbCl_3 \qquad\qquad (5a)$$

$$CH_3PbCl_3 \longrightarrow CH_3Cl + PbCl_2 \qquad\qquad (5b)$$

2. *The lability of methyllead linkages.* In comparative transmethylation studies, Chau et al. [62] reported that methyllead species would methylate Sn(II) and Sn(IV) but that
Pb(II) would not accept a methyl group from methyl derivatives of Sn, As, or Hg. Along with the earlier observations of Jewett et al. [63], this would suggest that the
order of transmethylating ability is Pb > Sn > Hg > As.

Lead(II) did accept methyl groups from a dimethylcobalt complex
[64], and lead dioxide did react with methylcobalamin under laboratory
conditions [58,65]; however, only trace quantities of methyllead
species were observed in both instances. Lead(II) salts underwent
methylation by biologically active sediments [66,67] and by marine
phytoplankton and the yeast *S. cerevisiae* [68]. Again, only low
quantities of methyllead compounds were observed.

Available evidence suggests that biomethylation of lead does
indeed occur but is considerably less extensive than for tin or
mercury. Only a portion of the initially formed methyllead species
apparently ends up as readily observable products. The importance
of biomethylation in the environmental mobility of lead remains
unclear.

2.7. Phosphorus

The only verified reports for biomethylation of phosphorus have
involved soil bacteria generating phosphinothricin [17], which will
be discussed in Sec. 3.3. Available evidence suggests that the
Challenger mechanism may be the route of biomethylation. Monitoring

the occurrence of methylphosphonic acid and its derivatives in
natural waters has been proposed as a method for detecting the
manufacture of nerve gases [69].

2.8. Tellurium

Biomethylation of tellurium, inasfar as it has been studied, appears
to parallel that of selenium [16,70,71]. Tellurium has been much
less investigated than selenium, and methylated derivatives appear
to play less of a role in its biochemistry and environmental
cycling. No organotellurium compounds have been reported in the
natural environment.

2.9. Germanium

Methylgermanium compounds occur widely in natural waters [72,73].
The fact that germanium compounds do not undergo methylation under
aerobic conditions but are methylated under anaerobic conditions
[72] suggests that reduction to a Ge(II) intermediate may be necessary, in which case methylation might occur via the Challenger
mechanism.

2.10. Antimony

Evidence for the environmental biomethylation of antimony is indirect. Methylstibonic and dimethylstibinic acids have been reported
in various natural waters [74-77] and biomethylation appears to be
their most likely source. Challenger found no evidence for biomethylation of antimony in most cases, although $KSbO_3$ did give small
quantities of an unidentified volatile Sb-containing species [15].
No traces of methylantimonials were found in phytoplankton [78] or
mollusk shells [79], although methylarsenicals were found in both

cases. A metabolic study on humans ingesting Sb(V) oxy compounds
showed no evidence for biomethylation [80].

There is clearly a strong need for more intensive and focused
work on whether, and to what extent, antimony might undergo biomethylation.

2.11. Thallium

No evidence for the biomethylation of thallium species in the natural
environment has been reported. However, thallous ion has been found
to undergo biomethylation under laboratory conditions [81,82].
Anaerobic bacteria formed $(CH_3)_2Tl^+$ from thallous ion in low yields,
which suggests biomethylation via the Challenger mechanism. The
authors also showed that monomethylthallium compounds (which have
been isolated) can act as methylating agents and also undergo disproportionation; the latter is abetted under reducing conditions [82].
No biomethylation of thallous ion occurred under aerobic conditions
[81]. Thallium is a prime candidate to be investigated for environmental biomethylation.

2.12. Halogens

The formation and environmental importance of methyl halides are
discussed elsewhere in this volume [7]. Suffice it to say here that
those compounds play a major role in the environmental mobility of
the halogens. Biomethylation is apparently limited to only one methyl
group per halogen atom. Dimethylhalonium hexafluoroantimonates have
been isolated and are thermally stable, but react rapidly with water
to form methanol [83]:

$$(CH_3)_2X^+ + H_2O \longrightarrow CH_3X + [CH_3OH_2]^+ \qquad (6a)$$

$$[CH_3OH_2^+] \longrightarrow CH_3OH + H^+ \qquad (6b)$$

Unlike other elements discussed in this chapter, more than one halogen atom may undergo attachment to a specific carbon atom through biochemical processes (the "biopolyhalogenation" of methane?). This has become a serious problem in purity control of drinking water, since algae in reservoirs can generate chloroform, bromoform, and mixed trihalomethanes [84,85] that will enter such water.

2.13. Other Representative Elements

2.13.1. Sulfur

The formation and biological roles of methylsulfur species such as methionine, $CH_3SCH_2CH_2CH(NH_2)CO_2H$, is amply covered in standard biochemistry textbooks. Methylsulfur compounds in nature (dimethylsulfide, dimethyldisulfide, methylthiol) are widespread and important in the biogeochemistry of the element [86].

2.13.2. Silicon

Except for one unconfirmed report [87], there is no evidence for biomethylation of silicon. The great bond strength of silicon-oxygen linkages and the difficulty of generating Si(II) under ambient biological conditions would seem to preclude biomethylation. Methylsilicone polymers have been found in marine sediments, where they have a long residency and little tendency to undergo bioaccumulation [88-92]. They have been proposed for use in investigations of sediments [93].

2.13.3. Bismuth

No evidence has appeared for the biomethylation of bismuth. No methylbismuth(V) compounds have ever been prepared, though methylbismuth(III) species are known. Both bismuth(III) oxide and sodium bismuthate reacted with methylcobalamin under laboratory conditions [58].

2.13.4. Indium

No evidence for the biomethylation of indium has appeared. Methyl-
indium compounds are stable to water. This element appears to be a
promising candidate for biomethylation, at least under laboratory
conditions.

2.14. Transition Metals

In general, the simple methyl derivatives of most transition metals
are unstable toward water (platinum and gold are two notable exceptions) and need special combinations of chelating agents attached to
the metal in order to exist in its presence. Methylcobalamin is the
best-known example, and its chemistry is well documented [4,26]. It
is stable enough to be used as a laboratory reagent, and occurs widely
in nature, both within organisms and without [1c]. The rhodium counterpart, methylrhodobalamin, has been synthesized in vitro and shows
biological activity [94]. Methylnickel species serve as important
intermediates in the metabolism of methanogenic bacteria [28].

3. BIOALKYLATION

3.1. Introduction

Compared to biomethylation, bioalkylation covers a wider range of
reactions and a smaller range of elements. Except for cobalt,
rhodium, and (possibly) antimony, bioalkylation is confined to the
metalloids and nonmetals. The mechanisms have little in common with
each other and will be discussed individually.

3.2. Arsenic

The number of papers on the bioalkylation of arsenic compounds has
increased dramatically during the last decade [13,14]. Many novel

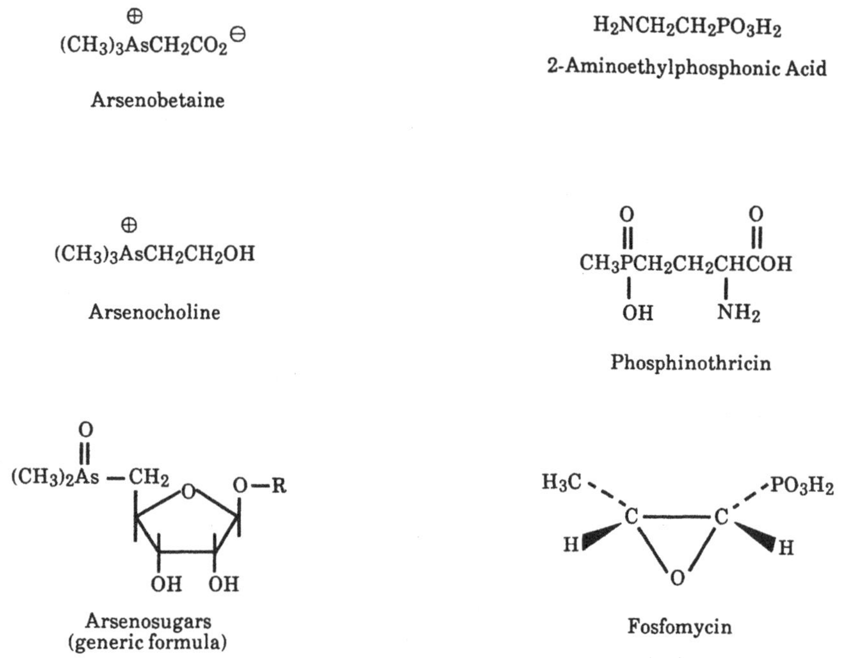

$$\overset{\oplus}{(CH_3)_3AsCH_2CO_2}\overset{\ominus}{}$$

Arsenobetaine

$$H_2NCH_2CH_2PO_3H_2$$

2-Aminoethylphosphonic Acid

$$\overset{\oplus}{(CH_3)_3AsCH_2CH_2OH}$$

Arsenocholine

Phosphinothricin

Arsenosugars
(generic formula)

Fosfomycin

FIG. 3. Some biogenic alkyl derivatives of arsenic and phosphorus.

organoarsenicals have been isolated from various organisms, the most widespread one being arsenobetaine (Fig. 3). While many of these compounds have been identified, many more have not. The complete metabolic pathways involving such arsenicals are still being worked out, but one crucial step almost certainly involves the attachment of a deoxyadenosyl group to an arsenic atom, with SAM as the alkylating agent.

If you examine the sulfur atom of SAM (Fig. 2), you see a sulfonium ion having three alkyl groups attached. Methyl is the smallest of these groups, and the most easily transferred, which is the reason that SAM plays a crucial role as a methyl transfer reagent (cf. Sec. 2.1). However, both the 5'-deoxyadenosyl and the 1-aminobutyryl groups can also undergo transfer; such transfer reactions are known in SAM biochemistry [24]. In marine organisms, the 5'-

deoxyadenosyl moiety is apparently transferred to cacodylic acid,
followed by hydrolysis of the adenine residue and esterification of
the resulting hydroxyl group to give the arsenosugar structure shown
in Fig. 3 [95]. Subsequent metabolism can convert such compounds to
arsenocholine and arsenobetaine, which themselves undergo further
conversion (Sec. 4).

3.3. Phosphorus

The first naturally occurring organophosphorus compound, 2-amino-
ethylphosphonic acid (AEP) (Fig. 3), was isolated in 1959 [96];
many biogenic alkylphosphorus compounds have since been reported in
a wide variety of organisms, including humans [17].

   The crucial precursor is phosphoenol pyruvate (PEP). This compound can undergo enzymatic rearrangement (Fig. 4) to phosphonopyruvate. The controlling enzyme, phosphoenolpyruvate mutase (EC 6.4.2.9),
has been isolated and studied [97-100]. Interestingly, this enzyme
will also catalyze the reverse reaction [98]! Phosphonopyruvate can
subsequently be metabolized to form AEP [97]. Alternatively, PEP can
undergo esterification with phosphonoformate (Fig. 4) to form carboxyphosphoenol pyruvate (CPEP) which is then enzymatically converted to
phosphinopyruvate [101]. This reaction differs from the preceding in
that loss of carbon dioxide apparently occurs simultaneously with the
formation of the P-C linkage, which may drive the reaction to comple-
tion [101]. In a comparative study of the two mutases [102], they
were found not to be interchangeable.

   Enzymatic formation of CPEP was studied as part of an investigation into the biosynthesis of phosphinothricin (Fig. 3). This molecule is the active portion of the phytotoxic tripeptide bialaphos
[17] and is itself commercially available as the herbicide glufo-
sinate. Phosphinothricin contains a methyl group bonded to phosphorus, and one author [17] has suggested the mechanism shown in Eq. (7).

   To date, the great majority of alkylphosphorus compounds found
in nature are phosphonic acid derivatives [16]; phosphinothricin is

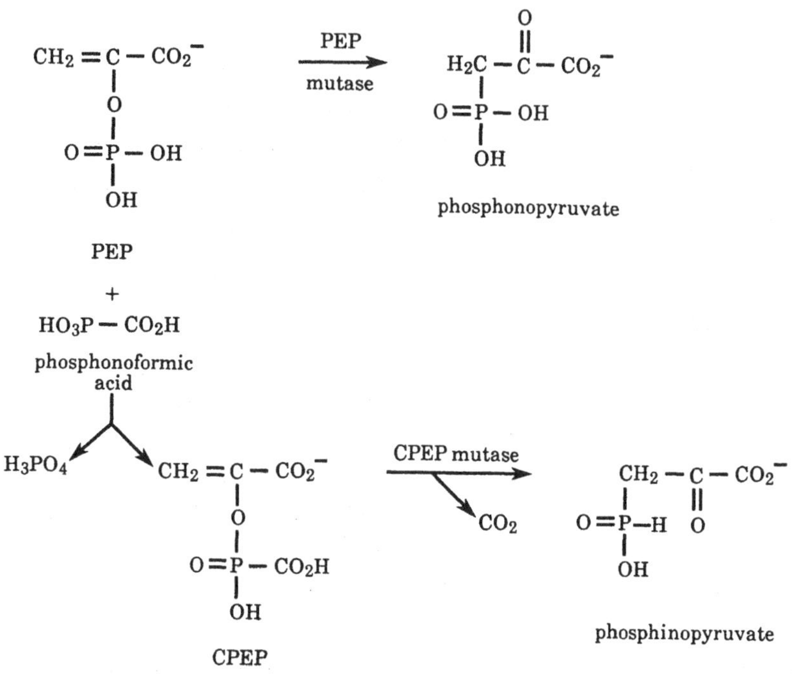

FIG. 4. Formation of phosphorus-carbon bonds from phosphoenol-pyruvate.

the major exception. Many phosphonates have been isolated from soil fungi, who use them as protective agents against bacteria. The best known is fosfomycin (Fig. 3), which has been utilized as an anti-biotic, and its biosynthesis investigated [103]. AEP is probably the

$$R-P-H \longrightarrow R-P: \tag{7a}$$

$$R-P: + SAM \longrightarrow SAH + R-P-CH_3 \tag{7b}$$

most widespread alkylphosphonate, generally occurring as esters in a wide variety of phosphonolipids [104,105]. Recently, alkylphospho-nates were discovered in marine sediments [106].

3.4. Selenium

Selenium biochemistry has many parallels with that of sulfur. Selenoproteins can be formed from various selenium-containing precursors [107]; the crucial precursor appears to be selenocysteine [108]. Selenoadenosylmethionine could be formed in vitro and, like its sulfur counterpart, can serve as a transmethylating agent [29]. Selenomethionine itself could replace methionine as a support for cell growth, but only at levels below 10 μM; at higher levels it became cytotoxic. While much has been discovered about selenium bioalkylation [16], much more remains unknown, and there is considerable active work in this area, especially as to the extent of selenium's biochemical resemblance to arsenic.

3.5. Tellurium

Much less is known about the biochemistry of tellurium than of selenium, and this is true for its bioalkylation as well. A recent report indicated that Te-tolerant fungal strains, grown on sulfur-free media with 0.2% sodium tellurite, could incorporate tellurium into proteins; tellurocysteine and telluromethionine were also reported [109]. A recent review suggests that research in this area might prove very rewarding to future investigators [71].

3.6. Antimony

Evidence for the bioalkylation of antimony is even more indirect than for its biomethylation. Two reports [110,111] claimed that [125]Sb-labeled antimony trichloride was taken up by algae to form a 5'-dimethylstibnoso-5'-deoxyribosyl derivative that behaved in close parallel fashion to the arsenic analog. These results suggest a parallel in the behavior of antimony to arsenic in marine organisms. There is a great potential for further work on this subject.

3.7. Halogens

So many concerns have been expressed about environmental pollution
arising from anthropogenic organohalogens (polychlorobiphenyls,
chlorofluorocarbons and the ozone layer, etc.) that it may come as
a surprise that many organohalogen compounds in addition to the
methyl halides are formed through biological processes! These were
recently reviewed [18,112a]. The active intermediates are hypo-
halites (either bound or as free acids) generated by haloperoxidase
enzymes [112a,113]. Numerous species of bacteria, fungi, and marine
organisms can generate such compounds. As with the methyl derivatives, each halogen is only bonded to a single carbon atom. No
organohalooxy derivatives (e.g., $RI(OH)_2$) have been reported. Most
naturally occurring derivatives have a single halogen per carbon,
although some have two or three. The great majority of reported
species are aliphatic. However, some aromatic halides have been
reported. The most intensely studied of these are the iodo species
thyroxine and its analogs or derivatives [112b,114,115].

3.8. Transition Metals

Formation of the 5'-deoxyadenosyl derivative of cobalamin (coenzyme
B_{12}) is the best known example of bioalkylation of a transition metal
[4,27]. In fact, the biological importance of vitamin B_{12} arises
primarily from its facile formation and cleavage of cobalt-carbon
bonds. The rhodium analog, 5'-deoxyadenosylrhodobalamin, was prepared in vivo by use of *Propionibacterium shermanii* cultures [94] and
was found to be a weak antimetabolite of its cobalt counterpart.
Other metal analogs of cobalamin—iron [116], copper [117], and zinc
[117]—were prepared by incorporation of the metal into metal-free
corrinoid, but no derivatives containing metal-carbon bonds were
reported.

4. BIOLOGICAL CLEAVAGE OF
METAL(LOID)-CARBON BONDS

Just as both biotic and abiotic processes are involved in the formation of metal(loid)-carbon bonds, they are also involved in the
corresponding cleavage reactions. Abiotic pathways would include
photolysis (especially important in the atmosphere), exchange reactions, and protolysis; see Eqs. (2) and (3).

Biodealkylation appears to be enzymatically controlled, and the
alkyl group usually is converted to the corresponding hydrocarbon.
This process has been studied most extensively for mercury [3,5], but
is also known in some detail for phosphorus [17], where investigations have been spurred by the increasing use of organophosphorus
compounds as herbicides. Some methylarsenicals will undergo loss of
methyl groups [11,14]. Hydrosulfide ions enhanced Hg-C bond cleavage
in CH_3HgCl by strains of S. cerevisiae [118]. Two separate groups of
aerobic microorganisms have been reported to convert arsenobetaine to
trimethylarsine oxide [119,120]. It is worth noting that, of the
four As-C bonds in arsenobetaine, the $As-CH_2CO_2^-$ bond is cleaved prefer-
entially. This contrasts to betaine itself, which is metabolized to
N,N-dimethylglycine and, ultimately, to glycine. Butyltin compounds
are known to undergo stepwise loss of butyl groups [1,2].

Almost all such processes seem to involve microorganisms and
occur in sediments, i.e., the same situation as for most methylations.
It would seem that any metal that can undergo biomethylation can also
undergo biodemethylation.

5. BIOMETH/ALKYLATION AND
BIOGEOCHEMICAL CYCLES

5.1. Introduction

Every element (with the probable exception of the noble gases) undergoes a variety of transformations in the natural environment. While
some of these transformations arise from physical processes (e.g.,

conversion of graphite to diamond), the great majority are due to
chemical reactions. Some of these compounds can be formed, transformed, and reformed in various ways; hence, the concept of a cycle.
Chemical reactions by definition involve at least two different elements. When such chemical reactions occur as part of a cycle, they
represent the interaction or overlap of the natural cycles of the
elements involved. Carrying this idea to its logical conclusion,
one might say that there is one single grand cycle, involving all
the physical and chemical transformations occurring on this planet.
Such a grand cycle would be difficult, perhaps impossible, to put
onto paper. Conceptually, then, scientists find it easier to draw
up cycles in which a single element (or, in certain cases, a compound, such as water) plays the predominant role. Nevertheless, it
is important to remember that the cycle of any one element does overlap with cycles of other elements. Anything that perturbs one cycle
will perturb those that interact with it [86,121,122]. Since alkyl
metal(loid)s contain carbon, they may be considered as part of the
global carbon cycle—probably a small part in comparison to methane
or carbon dioxide, but an important part nevertheless. Cycles have
been proposed for the various metal(loid)s that undergo biometh/
alkylation [1,2,11,14,86,121].

Everything that has been written about biogeochemical cycles
involves not only transformation (physical and chemical) but also
mobility. Atoms and molecules can move considerable distances, and
such movement is an integral, important part of these cycles. Dis-
persion through the constantly changing atmosphere is an important
route for volatile species. Dissolution in water and transport by
ceaseless currents (stream, river, estuary, ocean) provides another
means of dispersal. Movement in soils, sediments, and sludges, albeit
much slower, is likewise important. Living organisms play an important, and not fully appreciated, role in elemental cycles. The
majority of organisms travel during their lifetime, whether volun-
tarily or not. Often they travel great distances, especially migrat-
ing species. When they move, the chemical compounds within them also

move. When they die or are ingested by another organism, these
compounds are released or also ingested.

5.2. Biomethylation and Natural Cycles

As previously mentioned, attachment of a methyl or another alkyl
group to an atom of a heavy element changes both the physical and
the chemical properties of that element. This in turn will affect
the biogeochemical cycle of the element, usually by opening up new
pathways for transport. Inorganic compounds of most metals have
little or no volatility; once methylated, however, the resulting compounds gain markedly in volatility. Methylated derivatives of mercury, arsenic, selenium, lead, and probably other elements have been
detected in the atmosphere. Biomethylation then can generate otherwise impossible volatile molecules that will facilitate mobility.

Sediments used to be considered as a "final" resting place for
heavy elements introduced into environmental waters. The Minamata
tragedy proved this view to be erroneous and led to the realization
that biomethylation facilitated removal of mercury (and other metals)
from sediments [123]. Their enhanced solubility in lipids enables
methyl metals to be taken up by living organisms. If such an organism
is the base of a food chain, then the organometal may undergo biological accumulation ("bioaccumulation"). This was the cause of the
poisonings at Minamata and elsewhere in Japan. Bioaccumulation is
found for other elements in addition to mercury; two recent reports
[124,125] show that the proportion of methylarsenicals among all
arsenic compounds increases along a food chain. While the quantity
of any metal compound within a single organism may be extremely small,
the number of organisms involved can be quite large, making the total
quantity of metal or metalloid substantial. Add to that any distance
traveled by the organisms in question, and you have a mechanism for
environmental distribution. This aspect of biogeochemical cycling
has received much less attention than it deserves.

5.3. Bioalkylation and Natural Cycles

Much that has been written about biomethylation will also apply to
bioalkylation. The roles of the two processes, however, appear to
be quite different. For phosphorus, sulfur, selenium, and possibly
the halogens, formation of alkyl metalloid compounds is usually an
important part of organismal metabolism, and the compounds formed
have important biological functions. For arsenic, antimony (?), and
tellurium (?), formation of alkyl derivatives (usually complicated
ones) represents an important detoxification process. The major
difference between biomethylation and bioalkylation may be summarized by the following statement: methyl metal(loid) derivatives
are generated to be excreted; other alkyl metal(loid) derivatives
are generated to be sequestered.

From the extensive work reported to date on organoarsenicals in
living organisms [13,14], it appears that the numerous arsenosugars,
etc., have been found *only* within cells, while methylarsenicals appear
within cells *and* in the surrounding waters, atmosphere, sediments,
etc. The same appears to also be true for other metalloids. While
the organisms live, alkyl metalloids remain sequestered within them
and travel wherever they do. When the organism dies, the compounds
are released to the external environment, where they decompose. If
the organism is eaten, the compounds are ingested and metabolized or
released at a later time. In any event, they participate in the biogeochemical cycle.

6. CONCLUSIONS

Biomethylation and bioalkylation are processes that are more widespread than most people realize. As the chapters in this volume
indicate, research workers have generated an enormous literature,
which is growing rapidly. Research efforts tend to be unevenly distributed, and there is less interaction among research groups than

there might be. Many surprises await researchers in this area. The recent unexpected report of phosphonates in sediments [100] is one example.

Virtually all reported work on biometh/alkylation has focused on those compounds that can be detected, even if they cannot be immediately characterized. Such compounds do not necessarily represent the only products of these processes. Much of the biological role of vitamin B_{12} involves the transient existence of alkylcobalt intermediates. It may well be that other such intermediates occur in nature, both endo- and exocellularly. Their involvement in biological processes goes unrecognized. The formation/cleavage of transient metal-carbon bonds by abiotic alkyl transfer under environmental conditions deserves investigation. It has been proposed [126] that dialkylmercurials may undergo exchange with mercury (and possibly other metals) through formation of "organic calomel" intermediates:

$$R_2Hg + Hg^* \underset{\longleftarrow}{\longrightarrow} [R\text{-}Hg\text{-}Hg^*\text{-}R] \underset{\longleftarrow}{\longrightarrow} R_2Hg^* + Hg \qquad (8)$$

Similarly, the steady improvement of analytical techniques will increase available detection "windows", which will doubtlessly lead to the discovery of additional methyl and alkyl metal(loid) compounds in nature and expand our knowledge of the roles of these compounds in natural cycles.

For a note added in proof, see Ref. 127.

ACKNOWLEDGMENT

The author expresses appreciation to Mrs. Elaine Seliskar for typing this manuscript.

ABBREVIATIONS AND DEFINITIONS

AEP   2-aminoethylphosphonic acid
CPEP   carboxyphosphoenol pyruvate

PEP   phosphoenol pyruvate
SAH   *S*-adenosylhomocysteine
SAM   *S*-adenosylmethionine
Gosio-gas   an early name for biologically generated trimethylarsine

REFERENCES

1. P. J. Craig (ed.), *Organometallic Compounds in the Environment,*
    Longman, Burnt Mill, Harlow, UK, 1986, (a) 19-23; (b) 30-32;
    (c) 351-355.

2. J. S. Thayer, *Organometallic Compounds and Living Organisms,*
    Academic Press, New York, 1984, (a) 206-207; (b) 205-206; (c)
    39-60; (d) 7-9.

3. A. D. Ryabov, *Angew. Chem. Int. Ed., 30,* 931 (1991).

4. J. M. Pratt, Chap. 8, this volume.

5. Chap. 10, this volume.

6. J. S. Thayer, *Appl. Organomet. Chem., 3,* 123 (1989).

7. D. B. Harper, Chap. 11, this volume.

8. S. P. Wasik, in *Organometals and Organometalloids: Occurrence
    and Fate in the Environment* (F. E. Brinckman and J. M. Bellama,
    eds.), American Chemical Society, Washington, D.C., 1978, pp.
    314-326.

9. Y. Arakawa and O. Wada, Chap. 4, this volume.

10. Y. Yamamura and F. Arai, Chap. 5, this volume.

11. P. J. Craig, in *The Chemistry of the Metal-Carbon Bond* (F. R.
    Hartley, ed.), John Wiley and Sons, Chichester, 1989, Vol. 5,
    pp. 437-463.

12. D. A. Klein and J. S. Thayer, in *Soil Biochemistry* (J. M. Bollag
    and G. Stotzky, eds.), Marcel Dekker, New York, 1990, pp. 431-
    481.

13. M. Vahter and E. Marafante, Chap. 6, this volume.

14. W. R. Cullen and K. J. Reimer, *Chem. Rev., 89,* 713 (1989).

15. F. Challenger, in *Organometals and Organometalloids: Occurrence
    and Fate in the Environment* (F. E. Brinckman and J. M. Bellama,
    eds.), American Chemical Society, Washington, D.C., 1978, pp.
    1-20.

16. U. Karlson and W. T. Frankenberger, Chap. 7, this volume.

17. J. S. Thayer, *Appl. Organomet. Chem.*, *3*, 203 (1989).

18. D. J. Faulkner, in *The Handbook of Environmental Chemistry. 1A: The Natural Environment and the Biogeochemical Cycles* (O. Hutzinger, ed.), Springer-Verlag, Berlin, 1980, pp. 229-252.

19. D. Mennie and P. J. Craig, Chap. 2, this volume.

20. R. Holliday, M. Monk, and J. E. Pugh (eds.), *DNA Methylation and Gene Regulation*, Cambridge University Press, Cambridge, UK, 1991.

21. A. Razin, H. Cedar, and A. D. Riggs (eds.), *DNA Methylation: Biochemistry and Biological Significance*, Springer-Verlag, New York, 1984.

22. G. L. Cantoni and A. Razin (eds.), *Biochemistry and Biology of DNA Methylation*, Alan R. Liss, New York, 1985.

23. R. T. Borchardt, C. R. Creveling, and P. M. Ueland (eds.), *Biological Methylation and Drug Design*, Humana Press, Clifton, New Jersey, 1986.

24. F. Salvatore, E. Borek, V. Zappia, H. G. Williams-Ashman, and F. Schlenk, *The Biochemistry of Adenosylmethionine*, Columbia University Press, New York, 1977.

25. K. Takahashi, H. Yamauchi, M. Mashiko, and Y. Yamamura, *Nippon Eis. Zacchi*, *45*, 613 (1990); *Chem. Abs.*, *114*, 96350k (1990).

26. D. Dolphin (ed.), B_{12}, Wiley-Interscience, New York, 1982.

27. F. E. Brinckman, G. J. Olson, and J. S. Thayer, in *Marine and Estuarine Geochemistry* (A. C. Sigleo and A. Hattori, eds.), Lewis, Ann Arbor, Michigan, 1985, pp. 227-238.

28. B. Jaun, Chap. 9, this volume.

29. E. O. Kajander, R. J. Harvima, T. O. Eloranta, H. Martikainen, M. Kantola, S. O. Karenlampi, and K. Akerman, *Biol. Trace Elem. Res.*, *28*, 57 (1991).

30. W. Maher and E. Butler, *Appl. Organomet. Chem.*, *2*, 191 (1988).

31. K. D. Huysmans and W. T. Frankenberger, *Sci. Total Environ.*, *105*, 13 (1991).

32. O. Lindqvist, K. Johansson, M. Aastrup, A. Andersson, L. Bringmark, G. Hovsenius, L. Haakanson, A. Iverfeldt, M. Meili, and B. Timm, *Water Air Soil Poll.*, *55*, 1 (1991).

33. R. E. DeSimone, M. W. Penley, L. Charbonneau, S. G. Smith, J. M. Wood, H. A. O. Hill, J. M. Pratt, S. Ridsdale, and R. J. P. Williams, *Biochim. Biophys. Acta*, *304*, 851 (1973).

34. J. S. Thayer, *Inorg. Chem.*, *18*, 1171 (1979).

35. C. C. Gilmour and E. A. Henry, *Environ. Poll.*, *71*, 131 (1991).

36. M. R. Winfrey and J. W. M. Rudd, *Environ. Toxicol. Chem.*, *9*, 853 (1990).

37. G. C. Compeau and R. Bartha, *Appl. Environ. Microbiol., 50,* 498 (1985).

38. G. C. Compeau and R. Bartha, *Appl. Environ. Microbiol., 53,* 261 (1987).

39. A. Kerry, P. M. Welbourn, B. Prucha, and G. Mierle, *Water Air Soil Poll., 56,* 565 (1991).

40. T. Matilainen, M. Verta, M. Niemi, and A. Uusi-Rauva, *Water Air Soil Poll., 56,* 595 (1991).

40. T. A. Jackson, *Appl. Organomet. Chem., 3,* 1 (1989).

41. M. Berman, T. Chase, and R. Bartha, *Appl. Environ. Microbiol., 56,* 298 (1990).

42. P. H. Masscheleyn, R. D. Delaune, and W. H. Patrick, *J. Environ. Qual., 20,* 522 (1991).

44. D. Tanzer and K. G. Heumann, *Anal. Chem., 63,* 1984 (1991).

45. E. T. Thompson-Eagle, W. T. Frankenberger, and U. Karlson, *Appl. Environ. Microbiol., 55,* 1406 (1989).

46. Z. Wang, L. Zhao, and A. Peng, *Chem. Abs., 111,* 76941p (1989).

47. W. T. Frankenberger and U. Karlson, *Soil Sci. Soc. Am. J., 53,* 1435 (1989).

48. J. Yonemoto, *Chem. Abs., 111,* 168833h (1989).

49. R. J. Huggett, M. A. Unger, P. F. Seligman, and A. O. Valkirs, *Environ. Sci. Technol., 26,* 232 (1992).

50. E. A. Clark, R. M. Sterritt, and J. N. Lester, *Environ. Sci. Technol., 22,* 600 (1988).

51. M. D. Mueller, L. Renberg, and G. Rippen, *Chemosphere, 18,* 2043 (1989).

52. C. Stewart and S. J. DeMora, *Environ. Technol., 11,* 565 (1990).

53. C. C. Gilmour, J. H. Tuttle, and J. C. Means, *Microb. Ecol., 14,* 233 (1987).

54. N. S. Makkar and J. J. Cooney, *Geomicrobiol. J., 8,* 101 (1990).

55. R. J. Maguire, *Environ. Sci. Technol., 18,* 291 (1984).

56. J. Ashby and P. J. Craig, *Appl. Organomet. Chem., 1,* 275 (1987).

57. J. R. Ashby and P. J. Craig, *Sci. Tot. Environ., 100,* 337 (1991).

58. J. S. Thayer, *Appl. Organomet. Chem., 1,* 545 (1987).

59. D. Shugui, H. Guolan, and C. Yong, *Appl. Organomet. Chem. 3,* 437 (1989).

60. T. Hamasaki, H. Nagase, T. Sato, H. Kito, and Y. Ose, *Appl. Organomet. Chem., 5,* 83 (1991).

61. O. F. X. Donard and J. H. Weber, *Nature, 332,* 339 (1988).

62. Y. K. Chau, P. T. S. Wong, C. A. Mojesky, and A. J. Carty, *Appl. Organomet. Chem.*, *1*, 235 (1987).

63. K. L. Jewett, F. E. Brinckman, and J. M. Bellama, in *Marine Chemistry in the Coastal Environment* (T. M. Church, ed.), American Chemical Society Symposium Series 18, Washington, D.C., 1975, pp. 304-318.

64. S. Rapsomanikis, O. F. X. Donard, and J. H. Weber, *Appl. Organomet. Chem.*, *1*, 115 (1987).

65. J. S. Thayer, *J. Environ. Sci. Health—Environ. Sci. Eng.*, *A11*, 471 (1983).

66. I. Berdicevsky, M. Shachar, and S. Yannai, *Arch. Toxicol. Suppl.*, *6*, 285 (1983).

67. A. P. Walton, L. Ebdon, and G. E. Milward, *Appl. Organomet. Chem.*, *2*, 87 (1988).

68. R. M. Harrison and A. G. Allen, *Appl. Organomet. Chem.*, *3*, 49 (1989).

69. S. Black, B. Morel, and P. Zapf, *Nature*, *351*, 515 (1991).

70. T. G. Chasteen, *Chem. Abs.*, *115*, 173832u (1991).

71. T. Sadeh, in *The Chemistry of Organic Selenium and Tellurium Compounds* (S. Patai, ed.), John Wiley and Sons, Chichester, 1987, Vol. 2, pp. 367-376.

72. B. L. Lewis, M. O. Andreae, and P. N. Froelich, *Mar. Chem.*, *27*, 179 (1980).

73. B. L. Lewis and H. P. Mayer, Chap. 3, this volume.

74. M. O. Andreae, in *Trace Metals in Seawater* (C. S. Wong, E. Broule, K. W. Bruland, J. L. Burton, and E. D. Goldberg, eds.), Plenum Press, New York, 1983, pp. 1-19.

75. K. K. Bertine and D. S. Lee, in *Trace Metals in Seawater* (C. S. Wong, E. Broule, K. W. Bruland, J. L. Burton, and E. D. Goldberg, eds.), Plenum Press, New York, 1983, pp. 21-38.

76. M. O. Andreae and P. N. Froelich, *Tellus*, *36b*, 101 (1984).

77. M. O. Andreae, J. T. Byrd, and P. N. Froelich, *Environ. Sci. Technol.*, *71*, 731 (1983).

78. R. Kantin, *Limnol. Oceanogr.*, *28*, 165 (1983).

79. W. R. Cullen, M. Dodd, B. U. Nwata, D. A. Reimer, and K. J. Reimer, *Appl. Organomet. Chem.*, *3*, 351 (1989).

80. R. Bailly, R. Lauwerys, J. P. Buchet, P. Mahieu, and J. Konings, *Br. J. Industr. Med.*, *48*, 93 (1991).

81. F. Huber and H. Kirchmann, *Inorg. Chim. Acta*, *29*, L249 (1978).

82. F. Huber, U. Schmidt, and H. Kirchmann, in *Organometals and Organometalloids: Occurrence and Fate in the Environment* (F. E. Brinckman and J. M. Bellama, eds.), American Chemical Society, Washington, D.C., 1978, pp. 65-81.

83. G. A. Olah and J. R. DeMember, *J. Am. Chem. Soc., 92,* 2562 (1970).

84. G. Asplund and A. Grimvall, *Environ. Sci. Technol., 25,* 1347 (1991).

85. A. A. Karimi and P. C. Singer, *J.A.W.W.A., 83,* 84 (1991).

86. P. Brimblecombe and A. Y. Lein, *Evolution of the Global Biogeochemical Sulfur Cycle,* John Wiley and Sons, Chichester, 1989.

87. W. Heinen, *Arch. Biochem. Biophys., 110,* 137 (1965).

88. C. Guillemaut, J. Aubert, and H. Augier, *Chem. Abs., 109,* 106007s (1988).

89. C. L. Frye, *Sci. Tot. Environ., 73,* 17 (1988).

90. R. Firmin and A. L. J. Raum, *Rev. Int. Oceanogr. Med., 85/6,* 46 (1987).

91. A. Opperhuizen, G. M. Asyee, and J. R. Parsons, *Chem. Abs., 109,* 215488v (1988).

92. R. Firmin and A. L. J. Raum, *Stud. Environ. Sci., 34,* 43 (1988).

93. R. E. Pellenbarg and H. W. Carhart, *Appl. Organomet. Chem., 5,* 79 (1991).

94. V. B. Koppenhagen, B. Elsenhans, F. Wagner, and J. J. Pfiffner, *J. Biol. Chem., 249,* 6532 (1974).

95. K. A. Francesconi, R. V. Stick, and J. S. Edmonds, *J. Chem. Soc. Chem. Commun.,* 928 (1991).

96. M. Horiguchi and M. Kandatsu, *Nature, 184,* 901 (1959).

97. H. M. Seidel, S. Freeman, H. Seto, and J. R. Knowles, *Nature, 335,* 457 (1988).

98. T. Hidaka, M. Mori, S. Imai, O. Hara, K. Nagaoka, and H. Seto, *J. Antibiot., 42,* 491 (1989).

99. H. M. Seidel, S. Freeman, C. H. Schwalbe, and J. R. Knowles, *J. Am. Chem. Soc., 112,* 8149 (1990).

100. E. D. Bowman, M. S. McQueney, J. D. Scholten, and D. Dunaway-Mariano, *Biochemistry, 29,* 7059 (1990).

101. T. Hidaka and H. Seto, *J. Am. Chem. Soc., 111,* 8012 (1989).

102. T. Hidaka and H. Seto, *Agric. Biol. Chem., 54,* 2467 (1990).

103. F. Hammerschmidt and H. Kaehlig, *J. Org. Chem., 56,* 2364 (1991).

104. M. C. Moschidis, *Prog. Lipid Res.*, *23*, 223 (1985).

105. T. Hori and Y. Nozawa, in *Phospholipids* (J. N. Hawthorne and
      G. B. Ansell, eds.), Elsevier, Amsterdam, 1982, pp. 95-128.

106. E. D. Ingall, P. A. Schroeder, and R. A. Berner, *Geochim.
      Cosmochim. Acta*, *54*, 2617 (1990).

107. R. A. Sunde and J. K. Evenson, *J. Inorg. Biochem.*, *43*, 286
      (1991).

108. R. F. Burk, *FASEB J.*, *5*, 2274 (1991).

109. S. E. Ramadan, A. A. Razak, A. M. Ragab, and M. El-Meleigy,
      Biol. Trace Elem. Res., *20*, 225 (1989).

110. A. A. Benson, in *Progress in Clinical and Biological Research*,
      Alan R. Liss, New York, 1988, pp. 385-391.

111. A. A. Benson, in *The Biological Alkylation of Heavy Elements*
      (P. J. Craig and F. Glockling, eds.), Royal Society of Chem-
      istry Spec. Public. #66, London, 1989, pp. 135-137.

112. K. L. Kirk, *Biochemistry of the Elemental Halogens and Inorganic
      Halides*, Plenum Press, New York, 1991, (a) 155-190; (b) 135-154.

113. R. Wever, M. G. M. Tromp, B. E. Krenn, A. Marjani, and M. Van
      Tol, *Environ. Sci. Technol.*, *25*, 446 (1991).

114. J. H. Oppenheimer and H. H. Samuels (eds.), *Molecular Basis of
      Thyroid Hormone Action*, Academic Press, New York, 1983.

115. G. Hennemann (ed.), *Thyroid Hormone Metabolism*, Marcel Dekker,
      New York, 1986.

116. R. Bieganowski and W. Friedrich, *Z. Naturforsch.*, *36C*, 9 (1981).

117. B. Eisenhans and I. H. Rosenberg, *Biochemistry*, *23*, 805 (1984).

118. B. Ono, N. Ishii, S. Fujino, and I. Aoyama, *Appl. Environ.
      Microbiol.*, *57*, 3183 (1991).

119. K. Hanaoka, T. Motoya, S. Tagawa, and T. Kaise, *Appl. Organo-
      met. Chem.*, *5*, 427 (1991).

120. K. Hanaoka, S. Tagawa, and T. Kaise, *Appl. Organomet. Chem.*, *5*,
      435 (1991).

121. O. Hutzinger (ed.), *The Handbook of Environmental Chemistry*,
      Vol. 1, The Natural Environment and the Biogeochemical Cycles,
      Springer-Verlag, Berlin, Part A: 1980; Part B: 1982; Part C:
      1984; Part D: 1984; Part E: 1989.

122. W. H. Schlesinger, *Biogeochemistry: An Analysis of Global
      Change*, Academic Press, San Diego, 1991.

123. F. M. D'Itri, in *Sediments: Chemistry and Toxicity of In-Place
      Pollutants* (R. Baudo, J. Giesy, and H. Muntau, eds.), Lewis,
      Ann Arbor, Michigan, 1990, pp. 193-214.

124. S. Maeda, R. Inoue, T. Kezono, T. Tokuda, A. Ohki, and T. Takeshita, *Chemosphere, 20,* 101 (1990).

125. S. Maeda, A. Ohki, T. Tokuda, and M. Ohmine, *Appl. Organomet. Chem., 4,* 251 (1990).

126. O. A. Reutov and K. P. Butin, *J. Organomet. Chem., 413,* 1 (1991).

127. A recent report raises doubts about the occurrence of simple methylantimony compounds in natural waters [128]. Reviews have appeared on biomethylation [129], marine chemistry of organoarsenicals [130], and the general environmental chemistry of arsenic [131].

128. M. Dodd, S. L. Grundy, K. J. Reimer, and W. R. Cullen, *Appl. Organomet. Chem., 6,* 207 (1992).

129. S. Krishnamurthy, *J. Chem. Educ., 69,* 347 (1992).

130. K. Hanaoka, S. Tagawa, and T. Kaise, *Appl. Organomet. Chem., 6,* 139 (1992).

131. S. Tamaki and W. T. Frankenberger, *Rev. Environ. Contam. Toxicol., 124,* 79 (1992).

2

Analysis of Organometallic Compounds in the Environment

Darren Mennie and Peter J. Craig

Department of Chemistry
Leicester Polytechnic
P.O. Box 143
Leicester LE1 9BH, England

1. OCCURRENCE OF ORGANOMETALLIC COMPOUNDS

The occurrence of organometallic species in the environment is due to
two processes, anthropogenic input [1] and biogeochemical cycling,
which may involve the methylation of inorganic or organometallic pre-
cusors [2] (Sec. 2). Anthropogenic inputs of organometallics into
the environment fall into three major subdivisions:

1. Direct application in the natural environment of the
    organometallic, e.g., as a fungicide [3]

2. Indirect release, e.g., due to leaching of organometallics
    used as plastics stabilizers, i.e., organotins from poly-
    vinylchloride (PVC) material [4]

3. Controlled release of organometallics from suitably
    designed materials, such as the release of the biocidal
    tributyltin species from specially formulated self-
    polishing copolymer paints [5]

A comprehensive treatment of the uses of organometallic compounds is
beyond the scope of this chapter; however, uses of selected organometallic compounds are presented in Table 1 [6-9]. It should be
realized, in view of space limitations, that the reference citations
below serve as examples and are indicative only.

Table 2 presents an overview of the main organometallic species
detected in the environment. Some are formed there; some enter the
environment as such. Organolead compounds are discussed further in
Sec. 3. Knowledge of the amounts of organometallics manufactured
each year coupled with their method of utilization allows estimates

TABLE 1

Some Uses of Organometallic Compounds

Compound	Use	Ref.
C_2H_5HgX	Cereal seed treatment	6
$p\text{-}CH_3C_6H_4HgX$	Spermicide	6
$((C_4H_9)_3Sn)_2O$	Wood preservative, fungicides	7
$(C_4H_9)_2Sn(OCOC_{11}H_{23})_2$	Anthelmintics for poultry	6
$(C_4H_9)_2Sn(OCOR)_2$	Polymerization catalysts	7
R = e.g., CH_3, $i\text{-}C_8H_{17}$,		
$C_{11}H_{23}$, $C_{12}H_{25}$		
$RSn(SCH_2COO\text{-}i\text{-}C_8H_{17})_3$	Used in combination as heat and light stabilizers for rigid PVC	7
R = CH_3, C_4H_9, C_8H_{17}		
$R_2Sn(SCH_2COO\text{-}i\text{-}C_8H_{17})_2$		
$(CH_3)_4Pb$, $(C_2H_5)_4Pb$	Antiknock agents in petrol	9
Inorganic arsenic acid salts	Cotton defoliants	9
$(CH_3)_2AsO(OH)$	Cotton defoliant	8
$CH_3AsO(ONa)_2$	Grass herbicide	8

of the amount of anthropogenic organometallics reaching environmental matrices to be made.

Biogeochemical cycles for several organometallic elements have been studied [10], but presently quantitative knowledge of such cycles is practically incomplete; hence estimates of natural fluxes vary widely [11].

The toxicity of organometallic species is usually dependent on the degree and nature of the alkyl/aryl substitution [12]. Table 3 lists the toxicities as LD_{50} (mg/kg) of various alkyl/aryl organotins

TABLE 2

Some Organometallic Species Detected in the Environment

Formula	Generic name
CH_3SnX_3	Monomethyltin
$(CH_3)_2SnX_2$	Dimethyltin
$(CH_3)_3SnX$	Trimethyltin
SnH_4	Stannane
$C_4H_9SnX_3$	Monobutyltin
$(C_4H_9)_2SnX_2$	Dibutyltin
$(C_4H_9)_3SnX$	Tributyltin
$(C_4H_9)_2CH_3SnX$	Dibutylmethyltin
$(C_6H_5)_3SnX$	Triphenyltin
CH_3HgX	Methylmercury
$(CH_3)_2Hg$	Dimethylmercury
$(CH_3)_4Pb$	Tetramethyllead
$(C_2H_5)_4Pb$	Tetraethyllead
$(CH_3)_{4-n}Et_nPb$	Alkyllead
$CH_3AsO(OH)_2$	Monomethylarsenic acid
$(CH_3)_2AsO(OH)$	Dimethylarsenic acid
$(CH_3)_3As$	Trimethylarsenic
$(CH_3)_4As^+$	Tetramethylarsonium ion
$(CH_3)_3AsO$	Trimethylarsine oxide
$(CH_3)_2AsOCH_2R$	Arsenosugars (R = sugar, riboside grouping)

Note: X = anionic grouping considered for convenience as being singly charged.

Source: Data are compiled from Refs. 6–9.

TABLE 3

Toxicities of Alkyl/Aryl Organotins

Compound	LD_{50} (mg/kg)	Target species
$(C_4H_9SnO(OH))_n$	6000 (oral)	Mouse
$C_4H_9SnCl_3$	1400 (oral)	Mouse
$(C_4H_9)_2SnCl_2$	182 (oral)	Rat
$(C_4H_9)_2SnS$	145 (oral)	Rat
$(C_4H_9)_3SnOCOC_6H_5$	108 (oral)	Mouse
$(C_4H_9)_3SnCl$	117 (oral)	Mouse
$(C_6H_{11})_3SnOCOCH_3$	1000 (oral)	Mouse
$(C_6H_5)_3SnOCOCH_3$	136 (oral)	Rat
$(C_6H_5)_3SnCl$	80 (oral)	Rat
$(C_2H_5)_3SnOCOCH_3$	4 (oral)	Rat

Source: Adapted from Ref. 12.

and shows the wide variations between different species of the same element. The chemistry of arsenic reveals this tendency even more clearly.

Table 4 illustrates species specificity for some organic groups possible for R_3SnX. Table 5 illustrates variations in toxicity for different arsenic compounds.

Consequently detection systems employed in analytical environmental organometallic chemistry must be capable of speciating the different chemical forms in which a particular organometallic element is present within environmental matrices. This is discussed further in Sec. 6.4. In the natural environment, organometallic species will exhibit greater preference for some matrices compared to others [13]. In the particular case of the biocide tributyltin species there have

TABLE 4

Species Specificity of Triorganotin Compounds: R_3SnX

Species	R in most active R_3SnX compound
Insects	CH_3
Mammals	C_2H_5
Gram-negative bacteria	$n\text{-}C_3H_7$
Gram-positive bacteria,	
Fish, fungi, molluscs, plants	$n\text{-}C_4H_9$
Fish, fungi, molluscs	C_6H_5
Fish, mites	$c\text{-}C_6H_{11}$, $C_6H_5(CH_3)_2CCH_2$

Note: In general, maximum biological activity for R_nSnX_{n-4} occurs for n = 3.

TABLE 5

Toxicity of Selected Inorganic and Organometallic
Arsenic Compounds[a]

Compound	Toxicity
AsH_3	3
As_2I_3	20
$CH_3(AsOH)_2$	20
$CH_3AsO(OH)_2$	700–1800
$(CH_3)_2AsO(OH)$	700–2600
$(CH_3)_3AsCH_2COO$	High

[a]As LD_{50} in rats, mg/kg.

Source: Adapted from Ref. 8.

been reports of 3-4 orders of magnitude concentration enrichment factors in the surface microlayer and the sediments as compared to the water column [14]. Consequently Secs. 7, 8, and 9 of this chapter review the concentration of organometallics in sediments and suspended matter, dissolved state and in biota.

Clearly the affinity of a particular organometallic for an environmental matrix will be a function of the physical and chemical properties of both; hence this chapter will also seek to present an overview of the analysis of organometallic compounds.

2. FORMATION OF ORGANOMETALLIC COMPOUNDS IN THE ENVIRONMENT

The formation of organometallic compounds in the environment is important as most organometallic compounds generally exhibit greater toxicity than their inorganic counterparts [15]. This is not the case with arsenic, where the inorganic form displays higher mammalian toxicity than the organometallic forms [16].

The production of organometallic compounds in the environment may be the result of biological or nonbiological processes [17]. The predominant biological process producing organometallics is biomethylation [18].

Methylation of an inorganic precusor may also lead to the production of di-, tri-, and tetramethyl organometallics via rearrangement following the stepwise dismutation of monomethyl metals or by further addition of methyl groups. Methyl group transfer mechanisms may be by radical addition [Eq. (1)], carbonium ion transfer [Eq. (2)], or carbanion transfer [Eq. (3)] depending on the oxidation state of the metal.

$$CH_3CoB_{12} + Sn(II) \longrightarrow CH_3Sn(III)^0 + CoB_{12}^0 \qquad (1a)$$

$$CH_3Sn(III)^0 + O_2 \longrightarrow CH_3Sn(IV) + O_2^- \text{ (aerobic)} \qquad (1b)$$

$$CH_3Sn(III)^0 + H_2OCoB_{12}^+ \longrightarrow CH_3Sn(IV)^+ + H_2O + CoB_{12}^0 \text{ (anaerobic)} \qquad (1c)$$

$$CH_3^+ + As(III)(OH)_3 \longrightarrow CH_3As(OH)_3^+ \longrightarrow CH_3As(V)O(OH)_2 + H^+ \quad (2)$$

Metals in their highest oxidation states (i.e., having no lone pair of electrons available) will be methylated via carbanion transfer [Eq. (3)], e.g., mercury(II) possibly from the natural carbanion donor vitamin B_{12} coenzyme CH_3B_{12} [19].

$$CH_3CoB_{12} + Hg^{2+} + H_2O \rightleftharpoons CH_3Hg^+ + H_2OCoB_{12}^+ \quad (3)$$

Methylation of mercury(II) may also occur via an abiotic trans-methylation process, i.e., by Eqs. (4) and (5) [20]:

$$(CH_3)_3Sn^+ + Hg^{2+} \longrightarrow (CH_3)_2Sn^{2+} + CH_3Hg^+ \quad (4)$$

$$(CH_3)_3Pb^+ + Hg^{2+} \longrightarrow (CH_3)_2Pb^{2+} + CH_3Hg^+ \quad (5)$$

Metals with a lone pair of electrons that are capable of oxidation, e.g., tin(II), may be methylated by methyl carbonium ion transfer with consequent oxidation to tin(IV). Methyl carbonium ion donation in the natural environment is likely to be from the s-adenosyl-methionine (SAM) system [19] (Eq. 6).

$$CH_3SAM^+ + Sn(II) \longrightarrow CH_3Sn(IV)^+ + SAM \quad (6)$$

The methylation of the inorganic precusor affects not only its toxicity, but also its solubility, which may modify the resulting affinity for an environmental matrix [18]. Methylation or dismutation of a partially alkylated species [Eq. (7)]

$$2CH_3Hg^+ + S^{2-} \longrightarrow (CH_3Hg)_2S \longrightarrow (CH_3)_2Hg + HgS \quad (7)$$

may result in volatilization of the produced species [21]. This may lead to transport of the organometallic between environmental matrices.

Organometallic species may be produced chemically by methylating agents that are common near plant life. These may include, for example, 3-(dimethylsulfonio)propionate (DMSP) [22] and methyl iodide [23,24], which provide CH_3 as a carbo cation [25].

Methylation of mercury has been observed by numerous authors [26,27]. This process is linked to the level of microbiological activity [26] and is higher the lower the oxygen concentration [27].

Methylation of mercury by reducing bacteria may be envisaged as a detoxification process due to the more deleterious effect of inorganic Hg(II) as compared to methyl mercury, inside the cell of the bacteria [28]. The importance of biomethylation of mercury was dramatically demonstrated at Minamata, Japan, where it caused the death of over 100 subsistence fishermen who consumed fish containing methylmercury [29], i.e., detoxification for bacteria is not detoxification for all species.

In the pioneering work of Challenger [30-32] the toxic gaseous arsenic compound produced by molds growing on wallpapers containing arsenical pigments was identified as trimethylarsine. This trimethylarsine was shown to be produced from arsenic trioxide by the direct methylating action of the mold. Challenger found that methyl compounds could also be formed from oxides of selenium and tellurium, again by direct action of the molds on the inorganic precursors [30-32].

Germanium and antimony have also been suggested to undergo methylation in the environment [33-35]. The detection of methylgermanium species in natural waters in the absence of anthropogenic sources of methylgermanium is further evidence for the methylation of germanium in the natural environment [36]. Work by Challenger and Barnard [37] provides circumstantial evidence that methylated forms of antimony may be produced by mold cultures.

Methylation of tin in the environment has been postulated because methyltin compounds have been identified in unpolluted waters [38] and demonstrated experimentally by several authors [39].

Methylation of lead within the environment has been postulated by several groups [39,40]. However, studies have not always been consistent and so doubt exists here as the widespread use of methyl- and ethylleads as petrol additives clouds the issue. The finding of methyllead compounds in the environment is not usually evidence for their having been formed there [9].

In summary, it is accepted that chemical and biological pathways exist that may result in the formation of methyl derivatives of

arsenic, selenium, tellurium, mercury, germanium, and antimony from
inorganic precursors within the environment. The case for environmental methylation of tin, mercury, arsenic, and selenium appears
most convincing.

3. FATE OF ORGANOMETALLIC COMPOUNDS IN THE ENVIRONMENT

The fate of an organometallic compound in the environment may be
influenced by the factors that also initially caused its presence in
the environment. Most neutral organometallic compounds exhibit hydrophobic character, e.g., $(CH_3)_2Hg$, and consequently organometallic compounds tend to accumulate in environmental matrices with high organic
content, e.g., the surface microlayer, lipid tissue [41]. Once an
organometallic compound enters the environment, environmental factors
such as sulfur and oxygen concentration, pH, salinity, and intensity
of incident UV radiation will affect its fate [42].

   Organometallic species in the environment, whether due to
anthropogenic sources or environmental methylation, may bioaccumulate,
i.e., the concentration of the organometallic within an organism or
plant may become several orders of magnitude higher than the concentration of the organometallic within the immediate external environment [43]. This process has been shown to occur with mercury, lead,
tin, and arsenic and is covered in more detail in Sec. 9.

   In the case of biological methylation (Sec. 2) the organometallic species may then be subjected to demethylation as part of
the biogeochemical cycling of that element. This effect has been
studied for methylmercury [44] demonstrating that aerobic bacteria
predominate in demethylation in estuarine sediments. Mercury demethylation is enhanced under aerobic conditions [45] and so while an ecosystem is aerobic methylmercury will tend to be demethylated to inorganic mercury. Under aerobic conditions at pH values greater than 8
the predominant organometallic species is dimethylmercury, which is
volatile and which may be lost from aquatic systems by diffusion [46].

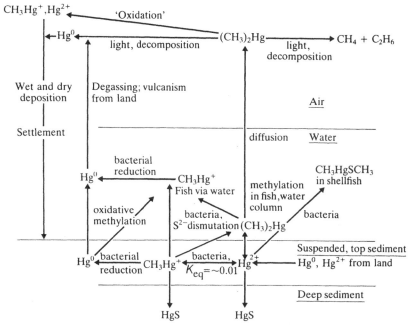

FIG. 1. Biogeochemical cycle for mercury. (Reprinted with permission from Ref. 6.)

This volatilization of the mercury species clearly results in transfer between environmental matrices.

The biogeochemical cycling of mercury species is shown in Fig. 1. Clearly the fate of organomercury compounds in the environment is a result of chemical and biological oxidations and reductions. For a more detailed treatment of the biogeochemistry of mercury, the reader is referred to Ref. 6.

With organolead compounds the primary route of entry into the environment is via their use as antiknock agents in petroleum, although this is decreasing. Lead antiknock additives are tetra-alkylleads (R_4Pb) and include tetraethyllead, tetramethyllead, and their mixed derivatives [47]. On combustion of the petroleum approximately 97-99.3% of the additive is converted to inorganic lead species,

TABLE 6

Alkyllead Compounds Detected in Environmental Samples

R_4Pb	R_3Pb^+	R_2Pb^{2+}
$(CH_3)_4Pb$	$(CH_3)_3Pb^+$	$(CH_3)_2Pb^{2+}$
$(CH_3)_3(C_2H_5)Pb$	$(CH_3)_2(C_2H_5)Pb^+$	$(CH_3)(C_2H_5)Pb^{2+}$
$(CH_3)_2(C_2H_5)_2Pb$	$(CH_3)(C_2H_5)_2Pb^+$	$(C_2H_5)_2Pb^{2+}$
$(CH_3)(C_2H_5)_3Pb$	$(C_2H_5)_3Pb^+$	
$(C_2H_5)_4Pb$		

Source: Compiled from Ref. 48.

while the remainder is emitted unchanged as the organolead compound [48]. Table 6 lists the alkyllead compounds which have been detected in environmental samples.

Organolead compounds may also enter the environment as a result of evaporative losses of fuel, from fuel tanks, or tankers, carburetors, and spillage during the production and transfer of antiknock compounds [9]. Once in the atmosphere the tetraalkylleads are subject to photolytic decomposition which may be rapid, 8% per hour for $(CH_3)_4Pb$ and 26% $(C_2H_5)_4Pb$ [49,50] depending on the conditions. This decomposition produces trialkyllead, dialkylleads and inorganic lead species as shown in Table 6.

Manufacture of tetraalkyllead has been shown to be responsible for high concentration of alkylleads in some aquatic environments [51]. These compounds and their partially dealkylated derivatives may contaminate water, accumulate in sediments, and have sufficiently long half lives to result in bioaccumulation of alkyllead species by benthos, fish, and plants. The distribution of alkyllead compounds in the aquatic environment suggests that chemical and biological transformation (i.e., hydrolysis, photolysis, biological dealkylation, methylation, and transmethylation) could occur after discharge. The predominance of trialkyl forms [48] is probably a reflection of the higher water solubility and lower volatility relative to the tetra-alkyl forms. Trialkylleads are the most toxic organoleads to mammals [52]; hence accumulation of trialkylleads in fish may represent a hazard to fish consumers.

A biogeochemical cycle for lead has been proposed and is shown in Fig. 2. Again for a more detailed treatment of organolead biogeochemistry the reader is referred to Ref. 9.

Owing to the multiplicity of uses of organotins there are numerous pathways by which they gain entry into the environment. In the case of the biocide tributyltin upon entering the aquatic environment,

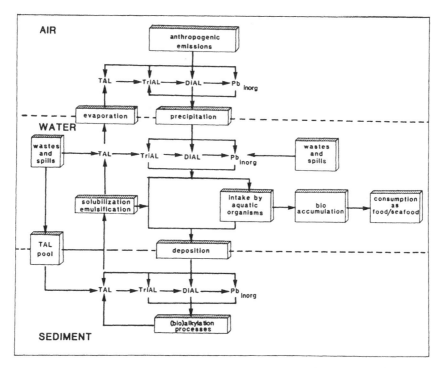

FIG. 2. Biogeochemical cycle for lead. (Reprinted with permission from Ref. 9.)

it tends to accumulate in the sediments and surface microlayer [53].
Once present in sediments tributyltin is believed to degrade via progressive debutylation steps, to dibutyl-, monobutyl-, and inorganic
tin [54].

Estimates of the half-life of tributyltin in sediment vary
widely, with reported half-lives of tributyltin in sediment ranging
from 4 months to 1.85 years in different sediments [55-57]. Presumably the nature of the sediment (i.e., anoxic or aerobic) affects the
distribution of the microflora which will have an impact on the stability of the tributyltin. Sediments having a high organic content
have a high number of natural ligands, which will reduce the bioavailability of the tributyltin species thereby increasing persistence.

Tributyltin species has been demonstrated to produce sublethal
effects against a number of nontarget organisms. This includes the
development of male characteristics in females, a phenomenon termed
imposex which has been observed in numerous species of molluscs
including dog whelks (Nucella lapillus) [58,59] and the common mud
snail (Nassarius obsoletus) [60]; these effects may occur with a
tributyltin concentration of 1 ng/liter. Shell thickening in the
oyster Crassorstrea gigas has also been demonstrated to be due to
tributyltin compounds [61], this effect being particularly marked at
1.6 µg/liter. These effects are assumed to be caused by bioaccumulation by the organisms from relatively low, i.e., ng/liter, levels
of tributyltin. The biogeochemistry of organotin is not fully understood, so that no quantitative biogeochemical cycle can be estimated.
A tentative cycle is presented in Fig. 3.

When organoarsenics are added to soil (e.g., in their use as
cotton defoliants; see Sec. 1) a portion will adsorb to the soil,
some will leach into the groundwater, and a further portion will be
volatilized, leading to transport between environmental matrices [16].

Arsenic compounds in biota form complex molecules, the so-called
hidden arsenicals, resistant to extraction for analysis, which include

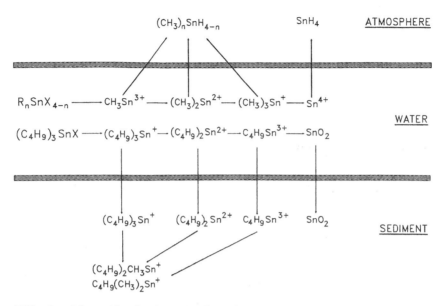

FIG. 3. Biogeochemical cycle for tin.

arsenobetaine, arsenocholine, and arsenosugars [62]. These organoarsenic species have only been detected as analytical chemistry has
improved. The biogeochemical cycling of arsenic involves not only
arsenate, the major species in seawater, but also the formation of
reduced and methylated species [63]. Arsenic intake by organisms
via food appears to be efficient; ingested organic arsenic is
retained in preference to inorganic arsenic, due to more effective
elimination of inorganic forms of arsenic. This results in a shift
in the ratio of arsenic in higher organisms toward organoarsenic
forms. Biomagnification of arsenic in food chains does not appear
to occur, with most evidence pointing to an actual biodiminution of
arsenic species [63]. The known biogeochemical cycle for arsenic
is presented in Fig. 4.

In summary, most organometallic compounds are polar and consequently in aquatic systems they accumulate in the sediments, bonded
by natural donor ligand molecules. Here they may bind to organic

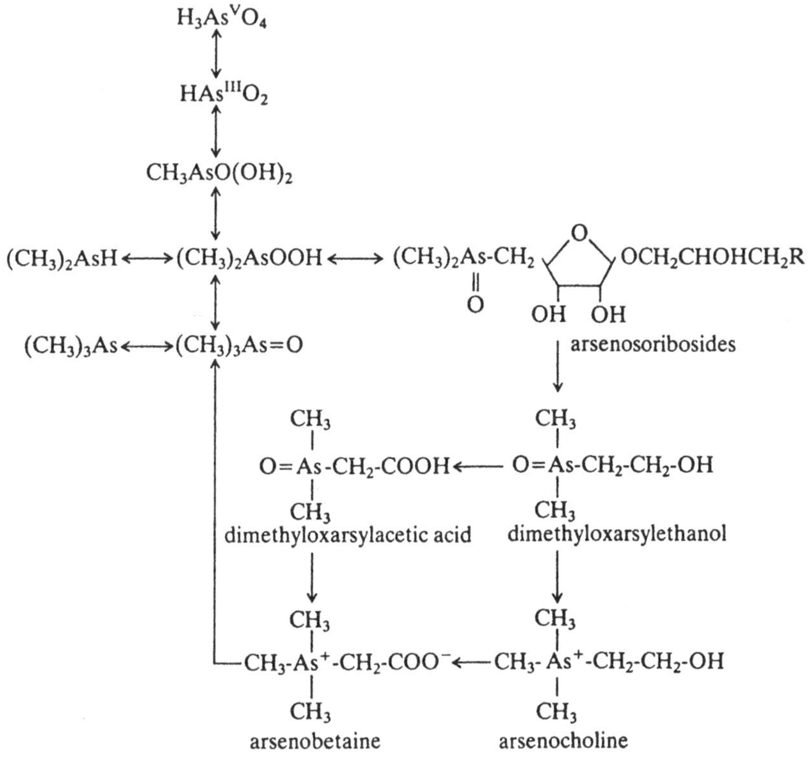

$$H_3As^VO_4$$

$$HAs^{III}O_2$$

$$CH_3AsO(OH)_2$$

$(CH_3)_2AsH \longleftrightarrow (CH_3)_2AsOOH \longleftrightarrow (CH_3)_2As\text{-}CH_2$ ⟍O⟍ $OCH_2CHOHCH_2R$

$(CH_3)_3As \longleftrightarrow (CH_3)_3As{=}O$

arsenosoribosides

$$O{=}As\text{-}CH_2\text{-}COOH \longleftarrow O{=}As\text{-}CH_2\text{-}CH_2\text{-}OH$$

dimethyloxarsylacetic acid   dimethyloxarsylethanol

$$CH_3\text{-}As^+\text{-}CH_2\text{-}COO^- \longleftarrow CH_3\text{-}As^+\text{-}CH_2\text{-}CH_2\text{-}OH$$

arsenobetaine   arsenocholine

FIG. 4. Biogeochemical cycle for arsenic. (Reprinted with permission from Ref. 8.)

substances and natural ligands, which results in reduced bioavailability until some process, e.g., disturbance, remobilizes the organometallic species. Once mobilized an organometallic species may become incorporated into an organism (Sec. 9) via the process of bioaccumulation, which is very important for filter feeders; then potentially organometallics may move through trophic levels, being bioconcentrated during the process due to their lipophillic nature.

Organometallics applied in terrestrial environments are usually chemically formulated to have susceptibility to UV radiation, which results in their degradation to inorganic forms over a period of time [7].

An understanding of the biogeochemical cycling of organometallics is well established for mercury and to some extent lead, while that of tin and arsenic is less well understood quantitatively. Little quantitative information is available on the biogeochemical cycling of other organometallic elements.

4. MAIN ORGANOMETALLIC SPECIES REQUIRING ANALYTICAL DETECTION FROM THE ENVIRONMENT

The main species requiring detection are clearly those added as anthropogenic pollutants. These include tetraalkylleads (Sec. 3), as despite new regulations limiting their use [47], substantial quantities continue to be emitted to the atmosphere. With organotins, the biocide tributyltin has also displayed decreasing use due to regulations. However, its use is continuing and the persistence of tributyltin and its degradation products in the environment imply that the quantitation of tributyltin will continue to be important. Organotin compounds are present (like other organometallic species) in significant quantities in plants and sediments in regions adjacent to use (Sec. 7) and so remobilization of this significant sink of organotin is possible, and could lead to continuing environmental problems even following a complete ban on the use of tributyltin as the biocide component of antifouling paints [57]. Organomercury compounds continue to be of interest due to their natural formation (Sec. 2), their ability to bioaccumulate, and the high toxicities displayed by these compounds [64].

The detection and speciation of other organometallic species, i.e., organoarsenic, organogermanium, and organoantimony, is also of great importance as the biogeochemical cycles of these elements have yet to be fully elucidated. Elucidation of biogeochemical cycles of elements is important as it is an essential precursor to assessing and quantifying the impact of anthropogenic pollution by metals and metalloids.

5. SAMPLING METHODOLOGIES

5.1. Introduction

The sampling methods employed will clearly be affected by the nature
of the sample. The sampling strategies used for the detection of
organometallic samples in air, water, sediment, and biological samples
will be discussed in Secs. 5.2-5.5.

5.2. Air Samples

In the case of airborne organolead compounds the concentrations
typically encountered are in the ng/m^3 range (expressed as lead)
with, e.g., 1 ng/m^3 observed at a rural location, with a maximum
observed concentration of 5400 ng/m^3 being observed in a multistorey
car park [9]. Environmental samples require preconcentration prior
to analysis (see Sec. 6).

   Arsines in air have been selectively and quantitatively sampled
using columns containing small glass beads coated with gold or silver.
The trapped arsenic species were eluted as their oxidation products
with potassium hydroxide solution [65]. The quantitative sampling of
arsine using coconut shell active carbon with a sodium carbonate-
impregnated filter [66] and with synthetic resin active carbon has
also been reported [67].

   Most of the mercury found in the atmosphere is mercury(0) or
methylmercury [68]. Sewage treatment facilities are an important
source of airborne mercury(0) and organomercurials [69]. Air sampling
for mercury generally involves trains of absorbents containing a
variety of solvents for the selective collection of the different
mercury species [70].

   Limited studies have been conducted on the analysis of airborne
organotins although the sampling of airborne dibutyltin and dioctyltin
chlorides by glass fiber filters has been achieved [71].

5.3. Water Samples

With water samples a critical point is that of contamination via the material used for sampling and/or storage. In the case of total mercury analysis Bloom [72] used borosilicate glass or Nalgene brand FEP, both of which have been cleaned by a hot oxidation technique [72].

In the analysis of germanium compounds contamination is not regarded as a serious problem due to the fact that germanium does not occur at high concentrations in everyday materials [73]. Samples are acidified to pH 2 (with 4 ml concentrated HCl per liter of sample) at the time of sampling or within an hour of collection, and have been stored for several years with no changes in germanium concentration. Samples have been collected and stored in polyethylene bottles rinsed with acid and water.

With organotin determinations, use of polyvinylchloride (PVC) containers may lead to contamination of the sample as the plastic contains a tin additive. Hence the use of PVC should be avoided. Samples may be collected with polycarbonate, polyethylene, or Teflon samplers. Ideally samples should be immediately filtered and acidified to pH 2 with nitric acid [74].

Water samples for analysis of organolead compounds have been taken using 4-liter Winchester bottles just below the surface [75]. No pretreatment of the samples was reported in this case.

Storage experiments conducted on μg/liter concentrations of arsenic species have shown that methylated arsenic species are stable for months if water samples containing them are frozen or preserved by the addition of 4 ml of concentrated hydrochloric acid per liter of sample. Preservation of arsenic(III) species is more difficult and can only be satisfactorily achieved by rapid freezing in liquid nitrogen followed by storage at >-80°C [76].

With selenium samples, immediate analysis is recommended due to the difficulties of storing samples containing methyl species of the element [77].

5.4. Sediment Samples

Sediment samples are usually removed from the surfical sediments
using a grab mechanism. Ideally these sediments will then be
filtered via a 63-μm filter and freeze-dried. This will arrest
biological activity and remove small stones, etc., from the sample,
thereby allowing a more realistic comparison to be achieved between
sites [78].

5.5. Biological Samples

These are divided into two major categories of organisms: (1) mobile
and (2) sedentary. The latter provides a biological integration of
concentrations of organometallics in the environment at a specific
location. Mobile species may also be used as biological indicators
of organometallic pollution, but here one is dealing with organometallic levels on the ecosystem level rather than at specific
locations.

Sedentary organisms are often collected by divers, whereas
fish samples may be collected using gill nets [75]. Freeze drying
is the preferred method of preservation of biological samples [78].

6.   DERIVATIZATION AND PRECONCENTRATION OF ORGANOMETALLIC COMPOUNDS

6.1. Derivatization

Most derivatization techniques allow in principle the quantitative
transformation of polar organometallic compounds to either increase
or decrease their volatilities and render them suitable for gas
chromatography (GC) or high-performance liquid chromatography (HPLC)
separation [74]. The general assumption is that the organometallic
group remains intact.

6.2. Derivatizing Agents

6.2.1. Sodium Borohydride (NaBH$_4$; Hydride Generation)

The reaction of organometallics with aqueous solutions of NaBH$_4$ to produce the hydride derivative of the organometallic has been extensively studied and utilized in analytical organometallic chemistry. Organometallic forms of arsenic, selenium, germanium, tin, mercury, antimony, and tellurium have been determined by hydride generation [79-85]. Hydride generation has the advantages of several orders of magnitude improvement in concentration sensitivity over conventional nebulizer sample introduction [83].

Hydride generation has been shown to be subject to several interferences, e.g., sulfides, oils. Other metals present in the sample have also been reported to interfere with the generation process [86,87]. In natural water samples this is unlikely to be a problem owing to the low concentrations of such species, but in polluted solid matrices such as sediments significant interference may occur [88]. Also in sediments interferences due to diesel oil and sulfur compounds have been implicated [89].

6.2.2. Sodium Tetraethylborate (NaBEt$_4$; Ethylation)

Owing to the various interferents in hydride generation, a number of workers have investigated the use of sodium tetraethylborate as a derivatizing agent [90,91]. Ashby and Craig [90] found that ethyl derivatives of butyltin extracts from sediments could be formed, e.g., Bu$_n$SnEt$_{4-n}$ ethylation of the extracted butyltin compounds does not appear to suffer as much from interferents present in sediment extracts. Ethylation is a relatively new method of derivatization, although it has also been shown to produce near-quantitative ethylated forms of inorganic and organic lead [91] and ethylated forms of inorganic and organic mercury [92].

6.2.3. *Grignard Reagents (Alkylation)*

Alkylation by Grignard reagents has been reported by numerous authors [93-99]. Here the extracted organometallic species are reacted with a suitable Grignard reagent RMgX (whre R is an alkyl group and X is a halogen) in a complex series of stages, to produce the alkylated organometallic species.

The reaction of Grignard reagents with organometallic species has been used to generate the methylated species [97,99], the ethyl-ated species [94,98], and the pentylated species [95,96]. It is therefore an alternative method of derivatization which, unlike those described in Secs. 6.2.1 and 6.2.2, is capable of generating a variety of products. Derivatization by Grignard reagents is useful when one wishes to determine structural information which the other derivatiza-tion techniques would not reveal. The derivatization reaction using Grignard reagents is a complex series of steps which may involve loss of analyte. The derivatization step used within a given laboratory is often determined by the ease with which it is accomplished there. Consequently hydride generation (Sec. 6.2.1) and ethylation (Sec. 6.2.2) tend to be the preferred methods of derivatization.

6.3. Sample Preconcentration

Air samples are usually preconcentrated by passing a known volume of air through a glass column (filled with a suitable chromatographic packing material) which is submerged in liquid nitrogen. This cryo-genic trapping is therefore achieved at -196°C and has been success-fully used in the determination of alkyllead compounds [9].

The use of cryogenic trapping with atomic absorption detection may also be described as a method of preconcentration when applied to aqueous samples. This is due to the fact that all the volatile organometallic analytes from a sample are preconcentrated onto the cryogenic column by the derivatization of the organometallic with either sodium borohydride or sodium tetraethylborate.

Solvent extraction of organometallic compounds [90] followed by reduction in solvent volume via evaporative methods is the other most common method for preconcentrating organometallic compounds.

6.4. Detection and the Interfacing of Detection and Delivery Systems

6.4.1. Introduction

The speciation of organometallic compounds is of great importance owing to the different toxicities shown by the different chemical forms of different organometallics of the same element found in the natural environment [12]. Analytical speciation is a two-stage process which involves separation (Sec. 6.4.2) followed by detection (Sec. 6.4.4) of the organometallic species. The apparatus used for separation and detection are linked by an interface (Sec. 6.4.3).

6.4.2. Separation of Organometallic Compounds

Numerous chromatographic methods have been used to separate organometallics; these are summarized in Table 7 [100-108].

Gas chromatographic separations usually rely on the use of an initial derivatization step. In the case of packed column and capillary column gas chromatography the volatile organometallic derivatives then show differing affinities for the mobile and stationary phases which results in their separation [100-102]. Separation is more efficient in capillary columns. With cryogenic gas chromatography the derivatized organometallics are initially swept into a liquid nitrogen-cooled U tube (i.e., -196°C) containing a suitable chromatographic phase where they are frozen onto the stationary phase. The organometallics are then desorbed by electrothermal heating of the U tube. This leads to elution of the species in order of boiling points [78]. Use of high performance liquid chromatography (HPLC) eliminates the need for a derivatization step as described in Sec. 6. The separation of the organometallics is again due to the different affinities that they show for the mobile and stationary phases [104].

TABLE 7

Common Separation Methodologies Used in Analytical Organometallic Chemistry

Method of separation	Typical stationary phase	Typical element/ carrier gas	Indicative references
Packed column gas chromatography (GC)	3% OV101 Chromosorb WHP, 80-100 mesh	Mercury, tin, lead/ Nitrogen	17,90,97,100,101
Capillary column gas chromatography (CGC)	Immobilized methylsilicone	Mercury, tin, lead/ Nitrogen	94,98,99,102,103
Cryogenic trap chromatography (Cry GC)	OV101 on 10% Chromosorb G AW-DMCS	Tin/Helium	40,48,55,62,78
High-performance liquid chromatography (HPLC)	Partisil-10-SCX (250X 4.6 mm id) column	Arsenic/ 80:20 methanol:water in 0.1 M NH_4OAc	104-108

6.4.3. Interfacing

The interfacing of certain detection systems, e.g., gas chromatography
(GC) with mass spectrometry (MS), GC-flame ionization detector (FID),
GC-flame photometric detector (FPD), and GC-electron capture detector
(ECD), is carried out commercially in the manufacture of the instru-
ment.

Other interfaces, e.g., GC-atomic absorption spectroscopy (AA),
high-performance liquid chromatography (HPLC), or graphite furnace
atomic absorption (GFAA), have been constructed within various laboratories. A typical interface between GC-AA is described by Clark and
Craig [100a]. Typically it consists of an insulated narrow bone
(0.76 mm), electrothermally heated stainless steel tube which conveys
the element from the GC into the quartz furnace of the AA. This is
not ideal as there is no stationary phase in the transfer line between
the GC and the AA. Consequently there is loss of chromatographic
equilibrium leading to zone broadening, resulting in a partial loss
of separation. A capillary GC-AA system has been built which utilizes
the electrothermally heated tube to heat a portion of the capillary
column which is uncoiled and threaded down the transfer line to maintain chromatographic integrity from injector to detector [100b].

An interface between HPLC and GFAA may be constructed by diverting aliquots of the eluent from the HPLC into an autosampler. These
aliquots are then analyzed by GFAA. The drawback of this type of
system is the use of a discrete sampling system to analyze a continuously flowing sample.

Interfaces between cryogenic trapping GC-AA systems have been
constructed by attaching the eluent from the U tube directly to the
quartz furnace in the AA [78].

6.4.4. Detection of Organometallic Species

Organometallic compounds are defined as possessing at least one
covalent metal-to-carbon bond; hence the species are a mixture of
metallic, carbonaceous, and (depending on the metallic valency and

TABLE 8

Main Detectors Used in Organometallic Detection

Detection system	Responds to ligands in R_nML_m	Responds to metal in R_nML_m	Separation method	Indicative references
Mass spectrometry	Yes	Yes	Capillary gas chromatography	94,100,101,103,109
Quartz furnace atomic absorption spectroscopy	No	Yes	Packed column gas chromatography Capillary column gas chromatography Cryogenic trap gas chromatography	90,100,101,103 40,48,55 62,91
Graphite furnace atomic absorption spectroscopy	No	Yes	Cryogenic trap gas chromatography High-performance liquid chromatography	85 105-108
Flame photometric detection	No	Yes	Capillary gas chromatography Packed column gas chromatography	94,98,99-110 17,97
Electron capture detection	Yes	No	Capillary gas chromatography Packed column chromatography	111 112-115
Flame ionization detection	Yes	No	Packed column gas chromatography	25,39,101
Atomic fluorescence	Yes	No	Cryogenic trap gas chromatography	92

degree of alkyl/aryl substitution) inorganic groupings. Consequently, a range of detectors are capable of organometallic detection, which may respond to one (or more) of the three groups that constitute an organometallic molecule (i.e., metal, organic groups, and inorganic groups).

The main detectors used in organometallic detection are listed in Table 8 [105-115]. The response of each detector to the metal and the ligand(s) is also presented as is the method of separation utilized with the detection system. The principles of each detector will now be briefly discussed.

Mass spectrometry separates a mixture of gaseous ions, according to their mass-charge (M/Z) ratios. A mass spectrum is a plot of relative pressure or concentration of the gaseous components as a function of the mass-charge ratios of the components.

Mass spectrometry may be used either qualitatively to determine the chemical structure of compounds, by comparing the mass spectrum of the sample with those of standards recorded under similar experimental conditions, or quantitatively by comparing the intensity of a single peak from the spectrum of the sample with that of a series of standards [116]. Unknown compounds may be identified a priori from the spectrum. Due to the expense of mass spectrometry the technique is generally used to confirm the chemical structure of compounds, which are usually determined only by retention times using any of the techniques listed in Table 8.

Atomic absorption spectrometers are conventionally used to detect solutions of metals in the elemental state via their characteristic atomic absorption spectra. Conventionally an atomic absorption spectrometer is used with a burner system which utilizes natural gas, propane, hydrogen, or acetylene as fuel and air, oxygen, or nitrous oxide as an oxidant. Liquid samples are aspirated into the flame as small droplets. The heat of the flame causes the solvent in the solution to evaporate rapidly, leaving behind the solid particles of solute. These particles then melt to form a liquid, evaporate to yield a gas, and dissociate into atoms. Radiation from the lamp

(which is specific to the element being determined) passes through
the flame and is partially absorbed by the sample atoms. The amount
of absorption is monitored with the detector.

Atomic absorption spectrometry in organometallic detection is
often used in conjunction with an electrothermally heated quartz
furnace, placed in the light path of the instrument. Both commercially available and laboratory-constructed quartz furnaces replace
the burner head of the instrument. Such furnaces are designed to fit
into the burner support in order to utilize the horizontal and lateral
optical adjustment of the instrument which facilitates optimum posi-
tioning of the furnace within the light beam of the instrument. The
quartz furnace is used to increase the residence time of the organometallic species in the light beam of the AA, which gives a significant increase in the sensitivity achieved when organometallics are
eluted (following chromatographic separation) into a conventional
atomic absorption spectrometry, utilizing traditional flame atomization. Atomization within a quartz furnace may be achieved at 1000°C
in a hydrogen-rich hydrogen/air flame [117]. GC-AA is a relatively
inexpensive, fairly robust technique and is the preferred choice for
many laboratories.

In GFAA spectroscopy the atomization is achieved in a graphite
furnace which is heated in a temperature program. This consists of
a drying stage at 110°C for 30 sec (to remove solvent) followed by
an ashing stage typically at a temperature of 350-1200°C for a period
of 45 sec. This stage removes volatile components and destroys the
chemical matrix of the sample. The final step is the atomization
stage in which the temperature is increased to a point at which discrete (gaseous) atoms of the analyte are formed. This atomization
stage occurs at a temperature, between 2000°C and 3000°C for a period
of approximately 5 sec. The absorption of the analyte is measured
during the atomization stage. Graphite furnace atomic absorption is
an extremely useful tool for elements, such as germanium, which form
extremely stable oxides in quartz furnaces and hence demonstrate poor
sensitivity via this technique. This is overcome by the use of the

graphite furnace method where the higher temperatures utilized permits atomization of the samples [85,105-108].

Flame photometric detection (FPD) uses a hydrogen air flame and a photomultiplier tube to measure the radiation emitted by an analyte within the flame. Generally the FPD is used to detect compounds containing sulfur or phosphorus. The FPD can be made element-selective by monitoring only radiation emitted at a characteristic wavelength of a particular element. This is achieved using the appropriate optical filter placed between the flame and the photo-multiplier tube. The use of such filters has enabled organo-metallic compounds which contain metallic atoms capable of being excited in a hydrogen air flame to be determined quantitatively by gas chromatography with flame photometric detection [99]. FPDs have excellent sensitivity, have been reported to be poisoned by various compounds, but are used extensively.

An electron capture detector (ECD) contains a radioactive β (e$^-$) particle emitting ^{63}Ni source and is commonly used with nitrogen carrier gas. The β particles ionize the nitrogen carrier gas [Eq. (8)]:

$$N_2 + 2e^- \longrightarrow 2N^+ + 4e^- \qquad (8)$$

A polarizing voltage is applied between the body of the detector and an insulated collector electrode. Under the influence of this polarizing voltage the electrons released by the ionization of the carrier gas are collected at the collector electrode and give rise to a current of the order of 1 nA, and this is amplified by the detector electronics.

In the simplest form of an ECD the polarizing voltage is provided by a steady direct current supply. The elution of an electron capturing analyte, i.e., nitro groups, phosphorus, oxygen, and halo-gens, into an ECD causes some electrons to be captured, resulting in a reduction in the standing current in the detector which is detected as "peak". More sophisticated ECDs utilize a system where the pulse frequency of the polarizing voltage is modulated to maintain a constant current in the detector (even during the elution of electron-

capturing species). Here the increase in frequency of pulse necessary to maintain the constant current, during the elution of an
electron capturing species, is recorded as a "peak". The ECD is a
highly sensitive detector. However, it is non-species- and non-
element-specific and has a limited linear range. Consequently, it
is not used as extensively as some other detectors [111-115].

A flame ionization detector (FID) typically burns 30-40 cm^3/min
of hydrogen in 250-400 cm^3/min of air as a small flame. A small jet
(to which a polarizing voltage of typically 160 V is applied) allows
the carrier gas to enter the flame. When only carrier gas enters the
flame, very little current (i.e., 1×10^{-11} A) flows between the
electrodes. When carbonaceous species elute into the flame they are
ionized, producing an increase in current which is then displayed as
a peak [101].

The FID is a non-species- and non-element-specific detector.
Burning an organometallic in the hydrogen/air flame causes deposition
of the metal oxide onto the detector. Continued use of an FID for
organometallic detection results in a gradual loss of sensitivity due
to the accumulation of metal oxide. Consequently, FID has not
achieved very widespread use as an organometallic detector.

The main application of atomic fluorescence has been the detection of inorganic and organic mercury samples [118]. Mercury vapor
(following a reduction of organometallic species) is fed into the
fluorescence detector using a chimney-type interface in a stream of
argon gas. An intense mercury source (a boosted discharge hollow
cathode lamp) is used to excite the mercury atoms in the vapor and
the fluorescence is measured with a photomultiplier tube [118].
Fluorescence offers several orders of magnitude improvement in
detection limits as compared to atomic absorption detection.

7. ORGANOMETALLIC COMPOUNDS IN SEDIMENTS AND SUSPENDED MATTER: CONCENTRATIONS AND CONSEQUENCES

The concentrations of organometallic compounds in sediments has
generally received less attention than the determination of concen-

TABLE 9

Some Concentration Values for Organometallic
Compounds in Sediments

Sample type			Ref.
Mercury Compounds	Inorganic	Methylmercury Organic	
Polluted sediments	1-20 µg/g	0.01-0.6 µg/g	6
Unpolluted sediments	0.2-0.4 µg/g	0	6
Organotin Compounds (polluted sediments)			
Tributyltin 1.4 µg/g	Dibutyltin 2.2 µg/g	Monobutyltin 0.059 µg/g	119-121
Triphenyltin 107 µg/g	Diphenyltin 171 µg/g	Monophenyltin 68 µg/g	122
Trimethyltin 0.027 µg/g	Dimethyltin 0.042 µg/g	Monomethyltin 0.069 µg/g	122

tration of organometallics in biota (Sec. 9) and in the dissolved
state (Sec. 8). This is presumably due to the difficulties of
extraction and the interferences reported with samples from this
environmental matrix. For typical descriptions of treatment of
organometallics in sediments, the reader is referred to Refs. 6-9,
47, 64, 74, and 76.

Concentrations of some organometallic compounds are given in
Table 9 [119-122]. The consequences of organometallic compounds in
the sediments depends on the ultimate fate of the compound (Sec. 3).
Sediments are, however, undeoubtedly a sink for organometallics under
some circumstances, e.g., following anthropogenic input. Sediments
are a source in other circumstances also, e.g., following mechanical
disturbance due to boating activity or biological mobilization, etc.,
leading to methylation [18] or bioaccumulation [43].

TABLE 10

Some Levels of Organometallic Compounds in the Dissolved State

Sample type			Ref.
Organogermanium Compounds			
	Monomethylgermanium	Dimethylgermanium	Trimethylgermanium	
Ocean:	0.4-12.3 ng/liter	0-11.5 ng/liter	0-1.0 ng/liter	85
Organotin Compounds			
	Monobutyltin	Dibutyltin	Tributyltin	
Harbor:	0-12.6 ng/liter	1.6-67 ng/liter	0-66 ng/liter	123
	—	240-280 ng/liter	27-93 ng/liter	124
Methylmercury			
Freshwater:	3.96-<0.002 ng/liter			92
Ocean:	0.088 ng/liter			6

There are few reports of organometallic concentrations in
suspended matter and as suspended matter ultimately becomes part
of the sediment no separate consideration will be given here.

8. ORGANOMETALLICS IN THE DISSOLVED STATE

The speciation of organometallics in the dissolved state has received
more attention than the speciation of organometallics in sediments,
presumably due to the relative simplicity of the sample matrix.
Detection of organometallics in the dissolved state often involves
preconcentration of the sample due to the low levels (i.e., ng/liter)
of organometallics that are encountered in environmental samples.
More comprehensive treatment of organometallic levels in the dissolved
state is given in Refs. 6-9, 47, 64, 74, and 76.

Table 10 lists typical concentrations of some organometallics
observed in aqueous samples [123,124]. Clearly the levels of organometallics observed in natural waters are 3-4 orders of magnitude less
than those observed in sediments (Sec. 7). This is to be expected
due to their generally lipophilic nature.

9. ORGANOMETALLICS IN BIOTA

The determination of organometallics in biota has received considerable attention over the years. This is presumably due to the human
consumption of biota contaminated with organometallics, the consequences of which were tragically demonstrated at Minamata (Japan)
[29] where over 100 people died as a result of eating fish containing methylmercury (Sec. 2). The effects of the organotin biocide
tributyltin on the commercially important oyster *Crassostrea gigas*
[61] also produced a wealth of data on the levels of organotins in
biota. Levels of organometallics observed in biota are given in
Table 11 [125-128]. The reader is referred to Refs. 6-9, 47, 64,

TABLE 11

Some Concentrations of Organometallics in Biota

Sample type				Ref.
Organotin Concentrations (ng/g)				
	Tributyltin	Dibutyltin	Monobutyltin	
Oyster samples	350-1.6	19	/	125
Fish liver samples	100	8.70	55	109
Organomercury Compounds (mg/kg)				
	Total mercury	Methylmercury		
Fish muscle	0.17-0.21	0.17-0.23		126
Seafood samples	1.37-0.07	0.535-0.11		127
Alkyllead Compounds (µg/kg)				
Fish samples	Muscle	Carcass		51
	128-3585	279-1716		
Organoarsenic Compounds (µg/g)				
	Monomethylarsenic	Dimethylarsenic (trimethylarsenic)		
Fish samples				128
Muscle	0.05	0.01	(2.01)	
Viscera	0.16	0.16	(2.43)	

74, and 76 for a more comprehensive treatment of levels of organometallics in typical biota.

10. CONCLUSIONS

This chapter has attempted to present an overview of a large field
of research in a narrow space, and this has forced concise discussion

of the subject. Organometallic species are continuing to be introduced into environmental matrices due to anthropogenic input (Secs.
1 and 4). There is also little doubt that certain elements, e.g.,
mercury, are methylated within the environment (Sec. 2). Consequently, there is a need for detection of organometallic species
in the environment. Sections 5 and 6 attempted to present an overview of the commonly used analytical techniques utilized in organometallic detection.

Speciation of organometallics will normally involve a chromatographic separation followed by detection with a range of detection of
varying specificity, which will be determined partly by the running
costs and the availability of analytical instrumentation within
laboratories. The use of HPLC with MS [128] will reveal more information about the chemical form of organometallics in the environment,
specifically the nature of the anionic group, which is not usually
detected following derivatization (Sec. 6). This technique may also
prove invaluable in the elucidation of biogeochemical cycles of
organometallic elements, which is an essential precursor to the
assessment of the impact of anthropogenic metal/metalloid pollutants.

ACKNOWLEDGMENTS

The authors are grateful to the following organizations for funding
in recent years: SERC and NERC (UK), Royal Society (UK), NAB (UK),
and the British Council (UK).

ABBREVIATIONS

AA	atomic absorption
CGC	capillary column gas chromatography
DC	direct current
DMSP	dimethylsulfoniopropionate
ECD	electron capture detector

FID	flame ionization detector
FPD	flame photometric detector
GC	gas chromatography
FGAA	graphite furnace atomic absorption spectroscopy
HPLC	high-performance liquid chromatography
LD_{50}	lethal dose to 50% of test species
MS	mass spectroscopy
M/Z	mass to charge ratio
PVC	polyvinylchloride
SAM	*S*-adenosylmethionine
UV	ultraviolet

REFERENCES

1. S. K. Bailey and I. M. Davies, *Sci. Tot. Environ., 76,* 185 (1988).

2. T. A. Jackson, *Appl. Organomet. Chem., 3,* 1 (1989).

3. A. G. Davies and P. J. Smith, in *Comprehensive Organometallic Chemistry* (G. Wilkinson, ed.), Pergamon Press, Oxford, 1982.

4. G. A. Senich, *Polymer, 23,* 1385 (1982).

5. M. A. Champ and W. L. Pugh, "Tributyltin Antifouling Paints: Introduction and Overview, in *Proc. Organotin Symposium of the Oceans '87 Conference,* Halifax NS, Canada, 1987, Vol. 4, IEEE, New York, p. 1297, 1987.

6. P. J. Craig, Organomercury Compounds in the Environment, in *Organometallic Compounds in the Environment* (P. J. Craig, ed.), Longman Group, Harlow, 1986, pp. 90-95.

7. S. J. Blunden and A. Chapman, Organotin Compounds in the Environment, in *Organometallic Compounds in the Environment* (P. J. Craig, ed.), Longman Group, Harlow, 1986, pp. 111-123.

8. M. O. Andreae, Organoarsenic Compounds in the Environment, in *Organometallic Compounds in the Environment* (P. J. Craig, ed.), Longman Group, Harlow, 1986, pp. 198-212.

9. C. N. Hewitt and R. M. Harrison, Organolead Compounds in the Environment, in *Organometallic Compounds in the Environment* (P. J. Craig, ed.), Longman Group, Harlow, 1986, pp. 160-197.

10. J. S. Thayer and F. E. Brinckman, in *Advances in Organometallic Chemistry,* Vol. 20 (F. G. A. Stone and R. West, eds.), Academic Press, New York, 1982, p. 313.

11. A. G. Howard and S. D. W. Comber, *Appl. Organomet. Chem.*, *3*, 509 (1989).

12. S. Nicklin and M. W. Robson, *Appl. Organomet. Chem.*, *2*, 487 (1988).

13. L. Randall and J. H. Weber, *Sci. Tot. Environ.*, *57*, 191 (1986).

14. J. J. Cleary and A. R. D. Stebbing, Organotins in the Water Column: Enhancement in the Surface Microlayer, in *Proc. Organotin Symposium of the Oceans '87 Conference*, Halifax, Nova Scotia, Canada, Vol. 4, IEEE, New York, 1987, pp. 1405.

15. J. J. Cooney and S. Wuertz, *J. Indust. Microbiol.*, *4*, 375 (1989).

16. W. R. Cullen and K. J. Reimer, *Chem. Rev.*, *89*, 713 (1989).

17. T. Hamasaki, H. Nagase, T. Sato, H. Kito, and Y. Ose, *Appl. Organomet. Chem.*, *5*, 83 (1991).

18. J. S. Thayer, *Appl. Organomet. Chem.*, *3*, 123 (1989).

19. J. R. Ashby and P. J. Craig, *Sci. Tot. Environ.*, *73*, 127 (1988).

20. Y. K. Chau, P. T. S. Wong, C. A. Mojesky, and A. J. Carty, *Appl. Organomet. Chem.*, *1*, 235 (1987).

21. G. E. Parris, W. R. Blair, and F. E. Brinckman, *Anal. Chem.*, *49*, 378 (1977).

22. R. H. Reed, *Mar. Biol. Lett.*, *4*, 173 (1983).

23. P. M. Gschwend, J. K. MacFarlane, and K. A. Newman, *Science*, *227*, 1033 (1985).

24. S. L. Manley and M. N. Dastor, *Limnol. Oceanogr.*, *32*, 709 (1986).

25. D. S. Lee and J. H. Weber, *Appl. Organomet. Chem.*, *2*, 435 (1988).

26. I. Andersson, H. Parkman, and A. Jernelöv, *Limnologica (Berlin)*, *20*, 347 (1990).

27. R. P. Mason and W. F. Fitzgerald, *Nature*, *347*, 457 (1990).

28. H. S. Pan-Hou and N. Imura, *Arch. Microbiol.*, *131*, 176 (1982).

29. A. Kudo and S. Miyahara, *Wat. Sci. Technol.*, *23*, 283 (1991).

30. F. Challenger, *J. Soc. Chem. Ind.*, *5*, 657 (1935).

31. F. Challenger, *Chem. Rev.*, *36*, 657 (1945).

32. F. Challenger, *Quart. Rev.*, *9*, 225 (1955).

33. L. Bertilsson and H. Y. Neujahr, *Biochemistry*, *10*, 2807 (1971).

34. H. Yamamoto and T. Yakoyama, *Bull. Chem. Soc. Jpn.*, *48*, 844 (1975).

35. G. Topping and I. M. Davies, *Nature*, *290*, 243 (1981).

36. J. S. Thayer, *Appl. Organomet. Chem.*, *1*, 227 (1987).

37. P. Barnard, Ph.D. thesis, University of Leeds, UK, 1947.

38. R. J. Maguire, R. J. Tkacz, Y. K. Chau, G. A. Bengert, and
     P. T. S. Wong, *Chemosphere*, *15*, 253 (1986).

39. S. Rapsomanikis, O. F. X. Donard, and J. H. Weber, *Appl. Organomet Chem.*, *1*, 115 (1987).

40. R. Harrison and A. G. Allen, *Appl. Organomet. Chem.*, *3*, 49 (1989).

41. J. J. Cleary and A. R. D. Stebbing, *Mar. Poll. Bull.*, *18*, 238 (1987).

42. G. Compeau and R. Bartha, *Appl. Environ. Microbiol.*, *48*, 1203 (1984).

43. G. E. Batley, C. Fuhua, C. I. Brockbank, and K. J. Flegg, *Aust. J. Mar. Freshwater Res.*, *40*, 49 (1989).

44. R. S. O. Oremland, C. W. Culbertson, and M. R. Winfrey, *Appl. Environ. Microbiol.*, *57*, 130 (1991).

45. O. Regnell and A. Tunlid, *Appl. Environ. Microbiol.*, *57*, 789 (1991).

46. T. R. Stolzenburg, R. R. Stanforth, and D. G. Nichols, *J. Am. Water Wks Assoc.*, *78*, 45 (1986).

47. M. Radojevic, in *Environmental Analysis Using Chromatography Interfaced with Atomic Spectroscopy* (R. M. Harrison and S. Rapsomanikis, eds.), Ellis Harwood, Chichester, 1989, p. 223.

48. C. N. Hewitt and M. Rashead, *Appl. Organomet. Chem.*, *2*, 95 (1988).

49. R. M. Harrison and D. P. H. Laxen, *Environ. Sci. Technol.*, *12*, 1385 (1978).

50. A. W. P. Jarvie, R. N. Markall, and H. R. Potter, *Environ. Res.*, *25*, 241 (1981).

51. P. T. S. Wong, Y. K. Chau, J. Yaromich, P. Hodson, and M. Whittle, *Appl. Organomet. Chem.*, *3*, 59 (1989).

52. P. Grandjean and T. Nielsen, *Residue Rev.*, *72*, 97 (1979).

53. L. W. Hall, *Mar. Poll. Bull.*, *19*, 431 (1988).

54. D. Aldeman, K. R. Hinga, and M. E. Q. Pilson, *Environ. Sci. Technol.*, *24*, 1027 (1990).

55. P. F. Seligman, J. G. Grouboug, A. O. Valkirs, P. M. Stang, R. Fransham, M. O. Stallard, B. Davidson, and R. F. Lees, *Appl. Organomet. Chem.*, *3*, 31 (1989).

56. R. J. Maguire and R. J. Tkacz, *J. Agric. Food Chem.*, *33*, 947 (1985).

57. S. J. deMora, N. G. King, and M. C. Miller, *Environ. Technol. Lett.*, *10*, 901 (1989).

58. G. W. Bryan, P. E. Gibbs, L. G. Hummerstone, and G. R. Burt, *J. Mar. Biol. Assoc. UK*, *66*, 611 (1986).

59. I. M. Davies, S. K. Bailey, and D. C. Moore, *Mar. Poll. Bull.*, *18*, 400 (1987).

60. B. S. Smith, *J. Appl. Toxicol.*, *1*, 22 (1981).

61. M. J. Waldock and J. E. Thain, *Mar. Poll. Bull.*, *14*, 411 (1983).

62. A. G. Howard and S. D. W. Comber, *Appl. Organomet. Chem.*, *3*, 509 (1989).

63. W. Maher and E. Butler, *Appl. Organomet. Chem.*, *2*, 191 (1988).

64. S. Rapsomanikis, in *Environmental Analysis Using Chromatography Interfaced with Atomic Spectroscopy* (R. M. Harrison and S. Rapsomanikis, eds.), Ellis Harwood, Chichester, 1989, p. 299.

65. R. S. Braman, in *Arsenical Pesticides* (E. A. Woolson, ed.), American Chemical Society, Washington, D.C., 1975, Chap. 8.

66. R. J. Corstello, P. M. Eller, and R. D. Hull, *Am. Industr. Hyg. Assoc. J.*, *44*, 21 (1983).

67. Y. Matsumura, M. Ono-Ogasawara, and M. Furuse, *Appl. Organomet. Chem.*, *5*, 71 (1991).

68. B. H. Belliveau and J. T. Trevors, *Appl. Organomet. Chem.*, *3*, 283 (1989).

69. A. O. Summers and S. Silver, *Ann. Rev. Microbiol.*, *32*, 637 (1978).

70. R. S. Braman and D. L. Johnson, *Environ. Sci. Technol.*, *8*, 996 (1974).

71. S. Vainiotalo and L. Hayri, *J. Chromatogr.*, *523*, 273 (1990).

72. N. S. Bloom and E. A. Crecelius, *Mar. Chem.*, *14*, 49 (1989).

73. M. O. Andreae and P. N. Froelich, *Anal. Chem.*, *53*, 287 (1981).

74. O. F. X. Donard and R. Pinel, in *Environmental Analysis Using Chromatography Interfaced with Atomic Spectroscopy* (R. M. Harrison and S. Rapsomanikis, eds.), Ellis Harwood, Chichester, 1989, p. 189.

75. P. T. S. Wong, Y. K. Chau, J. Yaromich, P. Hudson, and M. Whittle, *Appl. Organomet. Chem.*, *3*, 59 (1989).

76. E. A. Crecelius, N. S. Bloom, C. E. Cowan, and E. A. Jenne, *Speciation of Selenium and Arsenic in Natural Waters and Sediments, Vol. 2, Arsenic Speciation*, E.P.R.I. Battelle Northwest Laboratories, Washington, 1986.

77. G. A. Cutter, *Anal. Chim. Acta*, *98*, 59 (1978).

78. P. Quevauviller, R. L. Lavigne, R. Pinel, and M. Astruc, *Environ. Poll.*, *57*, 149 (1989).

79. M. Chamsaz, I. M. Khasawneh, and J. D. Winefordner, *Talanta*, *35*, 519 (1988).

80. Mu-Qing Yu, Gui-Qin Liu, and Qinhanjin, *Talanta*, *30*, 265 (1983).

81. J. Piwonka, G. Kaiser, and G. Tolg, *Fresenius Z. Anal. Chem.*,
     321, 225 (1985).

82. J. M. Rabadan, J. Galban, J. C. Vidal, and J. Aznarez, *J. Anal.
     Atom. Spectrosc.*, *5*, 45 (1990).

83. S. H. Vien and R. C. Fry, *Anal. Chem.*, *60*, 465 (1988).

84. T. Guo, W. Erler, H. Schulze, and S. McIntosh, *Atom. Spectrosc.*,
     11, 24 (1990).

85. G. A. Hambrick, P. N. Froelich, M. O. Andreae, and B. L. Lewis,
     Anal. Chem., *56*, 422 (1984).

86. J. Dedina, *Anal. Chem.*, *54*, 2097 (1982).

87. F. D. Pierce and H. R. Brown, *Anal. Chem.*, *49*, 1417 (1977).

88. O. F. X. Donard, L. Randall, S. Rapsomanikis, and J. H. Weber,
     Int. J. Environ. Anal. Chem., *27*, 55 (1986).

89. M. P. Stang and P. F. Seligman, *Chemosphere*, *15*, 253 (1986).

90. J. R. Ashby and P. J. Craig, *Sci. Tot. Environ.*, *78*, 219 (1989).

91. S. Rapsomanikis, O. F. X. Donard, and J. H. Weber, *Anal. Chem.*,
     58, 35 (1986).

92. N. Bloom, *Can. J. Fish Aquat. Sci.*, *46*, 1131 (1989).

93. H. H. Van den Broek, G. B. M. Hermes, and C. E. Goewie, *Analyst*,
     113, 1237 (1988).

94. M. D. Muller, *Anal. Chem.*, *59*, 617 (1987).

95. S. Ohhira and H. Matsui, *J. Chromatogr.*, *525*, 105 (1990).

96. R. J. Maguire and R. J. Tkacz, *J. Chromatogr.*, *268*, 99 (1983).

97. R. J. Maguire and H. Huneault, *J. Chromatogr.*, *209*, 458 (1981).

98. T. Ishizaka, S. Nemoto, K. Sasaki, T. Suzuki, and Y. Saito,
     J. Agric. Food Chem., *37*, 1523 (1989).

99. M. D. Muller, *Fresenius Z. Anal. Chem.*, *317*, 32 (1984).

100. (a) S. Clark and P. J. Craig, *Appl. Organomet. Chem.*, *2*, 33
     (1988). (b) P. J. Craig, D. Mennie, N. Ostah, O. F. Donard,
     and F. Martin, *Analyst*, *117*, 823-824 (1992).

101. J. R. Ashby and P. J. Craig, *Appl. Organomet. Chem.*, *1*, 275
     (1987).

102. R. C. Forster and A. G. Howard, *Anal. Proc.*, *26*, 34 (1989).

103. P. J. Craig and D. Mennie, *Micro Chemica Acta* (submitted).

104. L. Ebdon, S. J. Hill, and P. Jones, *Analyst*, *110*, 515 (1985).

105. M. Astruc, R. L. Lavigne, V. Desauziers, A. Astruc, R. Pinel,
     and O. F. X. Donard, *Heavy Met. Hydrol. Cycle*, *447* (1988).

106. K. L. Jewitt and F. E. Brinckman, *J. Chromatogr. Sci.*, *19*,
     583 (1981).

107. F. E. Brinckman, W. R. Blair, K. L. Jewitt, and W. P. Imerson, *J. Chromatogr. Sci., 15,* 493 (1977).

108. T. M. Vickney, H. E. Howell, G. V. Harrison, and G. R. Ramelow, *Anal. Chem., 52,* 1743 (1980).

109. C. A. Krone, D. W. Brown, D. G. Burrows, R. G. Bogar, S. L. Chan, and V. Varanase, *Mar. Environ. Res., 27,* 1 (1989).

110. J. J. Sulivan, J. D. Torkelson, M. M. Wekell, T. A. Hollingworth, W. L. Saxton, and G. A. Miller, *Anal. Chem., 60,* 626 (1988).

111. G. A. Junk and J. R. Richard, *Chemosphere, 16,* 61 (1987).

112. M. L. Schafer, U. Rhea, J. T. Peeler, C. H. Hamilton, and J. E. Campbell, *J. Agric. Food Chem., 23,* 1079 (1975).

113. M. Shariat, *J. Chromatogr. Sci., 17,* 527 (1979).

114. Y. K. Chau and H. Saitoh, *Int. J. Environ. Anal. Chem., 3,* 133 (1973).

115. L. Goolvard and H. Smith, *Analyst, 105,* 726 (1980).

116. R. D. Braun, *Introduction to Instrumental Analysis,* McGraw-Hill, New York, 1986.

117. B. Welz and M. Melcher, *Analyst, 108,* 213 (1983).

118. P. B. Stockwell and W. T. Corns, *Spectrosc. World, 4,* 14 (1992).

119. J. J. Cooney, A. J. Kronick, G. J. Olson, W. R. Blair, and F. E. Brinckman, *Chemosphere, 17,* 1795 (1988).

120. C. L. Matthias and J. M. Bellama, *Int. J. Environ. Anal. Chem., 35,* 61 (1989).

121. N. S. Makkar, A. J. Kronick, and J. J. Cooney, *Chemosphere, 18,* 2043 (1989).

122. L. Schebek, M. O. Andreae, and H. J. Tobschall, *Environ. Sci. Technol., 25,* 871 (1991).

123. M. A. Unger, W. G. MacIntyne, J. Greaves, and R. J. Huggert, *Chemosphere, 15,* 461 (1986).

124. M. Nagase, *Anal. Sci., 6,* 851 (1990).

125. G. E. Batley, C. Fuhua, C. I. Brockbank, and K. J. Flegg, *Aust. J. Mar. Freshwater Res., 40,* 49 (1989).

126. R. Chiojka, R. J. Williams, and S. Fredrickshon, *Mar. Poll. Bull., 21,* 570 (1989).

127. R. Buzina, K. Suboticanec, J. Vukusic, J. Sapunar, K. Antonic, and M. Zorica, *Sci. Tot. Environ., 78,* 45 (1989).

128. T. Kaise, K. Hanaoka, S. Tagua, T. Hirayama, and S. Fukui, *Appl. Organomet. Chem., 2,* 539 (1988).

3

Biogeochemistry of Methylgermanium Species in Natural Waters

Brent L. Lewis

College of Marine Studies
University of Delaware
Lewes, Delaware 19958, USA

H. Peter Mayer

Max-Planck Institut für Chemie
Saarstrasse 23
D-6500 Mainz, Germany

1. INTRODUCTION

Organic compounds of the group IVb elements (Si, Ge, Sn, Pb) are
among the most stable organometallics, with a gradation of properties
down the group from silicon to lead [1]. The organogermanium compounds resemble the very stable organosilicon compounds, while corresponding organotin and organolead compounds are generally much more
reactive. The 4+ valence state dominates in organometallic compounds
of all the group IVb elements, while the divalent state is rare.
Expansion of the coordination number of germanium to values greater
than 4 occurs rarely if at all [1]. Germanium compounds with less
than three germanium-carbon bonds and no bulky substituents are generally soluble in water, with hydroxide ions in the remaining coordination sites of the central Ge atom. Over the pH range of natural
waters (approx. 4-8), organogermanium compounds are expected to occur
as hydroxide species.

Three germanium species have been observed in the natural
environment: inorganic germanium (Ge_i), monomethylgermanium (MMGe),
and dimethylgermanium (DMGe). The geochemistry of Ge_i in low-
temperature environments is closely tied to that of silicon. In
estuaries and oceans, for example, Ge_i displays nutrient-type behavior
nearly identical to that of silicon. The organogermanium species, on
the other hand, do not appear to enter into the Ge_i/Si biogeochemical
cycle, nor do they display the active cycling observed for organometallic compounds of other metalloid elements, such as arsenic,
mercury, or selenium [2-6]. MMGe and DMGe display conservative
(nonreactive) behavior in estuaries and oceans, and are very stable

in seawater to photolytic and chemical oxidation, consistent with
their conservative behavior. They appear to be produced on the
continents, with transport to the oceans via rivers. Based on estimated fluvial inputs, MMGe and DMGe display oceanic residence times
on the order of several hundred thousand to a few million years.
Field observations and laboratory experiments suggest that methylgermanium may be produced by microbial methylation of Ge_i in anoxic
aquatic environments, perhaps in the sediments.

2. METHODS

2.1. Sampling

Although a "trace" element in seawater, germanium is not subject to
the severe contamination problems encountered for elements such as
iron or zinc. Seawater samples may be collected using standard
hydrographic methods, using either Niskin or Go-Flo bottles hung on
a standard hydrowire or rosette system. Filtration and acidification
of open-ocean samples does not appear to be necessary for the short-
term storage of samples for MMGe and DMGe. Sample preservation by
acidification to pH ~2 with hydrochloric acid is recommended, however,
for long-term sample storage and for the determination of Ge_i.
Samples should be stored in linear polyethylene (LPE) bottles. Glass
should be avoided, as this tends to leach Ge_i and Si.

Riverine and estuarine samples were collected by hand, filtered
through 0.4-μm Millipore filters, and stored unacidified at room temperature in LPE bottles. Rainwaters were collected with a nonmetallic
catch basin and stored unacidified and unfiltered.

2.2. Analysis

Inorganic and methylgermanium species in natural water samples can be
determined by graphite furnace atomic absorption with hydride generation and cryogenic chromatographic separation of the germanium species

[7,8]. The germanium species are reduced by $NaBH_4$ to the corresponding germane and methylgermane hydrides, stripped from solution by a helium gas stream, and trapped at liquid nitrogen temperature on a chromatographic column packed with 15% OV3 on Chromosorb W-AW-DMCS. The germanes are sequentially released into the graphite furnace by controlled heating of the column, taking advantage of differences in boiling point between the species. Absolute detection limits of this method are on the order of 150 pg for Ge_i, MMGe, and DMGe [8]. Analytical precision of the method has been improved to better than ±2% at the 300 pM level by use of high-precision flow meters to regulate gas flows and by maintenance of a constant gas flow at the outlet of the cryogenic chromatographic column [9].

For analysis of methylgermanium in anaerobic microbial cultures, analytical interferences from carbonate and sulfide have been encountered. These problems can be overcome by acidification of sample aliquots to pH 2.5 with H_3PO_4 and sparging with helium prior to analysis. The analytical sensitivity for the DMGe species can be further enhanced in these solutions (up to 300% increase) by addition of cysteine-HCl as a secondary reducing agent.

Sohrin [10] described a potential alternative method for the preconcentration and physical separation and collection of water-soluble organogermanium compounds, based on solvent extraction of their chlorides from aqueous hydrochloric acid solutions into CCl_4 and back extraction from the organic phase into an aqueous 1 M NaOH solution. Although the method was not tested at natural concentrations, it should in principle allow separation and preconcentration of organogermanium species from natural water samples.

3. METHYLGERMANIUM IN NATURAL WATERS

3.1. Marine Systems

3.1.1. Oceans

In estuaries and the oceans, Ge_i displays nutrient-like behavior, acting as a near perfect tracer of Si [11-14]. In contrast to Ge_i

and Si, MMGe and DMGe display conservative (nonreactive behavior)
in marine waters, with constant Ge/salinity ratios of approximately
9.4×10^{-12} mol Ge and 2.8×10^{-12} mol Ge per salinity unit, respectively [15,16]. MMGe and DMGe are uniformly distributed throughout
the oceans at concentrations of approximately 330 and 100 pM, respectively, accounting for >70% of the total known germanium in seawater.
No other organogermanium species have been observed in seawater.

Figure 1 shows oceanic profiles for MMGe and DMGe at sites in
the Bering Sea, the northwest Pacific Ocean, the Antarctic Ocean, and
the Sargasso Sea (northwest Atlantic Ocean). The Geosecs stations
had strong vertical gradients in Ge_i and Si [11], but no discernible
gradients in MMGe or DMGe. MMGe and DMGe behave conservatively
throughout the water column, with no surface-to-deep or interocean
differences other than those due to variations in salinity.

Detailed and precise profiles of MMGe and DMGe through the
upper 800 m of the water column in the Sargasso Sea show the same
nonreactive behavior [9]. There is no apparent production or removal
of methylgermanium in the euphotic zone or upper thermocline, implying
that marine plankton do not biomethylate or demethylate germanium and
consistent with the hypothesis that the methylgermanium species do not
enter into the marine Ge_i/Si biogeochemical cycle.

3.1.2. Estuaries

Methylgermanium species have been observed to behave conservatively
in a wide variety of estuaries, with conditions ranging from pristine
to severely polluted and from very low to very high productivity [15].
Plots of MMGe and DMGe concentration vs. salinity in axial transects
of these estuaries are linear in all cases, with barely detectable
concentrations in the river end member and high concentrations at the
seawater end. Figure 2 shows a composite plot of the data from four
estuaries, with the seawater data also plotted as cross-hatched fields.
The Suwannee River estuary (Florida) is relatively pristine with moderate productivity, while Delaware Bay is polluted with industrial,

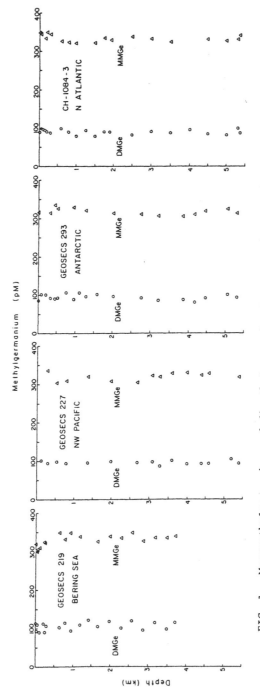

FIG. 1. Monomethylgermanium and dimethylgermanium concentrations at four ocean stations. (Reproduced by permission from *Nature*, Vol. 313, pp. 303-305, copyright 1985, Macmillan Magazines Ltd.; Ref. 15.)

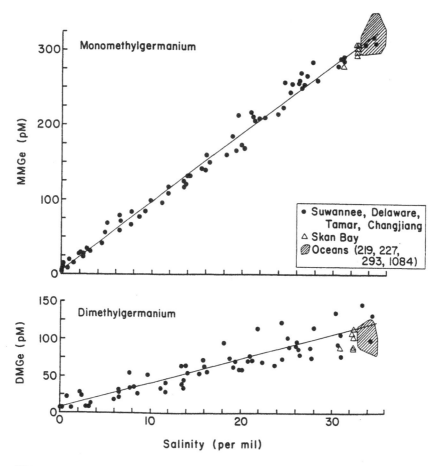

FIG, 2. Monomethylgermanium and dimethylgermanium concentrations versus salinity in estuaries and oceans. (Reproduced by permission from *Nature*, Vol. 313, pp. 303-305, copyright 1985, Macmillan Magazines Ltd.; Ref. 15.)

municipal, and agricultural wastes [17] and the Tamar (southern England) with high levels of arsenic and manganese from abandoned mining operations [18-20]. The Changjiang River estuary has a high suspended sediment load, which suppresses primary productivity at salinities <20 per mil, making it an ideal site for the study of nonbiological adsorption, desorption, and flocculation processes

[21]. Linear methylgermanium vs. salinity plots in these estuaries
suggest that biological and abiological estuarine processes neither
produce nor remove methylgermanium. The 35 per mil salinity end
members based on the estuarine data are 330 ± 15 pM MMGe and 120 ± 20
pM DMGe, in excellent agreement with the seawater values.

The conservative behavior displayed by the methylgermanium
species in estuarine and oceanic profiles suggests that the organogermanium species do not enter into the Ge_i/Si biogeochemical cycle
[9,12,15].

3.1.3. *Marine Anoxic Basins*

Based on estimated fluvial input fluxes and the apparent inert
behavior of MMGe and DMGe in the oceans, the estimated residence
time of methylgermanium in the oceans is on the order of ≥ 1 Ma [9,
15]. Assuming steady state, this translates to an oceanic removal
rate of $<10^{-4}$% of the total oceanic stock of methylgermanium per
annum. Preliminary data appeared to suggest that marine anoxic
basins may be a sink for methylgermanium [9]. Depleted levels of
MMGe and DMGe and excess Ge_i relative to the oceanic Ge_i/Si ratio
in the Framvaren Fjord (Norway) and the Black Sea were cited as
evidence of demethylation of methylgermanium in anoxic environments.
We now believe these data to be incorrect, probably due to a sample
storage artefact. Subsequent analyses in the Black Sea have shown
that Ge_i in this basin displays the same silicon-like behavior
observed in estuaries and the oceans. Both MMGe and DMGe appear to
behave nearly conservatively (P. N. Froelich, personal communication,
1991). MMGe concentrations were as predicted from the oceanic MMGe/
salinity ratio, even in the anoxic bottom waters. Although DMGe concentrations showed some depletion (~25%) in the deep waters, suggesting that a degree of demethylation may occur in the anoxic zone, it
does not appear that conversion of methylgermanium to Ge_i is a significant sink for methylgermanium from the marine environment.

A number of other potential removal processes, including photolytic decomposition, volatile and sea salt aerosol losses, thermal

destruction during hydrothermal circulation through the mid-ocean ridge axes, and removal by burial of interstitial waters or evaporite formation are also considered unlikely based on laboratory experiments and/or residence time calculations [9,22].

3.2. Pristine Rivers

Relatively pristine rivers which drain swamps and remote regions contain barely detectable levels of methylgermanium, ranging from <2 to ~3 pM for MMGe and from <3 to ~7 pM for DMGe [9]. This suggests either a natural terrestrial or freshwater source of methylgermanium on the continents or the recycling of sea salt inputs from atmospheric transport of marine aerosols and/or dissolution of marine evaporites. Evidence to date supports a continental source, probably biomethylation of Ge_i, although attempts to locate a natural source of methylgermanium in methanogenic swamps and their drainage basins have been unsuccessful, showing only very low MMGe and DMGe concentrations typical of lakes and pristine rivers [9]. The predicted methylgermanium concentrations in rivers due to sea salt aerosols are <0.7 pM MMGe and ~0.3 pM DMGe, much lower than observed in several pristine rivers [9]. Even if all the chloride in rivers were cyclic (from marine aerosols and evaporites), this could not account for the observed fluvial methylgermanium concentrations. Further, marine and continental rains contained no detectable methylgermanium. Vapor phase transport of marine methylgermanium through the atmosphere is also thought to be unimportant. Finally, the ratio of MMGe to DMGe in clean rivers with detectable methylgermanium is generally <0.5, compared to a seawater ratio of 3.1. Photolytic destruction of methylgermanium during atmospheric transport would tend to increase the MMGe/DMGe ratio, as conversion of methylgermanium compounds to Ge_i proceeds by successive removal of each methyl group, with conversion of MMGe to Ge_i as the last and rate-limiting step [9,15].

4. EVIDENCE FOR BIOLOGICAL METHYLATION
OF GERMANIUM

The transformation of metallic and metalloid elements from inorganic
to organic forms is often biologically mediated. This ability to
"biomethylate" metallic elements has been demonstrated for algae,
bacteria, and fungi [23-26]. Naturally occurring methylated species
have been reported in the environment for many of the metalloid elements, including antimony [27-29], arsenic [2,3,29], germanium [9,15,
16,29], lead [30], mercury [5], and tin [31]. Analogous with the
methylated species of other metalloids, one might expect naturally
occurring organogermanium species to be produced by the biological
methylation of inorganic germanium.

4.1. Anaerobic Sewage Digester

Conversion of Ge_i to MMGe, DMGe, and TMGe (trimethylgermanium) has
been observed in an anaerobic sewage digester (Table 1). An excess
of methylgermanium in the digester relative to that entering the
treatment plant, with an equimolar disappearance of Ge_i, indicates
in situ methylation of Ge_i, probably via a microbial biomethylation
pathway. Production in the digester of methylarsenical compounds,
known to be produced in sewage sludge by methanogenic bacteria [32],
supports this hypothesis [9]. The small quantity of TMGe observed
in the aerobic digesters is probably introduced when sludge from the
anaerobic system is recirculated into the aerobic tanks. During the
course of this study, the population of methanogenic bacteria in the
digester was inadvertently killed by a fluctuation in the digester
temperature. A concomitant decrease in the concentrations of the
methylgermanium species was observed (Table 1, Anaerobic-3). Rees-
tablishment of the methanogenic microbial population was accompanied
by an increase in methylgermanium levels (Anaerobic-4), further supporting methanogenesis as a mechanism for biomethylation of germanium.

TABLE 1

Germanium Speciation in the T. P. Smith Wastewater
Treatment Facility, Tallahassee, Florida

Sample	Ge_i (pM)	MMGe (pM)	DMGe (pM)	TMGe (pM)	ΣGe (pM)	$\frac{\Sigma MeGe}{\Sigma Ge}$
Raw	204	14	<10	<4	218	0.08
Aerobic	185	<7	<10	18	203	0.09
Anaerobic-1	103	<7	32	74	209	0.51
Anaerobic-2	128	36	13	18	195	0.34
Anaerobic-3	166	<7	<10	<4	166	0.02
Anaerobic-4	128	7	24	55	214	0.40

Raw = untreated sewage entering the plant. Aerobic = effluent from
the aerobic digesters. Anaerobic = effluent from the anaerobic
digesters. Samples were collected in Feb. 1985 (Raw, Aerobic,
Anaerobic-1,2), in March 1985 (Anaerobic-3), and in March 1986
(Anaerobic-4). ΣGe = sum of inorganic plus methylated species.
$\Sigma MeGe$ = sum of methylated species. "<" = values below detection
limits.
Source: Ref. 9.

4.2. Contaminated Rivers

Rivers in areas with coal-fired power plants or steel mills display
dramatic enrichments in Ge_i due to deposition of coal fly ash and
runoff from ash ponds [11,29]. In densely populated areas, these
rivers also contain elevated levels of methylgermanium, ranging from
3 to 100 times concentrations in pristine waters. Average methylgermanium concentrations in the polluted rivers sampled were approximately 17 ± 8 pM MMGe and 15 ± 6 pM DMGe [9]. In addition, MMGe/DMGe
ratios are higher than in clean rivers. These elevated methylgermanium concentrations may result from the use of Ge_i-rich river
waters as feedwaters for large municipal sewage treatment facilities,
with biomethylation of Ge_i and release of MMGe and DMGe back to the
rivers [9,16].

Two of the rivers sampled, the Cuyahoga (OH) and the Zenne
(Belgium), have methylgermanium concentrations much higher than the
other rivers. The Cuyahoga at Cleveland has inorganic and methylgermanium concentrations an order of magnitude higher than at Inde-
pendence only a few miles upstream, implying point contamination
sources for germanium within the city of Cleveland. The Zenne
receives large quantities of untreated sewage from the city of
Brussels, leading to high BOD (biological oxygen demand) and low
oxygen conditions analogous to those found in an anaerobic sewage
digester [33]. Methylgermanium concentrations in the Zenne are the
highest measured anywhere (~350 pM MMGe, ~250 pM DMGe) and may result
from biomethylation of Ge_i in the sediments (see below).

4.3. Culture Experiments

4.3.1. *Marine Algal Cultures*

A variety of marine algal species have been shown to biomethylate
metals. Arsenic(V), for example, may be converted by algae to
organoarsenicals or to As(III), probably as a detoxification mechanism [34,35]. A number of marine species have been tested for
their potential ability to methylate germanium, including diatoms,
a dinoflagellate, a blue-green alga, and a coccolithophorid. The
latter was the species *Hymenomonas carterae*, known to produce large
quantities of β-dimethylsulfoniopropionate and dimethylsulfide in
axenic culture [36]. Biomethylation of Ge_i was not observed in any
of these trials. This result is in keeping with the conservative
behavior and apparent stability of MMGe and DMGe in seawater, even
in highly productive waters.

4.3.2. *Zenne River Sediment Slurries*

Sediment cores were collected in March 1990 from the eastern arm
of the Zenne River, downstream from Brussels. Although the high
methylgermanium concentrations reported by Lewis et al. [9] for the
water column were for the western arm of the river, the eastern arm

was selected due to the presence of thick, black, organic-rich sediments. The sediments at this site are devoid of oxygen and display active bubble ebullition (probably methane), indicative of high microbial activity.

For the initial experiment, a single core was collected and five subcores taken from the main core over the depth interval 8-28 cm. To each subsample was added a complex medium to enhance bacterial growth and Ge_i (initial GeO_2 concentration ~100 nM) to induce methylgermanium production. Samples were incubated under anaerobic conditions in the dark at 30°C and ambient pH (7.7) for 10 months. The slurries were then centrifuged and the supernatant analyzed for the presence of methylgermanium. Three methylgermanium species—MMGe, DMGe and TMGe—were found in the supernatant, with DMGe and TMGe the main products (Table 2). The conversion of Ge_i to methylgermanium decreased with depth in the core, from a maximum of about 29% at 13 cm to 4.5% at 28 cm. No production of methylgermanium was seen in a sterile control, indicating a biomethylation pathway.

A second sediment slurry experiment was conducted with fresh Zenne sediment, collected in spring of the following year. For this experiment, the core was homogenized and separate aliquots of homogenized sediment were spiked with GeO_2 and/or MMGe, DMGe, and TMGe. The latter were included to test for adsorption of the germanium species onto sediments. After 3 months incubation, slurries were centrifuged and the supernatant analyzed for the germanium species. In contrast to the first experiment, no production of methylgermanium was observed in any of the aliquots. Approximately 90% of the DMGe and TMGe remained in the supernatant, while nearly 50% of the MMGe and >99% of the Ge_i had been lost by adsorption to the sediments. An aqueous sediment leach at pH 2 (acidified with H_3PO_4) recovered 90-100% of the methylgermanium. Ge_i remained on the sediments. These results indicate that the methylgermanium was not strongly adsorbed on the sediment or degraded during the incubation. It is not clear why no methylation of Ge_i was observed in this experiment, but it may be that 3 months was too short a time for the organo-

TABLE 2

Microbial Production of Methylated Germanium Compounds
in Zenne River Sediment Slurries

Depth (cm)	Dry wt. sediment (g)	%MMGe	%DMGe	%TMGe
8.00	7.55	0.9	11.7	8.2
13.00	9.77	3.0	16.3	10.2
18.00	9.34	2.4	14.2	7.9
23.00	7.40	1.7	8.5	4.8
28.00	7.40	0.7	2.4	1.4
Sterile control	2.75	0.0	0.0	0.0

Depth = depth in sediment. % = percentage of Ge_i transformed to
methylated species. Initial $[Ge_i]$ = 100 nM.

germanium species to increase to measurable levels. Alternatively,
the biological and/or chemical composition of the sediments may have
been different from that in the first trial, resulting in conditions
unfavorable for methylation. The latter is supported by a decrease
in pH in the second experiment from near 7.7 to 6.8 after only 3
months, whereas in the first trial the pH remained unchanged after
10 months.

4.3.3. *Microbial Enrichment Studies*

From the Zenne sediment slurries, microbial enrichments were prepared under anaerobic conditions using substrates selective for
methanogens or nitrate-, sulfate-, or iron-reducing bacteria.
Enrichments were grown in the presence of 20 nM or 10 μM GeO_2.

For methanogens, the substrates H_2/CO_2 (80:20) and acetate
(10 mM) were selected. With the acetate-containing medium, no
dominant microbial species developed after three transfers to fresh
media, and no methylgermanium production was observed. In the H_2/CO_2-
containing enrichment, a short rod-shaped organism, probably a homo-

acetogenic bacterium [37], was the dominant species after 6 days. Cell forms resembling *Methanospirillum* and *Methanosarcina* sp. [38] appeared after 11 days. After one transfer, the 20 nM GeO_2 enrichment contained 8.8 nM·MMGe and 2.1 nM DMGe. Subsequent transfers to fresh medium produced the same dominant species, but with varying abundance ratios. These cultures yielded mixed results, with some producing methylgermanium and some not. The 10 µM GeO_2 enrichments contained no measurable methylgermanium with a detection limit of ~20 nM. Although some methylgermanium may have been produced in these cultures, simply increasing the GeO_2 concentration did not stimulate a proportional increase in the amount of methylgermanium produced, perhaps due to toxic effects of germanium at high concentrations. Pure cultures of methanogenic bacteria species, selected for their reported presence in freshwater sediments and sewage sludge [39], failed to produce methylgermanium when grown with either 20 nM or 10 µM GeO_2 (pure strains obtained from the German Collection of Microorganisms and Cell Cultures, Braunschweig, Germany).

To select for anaerobic bacteria other than methanogens, Zenne sediment enrichments were prepared in basic medium supplemented with 5 mM glucose as a general substrate, 100 nM GeO_2, and nitrate, sulfate, or ferric iron as the electron acceptor. Although significant methane production and the reduction of nitrate, sulfate, or Fe(III) were observed in each culture, no methylgermanium was produced under these conditions. In a microbial enrichment containing 10 mM Na_2SO_4 and 20 nM GeO_2, the dominant organism was a lemon-shaped microbe, probably a *Desulfobulbus* species [40]. Methylgermanium production was observed in the second transfer from this enrichment (11.5 nM MMGe, 1.6 nM DMGe). Neither subsequent transfers from this enrichment nor a pure isolation of the lemon-shaped species displayed any methylgermanium production.

The results of these experiments suggest that the microbial methylation of germanium is not a simple, species-dependent process. Rather, it is likely influenced by the complex interaction of a number of variables, including the nature of the microbial community and various environmental parameters, such as the available substrate(s), oxygen concentration, pH, temperature, etc. The trimethylated germanium species (TMGe) produced in the sediment slurry experiments (and in the sewage digester) has not been observed in natural freshwater or seawater samples. Our experience with these compounds indicates that the TMGe species is less stable and has a higher vapor pressure than either MMGe or DMGe. This suggests that TMGe, if formed in the environment, may be rapidly removed from natural waters by degassing to the atmosphere and/or by demethylation to DMGe.

4.3.4. Chemical Methylation Experiments

Reactions of methylcobalamin and methyl iodide with metal or metalloid elements have been used as a chemical analog of in vivo biomethylation reactions [41-46]. Previous attempts to methylate Ge(IV) by reaction with methylcobalamin at pH 2 and 7 were unsuccessful under both aerobic and anaerobic conditions [16]. Subsequent experiments have demonstrated the methylation of Ge(II) by methylcobalamin and methyl iodide [47] using reaction conditions known to methylate Sn(II) [43, 48,49]. Ge(II) was methylated by methylcobalamin at pH 1. No methylation was observed at neutral pH, indicating that this reaction is unlikely to occur in natural waters. Methyl iodide, on the other hand, was found to methylate Ge(II) to MMGe at pH 7 (maximum yield ~6%). Neither DMGe nor TMGe were produced. The potential for methylation of germanium by methyl iodide in natural environments is being investigated. Methyl iodide is produced in the oceans by marine algae, especially *Laminaria* species in kelp beds (concentrations up to 3 μM) [50]. While these experiments suggest that kelp beds might be a potential marine source of methylgermanium, oceanic germanium is likely present in the 4+ rather than the 2+ state [29]. Although germanium reduction to Ge(II) might occur in anoxic marine sediments, it is unclear whether methyl iodide would also be available.

5. SUMMARY

Three methylgermanium species have been identified in natural waters:
inorganic, monomethyl-, and dimethylgermanium. Inorganic germanium
behaves as a trace element analog for silicon in low-temperature
environments. In contrast, the methylgermanium species display conservative (nonreactive) behavior in estuaries and oceans. They are
biologically unreactive, even in highly productive waters, and do not
appear to enter into the Ge_i/Si biogeochemical cycle. Together, MMGe
and DMGe account for >70% of the total known germanium in seawater
and are uniformly distributed throughout the oceans. In pristine
rivers, the methylgermanium species are present at barely detectable
levels. Based on estimated fluvial inputs, the oceanic residence
times of MMGe and DMGe are on the order of several hundred thousand
to a few million years.

The natural sources and sinks of methylgermanium have not yet
been positively identified. In contrast to our earlier supposition,
new data from the Black Sea indicate that demethylation in marine
anoxic basins is probably not a significant sink for methylgermanium
from seawater. Attempts to locate a natural freshwater source of
methylgermanium in methanogenic swamps and their drainage basins have
been unsuccessful. However, conversion of inorganic to methylgermanium in an anaerobic sewage digester and the presence of elevated
concentrations of MMGe and DMGe in polluted rivers suggest a microbial
biomethylation pathway, perhaps associated with freshwater methanogenesis. This hypothesis is supported by observation of extremely
high MMGe and DMGe concentrations in the Zenne River and by microbial
enrichment studies utilizing Zenne sediment samples. These experiments indicate microbial transformation of inorganic to methylgermanium in anoxic sediments. Chemical methylation of Ge(II) by
methylcobalamin and methyl iodide has also been demonstrated in
laboratory experiments. The ability to methylate germanium does
not appear to be solely dependent on the type of bacteria present.

Rather, methylation of Ge_i is likely controlled by a complex interaction of the microbial community and various chemical and physical
parameters, such as the organic content of the sediments, the available substrates for microbial growth, and temperature, pH, salinity,
etc. Coupled with the apparently weak source strength, this accounts
for the difficulty in locating a site of active methylgermanium production in the natural environment. It does appear that the necessary conditions are often enhanced in polluted systems.

ABBREVIATIONS

BOD	biological oxygen demand
Ge_i	inorganic germanium
LPE	linear polyethylene
MMGe	monomethylgermanium
DMGe	dimethylgermanium
TMGe	trimethylgermanium

REFERENCES

1. R. C. Poller, in *Comprehensive Organic Chemistry: The Synthesis
    and Reactions of Organic Compounds*, Vol. 3 (D. N. Jones, ed.),
    Pergamon Press, New York, 1979, pp. 1061-1109.

2. M. O. Andreae, *Deep-Sea Res., 25,* 391-402 (1978).

3. M. O. Andreae, *Limnol. Oceanogr., 24,* 440-452 (1979).

4. M. Berman, T. Chase, Jr., and R. Bartha, *Appl. Environ.
    Microbiol., 56,* 298-300 (1990).

5. R. P. Mason and W. F. Fitzgerald, *Nature, 347,* 457-459 (1990).

6. D. Tanzer and K. G. Heumann, *Atmos. Environ., 24,* 3099-3102
    (1990).

7. M. O. Andreae and P. N. Froelich, *Anal. Chem., 53,* 1766-1771
    (1981).

8. G. A. Hambrick, P. N. Froelich, Jr., M. O. Andreae, and B. L.
    Lewis, *Anal. Chem., 56,* 421-424 (1984).

9. B. L. Lewis, M. O. Andreae, and P. N. Froelich, *Mar. Chem., 27,*
    179-200 (1989).

10. Y. Sohrin, *Anal. Chem., 63,* 811-814 (1991).

11. P. N. Froelich, G. A. Hambrick, M. O. Andreae, R. A. Mortlock, and J. M. Edmond, *J. Geophys. Res., 90,* 1122-1141 (1985).

12. P. N. Froelich, L. W. Kaul, J. T. Byrd, M. O. Andreae, and K. K. Roe, *Estuar. Coast. Shelf Sci., 20,* 239-264 (1985).

13. P. N. Froelich, G. A. Hambrick, L. W. Kaul, J. T. Byrd, and O. Lecointe, *Geochim. Cosmochim. Acta, 48,* 1417-1433 (1985).

14. P. N. Froelich, R. A. Mortlock, and A. Shemesh, *Global Biogeo-chem. Cycles, 3,* 79-88 (1989).

15. B. L. Lewis, P. N. Froelich, and M. O. Andreae, *Nature, 313,* 303-305 (1985).

16. B. L. Lewis, M. O. Andreae, P. N. Froelich, and R. A. Mortlock, *Sci. Tot. Environ., 73,* 107-120 (1988).

17. J. H. Sharp, C. H. Culberson, and T. M. Church, *Limnol. Oceanogr., 27,* 1015-1028 (1982).

18. S. Knox, D. R. Turner, A. G. Dickson, M. I. Liddicoat, M. Whitfield, and E. I. Butler, *Estuar. Coast. Shelf Sci., 13,* 357-371 (1981).

19. S. Knox, W. J. Langston, M. Whitfield, D. R. Turner, and M. I. Liddicoat, *Estuar. Coast. Shelf Sci., 18,* 623-638 (1984).

20. W. J. Langston, *Can. J. Fish. Aquat. Sci., 40,* 143-151 (1983).

21. J. M. Edmond, C. Measures, R. E. McDuff, L. H. Chan, R. Collier, B. Grant, L. I. Gordon, and J. B. Corliss, *Earth Planet. Sci. Lett., 46,* 1-18 (1983).

22. B. L. Lewis, *A Survey of Methylgermanium Compounds in Natural Waters,* Master's thesis, Florida State University, 1985.

23. F. Challenger, *Chem. Rev., 36,* 315-361 (1945).

24. F. Challenger, in *Organometals and Organometalloids: Occurrence and Fate in the Environment* (F. E. Brinkman and J. M. Bellama, eds.), American Chemical Society, Washington, D.C., 1978, pp. 1-22.

25. A. O. Summers and S. Silver, *Ann. Rev. Microbiol., 32,* 637-672 (1978).

26. J. S. Thayer and F. E. Brinckman, in *Advances in Organometallic Chemistry,* Vol. 20 (F. G. A. Stone and R. West, eds.), Academic Press, New York, 1982, pp. 313-356.

27. M. O. Andreae, J.-F. Asmodé, P. Foster, and L. Van't dack, *Anal. Chem., 53,* 1766-1771 (1981).

28. M. O. Andreae, in *Trace Metals in Seawater* (C. S. Wong, E. Boyle, K. W. Bruland, J. D. Burton, and E. Goldberg, eds.), Plenum Press, New York, 1983, pp. 1-19.

29. M. O. Andreae and P. N. Froelich, *Tellus, 36B,* 101-117 (1984).

30. A. W. P. Jarvie, *Sci. Tot. Environ.*, *73*, 121-126 (1988).

31. J. R. Ashby and P. J. Craig, *Sci. Tot. Environ.*, *73*, 127-133 (1988).

32. B. C. McBride and R. S. Wolfe, *Biochemistry*, 4312-4317 (1971).

33. A. Lafontaine, J. Bouquiaux, D. De Brabander, J. Dierickx, J. Deleener, P. Dehavay, H. De Schepper, and G. Greven-Brauns, *Kwalteitoverzicht van een Aantal Belgische Oppervlakte Wateren in 1982 en 1983*, Inst. D'Hygiene et d'Epidem., Brussels, 1983.

34. M. O. Andreae and D. Klumpp, *Environ. Sci. Technol.*, *13*, 738-741 (1979).

35. J. G. Sanders and H. L. Windom, *Coast. Mar. Sci.*, *10*, 555-567 (1980).

36. A. Vairavamurthy, M. O. Andreae, and R. L. Iverson, *Limnol. Oceanogr.*, *30*, 59-70 (1985).

37. W. E. Balch, S. Schobert, R. S. Tanner, and R. S. Wolfe, *Int. J. Syst. Bacteriol.*, *27*, 355-361 (1977).

38. R. A. Mah and M. R. Smith, in *The Prokaryotes*, Vol. 1 (M. P. Starr, H. Stolp, H. G. Trüper, A. Balows, and H. G. Shlegel, eds.), Springer-Verlag, Berlin, 1981, pp. 948-977.

39. DSM, German Collection of Microorganisms and Cell Cultures, in *Catalogue of Strains*, Braunschweig, Germany, 1989.

40. F. Widdel, in *Biology of Anaerobic Microorganisms* (A. J. B. Zehnder, ed.), John Wiley and Sons, New York, 1988, pp. 469-585.

41. Y.-T. Fanchiang, W. P. Ridley, and J. M. Wood, *J. Am. Chem. Soc.*, *100*, 1010-1012 (1978).

42. Y.-T. Fanchiang, W. P. Ridley, and J. M. Wood, *J. Am. Chem. Soc.*, *101*, 1442-1447 (1979).

43. P. J. Craig and S. Rapsomanikis, *Environ. Sci. Technol.*, *19*, 726-730 (1985).

44. S. Rapsomanikis and J. H. Weber, in *Organometallic Compounds in the Environment* (P. J. Craig, ed.), Longman, Essex, UK, 1986, pp. 279-307.

45. B. Kräutler, in *The Biological Alkylation of Heavy Elements* (P. J. Craig and F. Glocking, eds.), Royal Society of Chemistry, Spec. Pub. 66, 1988, pp. 31-45.

46. R. J. P. Williams, in *The Biological Alkylation of Heavy Elements* (P. J. Craig and F. Glocking, eds.), Royal Society of Chemistry, Spec. Pub. 66, 1988, pp. 5-19.

47. H. P. Mayer and S. Rapsomanikis, *Appl. Organomet. Chem.*, *6*, 173-178 (1992).

48. Y.-T. Fanchiang and J. M. Wood, *J. Am. Chem. Soc.*, *103*, 5100-
     5103 (1981).

49. S. Rapsomanikis and J. H. Weber, *Environ. Sci. Technol.*, *19*,
     352-356 (1985).

50. J. E. Lovelock, *Nature, 256*, 193-194 (1975).

4

Biological Properties of Alkyltin Compounds

Yasuaki Arakawa

Department of Hygiene and Preventive Medicine
Faculty of Health Sciences
University of Shizuoka
52-1 Yada, Shizuoka-shi
Shizuoka 422, Japan

and

Osamu Wada

Department of Hygiene and Preventive Medicine
Faculty of Medicine
University of Tokyo
7-3-1 Hongo, Bunkyo-ku
Tokyo 113, Japan

1. INTRODUCTION

Alkyltin compounds are characterized by the presence of at least
one covalent carbon-tin bond. The compounds have a tetravalent
structure and are classified as mono-, di-, tri-, and tetraalkyltins
depending on the number of alkyl moieties. The anion is usually
chloride, fluoride, oxide, hydroxide, carboxylate, or thiolate.

Originally, organotin compounds were developed as thermal
stabilizers to prevent the thermal degradation of many chlorinated
compounds such as certain types of transformer oils, polyvinylchloride (PVC), chlorinated rubbers, paraffins, and modified plastics.
However, as the chemistry of organotin compounds became better
understood, their application expanded to catalytic and biologically
active agents. As a consequence of this rapid expansion, the concern about their environmental and health effects is increasing.
The purpose of this chapter is to present the current understanding
of the biological properties of alkyltin compounds.

2. BIOLOGICAL ACTIVITY OF ALKYLTINS

2.1. General Aspects

The biological activity of alkyltins is essentially determined by
the number and nature of the alkyl group bound to tin, whereas the
nature of the anionic group is only of secondary importance. The
tetra-, tri-, di-, and monoalkyltins display remarkable differences
in biological activities. Trialkyltin derivatives exert a powerful
toxic action on the central nervous system. Within the series of
trialkyltin derivatives, the lower homologs trimethyltin and triethyltin are the most toxic and the oral toxicity diminishes progressively from tri-*n*-propyltin to tri-*n*-hexyltin with tri-*n*-
octyltin being nontoxic. Tetraalkyltins resemble trialkyltins in
their biological activities, but their effects are often less and
delayed. This has been explained by a conversion of tetra- into
trialkyltin compounds in vivo [1]. Monoalkyltins do not have any
important toxic action [2]. The dialkyltin compounds from methyl
to hexyl induce bile duct lesions in rats and mice, but not in
rabbits or guinea pigs [3]. Further, dialkyltins such as dibutyltin and dioctyltin induce thymus atrophy, and their effects on the
immune system are the most sensitive criteria of their toxicity
[4-9].

In general, an alkyltin compound is much more toxic when it
is given interperitoneally than when it is given orally. However,
triethyltin and trimethyltin are exceptions to this rule. This
difference in toxicity results from the fact that except for the
dimethyl-, trimethyl-, and triethyltin compounds, the alkyltin
compounds are not very well absorbed from the gastrointestinal
tract.

2.2. Neurotoxicity

Interest in the neurotoxicity of alkyltin compounds increased after
the Stalinon affair, which happened in France in 1954 and resulted
in the death of about 100 people. The triethyltin derivatives were
identified as the toxic contaminant in Stalinon preparations. The
most constant symptom of the patients was severe and persistent
headaches. Other common symptoms were vomiting, retention of urine,
vertigo, abdominal pain, photophobia, weight loss, psychic disturb-
ances, and several cases of hypothermia (35°C). At autopsy, marked
interstitial edema of the white matter of the brain was seen.

Of all alkyltin derivatives, neurotoxic effects are limited
to trialkyltin derivatives such as trimethyltin and triethyltin.
No signs of neuronal damage or edema are observed in animals treated
with dialkyltin derivatives or any of the higher trialkyltin homologs
[10-12].

Triethyltin derivatives exert a powerful toxic action on the
central nervous system. The compounds produce a diffuse hemorrhagic
encephalopathy [13] and a generalized progressive weakness that
ended in death [14] as the acute symptoms, and further an interstitial edema confined to the white matter of the brain [13] as the
chronic symptoms. The basic pathological lesion of this edema is
limited to myelin. The myelin sheaths are dilated and filled with
fluid. Neonatal exposure to triethyltin causes a reduction of brain
weight and delayed myelinogenesis, but no cerebral edema is observed
[15]. Consequently, adult electrophysiological alterations [16],
deviation from neurobehavioral function [17], and neuromorphological
alterations [18] are observed.

Trimethyltin derivatives have the same order of neurotoxicity
as the triethyltin derivatives. The neurological effects of trimethyltin are tremor [2,19], hyperexcitability [2,19], aggressive-
ness [2,19], necrosis of neurons in the hippocampus and the pyriform
cortex [10,20], neuronal damages in the amygdaloid nucleus, brainstem
neurons, neocortex, spinal cord, sensory neurons, and retina [21-26],

selective vulnerability of the hippocampus [21-26], behavioral and
neuropathological alterations [27,28], but no cerebral edema is
found in contrast to triethyltin. Particularly in humans, a variety
of psychomotor changes, including irritability, depression, aggresive-
ness, headaches, tremors, convulsions, and changes in libido [29,30]
are observed.

Although tripropyltin and tributyltin derivatives produce
brain edema, their neurological effects seem to be much less as compared with triethyltin derivatives [19,31]. This may only be a
reflection of their poor absorption. Lower alkyltin compounds may
be readily absorbed from the gut but there are species differences
in the degree of absorption.

2.3. Hepatotoxicity

Some monobutyltin compounds cause steatosis of hepatocytes and
enlargement of the liver [32]. Dibutyltin compounds cause a specific lesion in bile ducts [19]. This lesion consists of an inflam-
matory reaction in the wall of the bile duct. Of the dialkyltin
compounds, bile duct damage is primarily produced by dibutyltin but
to a lesser degree also by diethyltin, dipropyltin, dipentyltin,
and dihexyltin compounds. Moreover, intrahepatic and extrahepatic
cholangitis is also produced by tributyltin compounds [11,33] and
tricyclohexyltin [34].

2.4. Cutaneous Toxicity

Various triorganotin compounds exert irritating effects on skin or
eyes. After a single application, tributyltin causes a severe epidermal damage, i.e., almost total epidermal necrosis and marked
dermal inflammation in the skin [35]. Triphenyltin does not appear
to have any action on the skin [36,37], but it irritates extremely
the eyes, resulting in corneal opacity. Tricyclohexyltin is an
irritant to both skin and eyes [34].

TABLE 1

Relative Organ Weights of Rats Fed Various
Organotin Compounds (% of Control)

	Body weight	Thymus	Adrenal	Liver	Spleen	Kidney	N
Bu_2SnCl_2	97	48[***]	125	93	98	96	6
Bu_3SnCl	90	58[***]	132[*]	93	75[*]	86[*]	7
$BuSnCl_3$	95	106	116	91	89	106	6
Bu_4Sn	101	88	119	89	79[*]	96	6
Pr_2SnCl_2	84	89	98	82[**]	88	97	7
Me_2SnCl_2	96	92	97	85[*]	112	99	6
$MeSnCl_3$	105	85	103	101	92	100	6
Oc_2SnCl_2	101	39[****]	100	104	116	100	7
Ph_3SnCl	91	100	113	89[*]	87	99	6
$SnCl_4$	88	83	110	92	73[**]	103	6

Note: Wistar-derived weanling rats (male, 40-45 g) fed independently 100 ppm organotin for 10 days. All values are the mean of 6-7 animals; those marked with asterisks differ significantly (Student's t test) from the corresponding control value: [*]$p < 0.05$, [**]$p < 0.02$, [***]$p < 0.01$, [****]$p < 0.001$.

2.5. Immunotoxicity

2.5.1. Thymus Atrophy

Dibutyltin, dioctyltin, and tributyltin compounds cause a severe thymus atrophy (Table 1) [4-9,38,39]. This atrophy is reversible (Fig. 1), and its degree correlates to the concentration of organotin in the thymus (Fig. 2) [7,38,39]. At histopathological examination, there is marked depletion of lymphocytes in the thymic cortex rather than in the thymic medulla (Fig. 3). Further, cortisone pretreatment studies reveal that immature cells in the thymic cortex are far more sensitive to organotin compounds than mature cells [34,36]. DNA synthesis of the thymic lymphocytes is significantly inhibited by dibutyltin and dioctyltin compounds at the concentration of 10^{-7} M

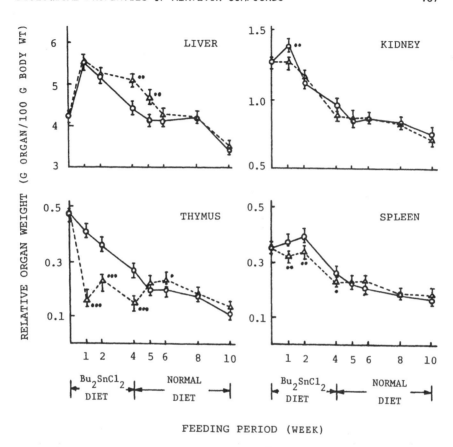

FIG. 1. Relative organ weights of rats fed 0 or 100 ppm Bu_2SnCl_2 for 4 weeks and after that period a normal diet for 6 weeks. Vertical bars denote SE of the mean for 10 determinations; those marked with asterisks differ significantly (Student's t test) from the corresponding control value: • $p < 0.05$, •• $p < 0.01$, ••• $p < 0.001$. (o) control (0 ppm) group, (△) 100 ppm group.

at which cell viability is not yet impaired (Fig. 4). Moreover, a parallelism between dose-response curves of dialkyltin for DNA synthesis and cell viability is found. These results indicate that organotins primarily induce an inhibition of cell proliferation and secondarily cause cell death [8,9,38-41].

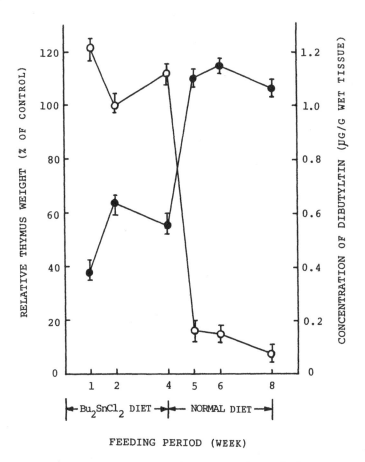

FIG. 2. Relationship between thymus atrophy and the concentration
of dibutyltin in the thymus of rats fed 100 ppm Bu_2SnCl_2 for 4 weeks
and after that period a normal diet for 6 weeks. (o) Concentration
of Bu_2SnCl_2 in rat thymus ($\mu g/g$ tissue), (●) relative thymus weights
are given as percentages of control values. Vertical bars indicate
SE of the mean for seven determinations.

2.5.2. *Immunosuppression*

Various immune function studies indicate that the cell-mediated
immunity and the T-cell-dependent humoral immunity are suppressed
by dialkyltin exposure [42,43]. A dose-related suppression of cell-
mediated immunity occurs in such manifestations as tuberculin hyper-
sensitivity, skin graft rejection, graft-vs.-host reactivity, and

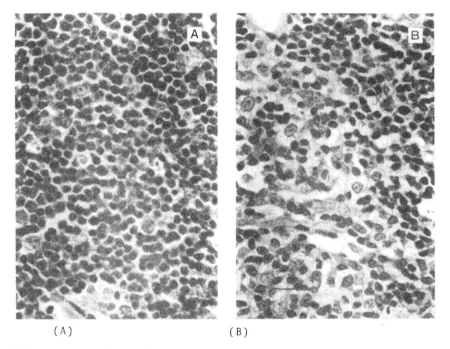

(A)   (B)

FIG. 3. Part of thymic cortex from a control rat (A) and a rat (B) fed 100 ppm Bu_2SnCl_2 for 2 weeks. Note the complete lymphocyte depletion of the cortex (B). Hematoxylin and eosin; ×400.

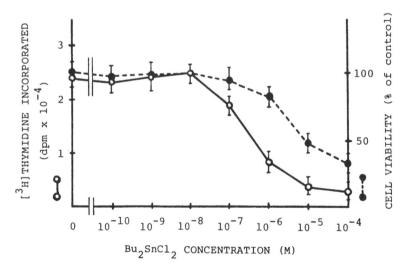

FIG. 4. Effect of Bu_2SnCl_2 concentrations on DNA synthesis (O) and viability (●) of rat thymocytes. Cells (10^6 cells/ml) were cultured in triplicate during 24 hr and [^{3}H]tymidine was present during the last 4 hr of the culture period. Vertical bars denote SE of the mean for seven determinations.

lymphocyte transformation by the T-cell mitogens phytohemagglutinin
(PHA) and concanavalin A (ConA). The resistance against *Listeria
monocytogenes*, also a T-cell-dependent phenomenon, is dose relatedly
decreased.

Triorganotin compounds such as tributyltin and triphenyltin
also suppress the thymus-dependent immune responses. However, possible immunotoxic properties of trimethyltin or triethyltin compounds
may be overshadowed by their neurotoxicity [12]. For the higher
trialkyltin homologs only some (trihexyltin) or no effect (trioctyltin) on the thymus is observed.

2.6. Antitumor Activity

Dialkyltin compounds, particularly dibutyltin, exhibit antitumor
activity toward various tumor systems such as Ehrlich ascites tumor,
IMC carcinoma, p-388 lymphocytic leukemia, and sarcoma 180 in descending order of activity in mice. As shown in Table 2, dibutyltin exhibits the highest activity against the Ehrlich ascites tumor system
[41,44-57]. With dosage ranged from 0.1 to 3.0 mg/kg, life span
values (T/C values, a compound is considered active if T/C > 115%)
ranges from 98% to 186%. Particularly, the maximum activity is
observed at a single and high dose (3 mg/kg).

In a two-stage mouse skin carcinogenesis system of initiation
and promotion, the optimal dose of dibutyltin inhibits the promotion
stage, especially the first phase 12-o-tetradecanoylphorbol-13-acetate
(TPA) promotion of the two-stage promotion system, more strongly than
the initiation stage (Fig. 5) [44-57].

On the other hand, in vitro studies show consistently that
dibutyltin dramatically inhibits the proliferation of malignant cells
such as thymic lymphosarcoma cells and HeLa cells (Fig. 6) [39-41],
and further it inhibits the initiation stage of the two-stage transformation of BALB/c 3T3 cells (Fig. 7) [39,44-56].

TABLE 2

Screening Data for Antitumor Activity
of Dibutyltin Compound

Tumor	Dose[a]		Life span[b]	
	mg/kg i.p.	Injection time	Survival time (day)	T/C (%)
Sarcoma 180	0	5	15.1	100
	0.1	5	17.6	117
	0.3	5	17.8	118
	1.0	5	15.8	105
	2.0	2	15.0	100
	2.0	4	15.0	100
	3.0	1	13.2	87
	3.0	2	13.0	86
IMC carcinoma	0	5	16.5	100
	0.1	5	18.6	113
	0.3	5	19.1	116
	1.0	5	19.8	120
	2.0	2	20.8	126
	2.0	4	21.3	129
	3.0	1	21.0	127
	3.0	2	22.0	133
Lymphocyte leukemia P-388	0	5	10.4	100
	0.1	5	10.6	102
	0.3	5	12.1	116
	1.0	5	11.8	114
	2.0	2	12.3	118
	2.0	4	12.2	117
	3.0	1	12.1	116
	3.0	2	12.5	120
Ehrlich ascites tumor	0	5	21.1	100
	0.1	5	20.7	98
	0.3	5	21.8	104
	1.0	5	22.7	108
	2.0	2	29.1	138
	2.0	4	28.1	133
	3.0	1	39.3	186
	3.0	2	37.6	178

[a]A total of one to five injections were given at daily intervals in one experiment.
[b]The increase in survival of treated animals over control is expressed as T/C (%). The values are the means of 10 animals per group.

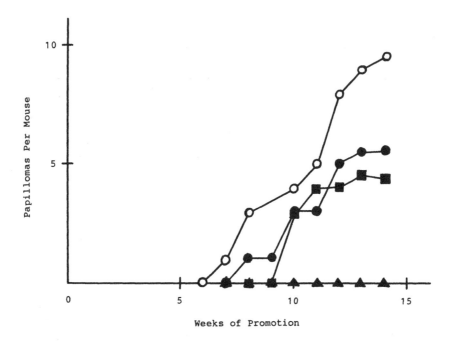

FIG. 5. Inhibitory effect of Bu_2SnCl_2 on two-stage mouse skin
carcinogenesis system of initiation and promotion. The mice were
initiated with 100 nmol of DMBA and promoted with 5 μg of TPA and
2.5 μg of mezerein. Bu_2SnCl_2 (5 μg) was applied 30 min before
treatment with the initiater or promoters. (o) Control, (●) DMBA-
Bu_2SnCl_2 (initiation stage), (▲) TPA-Bu_2SnCl_2 (promotion stage I),
(■) Mezerein-Bu_2SnCl_2 (promotion stage II).

In contrast to this result of in vitro two-stage transformation studies, in vivo studies of two-stage mouse skin carcinogenesis
systems show consistently that dibutyltin effectively inhibits the
TPA promotion stage. The reason for this apparent discrepancy
between results in vivo and in vitro is unknown. Further in vivo
and in vitro studies on its effects are needed. Other alkyltin
compounds such as monoalkyltin and trialkyltin are inactive against
in vivo carcinogenesis system.

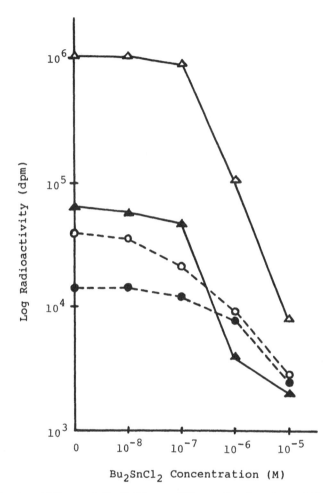

FIG. 6. Effect of Bu_2SnCl_2 on DNA synthesis of proliferating cells.
Cells (each, 1 x 10^6 cells/ml) were cultured with varying amounts of
Bu_2SnCl_2 in octuple during 24 hr and [^{3}H]thymidine was present during
the last 4 hr of the culture period. (○) ConA-stimulated thymocytes,
(●) nonstimulated thymocytes, (▲) thymic lymphosarcoma cells, (■)
HeLa cells.

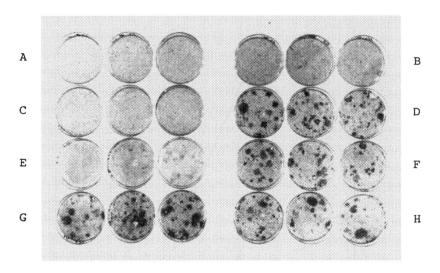

FIG. 7. Effect of Bu_2SnCl_2 on two-stage in vitro transformation of
BALB/c 3T3 cells. The cells (1×10^4) were initiated with MCA
(0.1 µg/ml) and promoted with TPA (0.1 µg/ml). Bu_2SnCl_2 was added
simultaneously with either MCA or TPA. A: whole negative control,
B: MCA alone (promotion blank), C: TPA alone (initiation blank),
D: MCA-TPA (positive control), E: [MCA + $Bu_2Sn(3 \times 10^{-7}$ M)]-TPA,
F: MCA-[TPA + $Bu_2Sn(3 \times 10^{-7}$ M)], G: [MCA + $Bu_2Sn(3 \times 10^{-8}$M)]-TPA,
H: MCA-[TPA + $Bu_2Sn(3 \times 10^{-8}$ M)].

2.7. Anti-inflammatory Action

Organotin compounds such as dibutyltin, tributyltin, and triphenyltin
significantly suppress not only chemotactic response of neutrophils
[58-60] but also the release of arachidonic acid [58-60] and a lysosomal enzyme such as β-glucuronidase in neutrophils. Moreover, each
of these suppressions is dose-dependent and their dose-response
curves are parallel to each other [58-60]. These results let our
interest focus on the anti-inflammatory action of these organotin
compounds because the steroids such as glucocorticoids possibly exert
their anti-inflammatory action by preventing the release of arachidonic acid from phospholipids, chemotaxis, and release of lysosomal
enzymes.

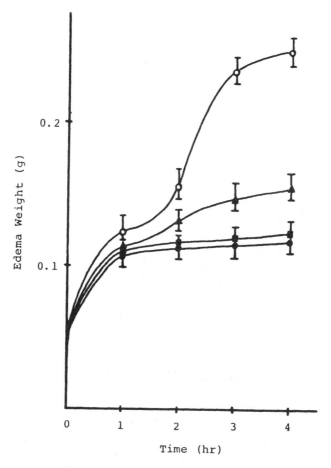

FIG. 8. Effect of organotin compounds on the development of edema
after subplantar injection of 0.5 mg of carrageenan in 0.05 ml of
pyrogen-free saline. The drugs were given p.o. at a dose of 10 mg/kg
1 hr before the irritant. Each point is corrected for the reading
taken at each time after the saline injection as blank control.
Vertical bars denote SE of the mean for eight determinations.
(o) Positive control, (●) Bu_2SnCl_2, (▲) Ph_3SnCl, (■) hydrocortisone.

Consequently, the anti-inflammatory action of organotin compounds was assessed on the carrageenan-induced foot edema as compared with that of glucocorticoids (Fig. 8) [61,62]. Dibutyltin
produces strong dose-dependent inhibition of the second phase which

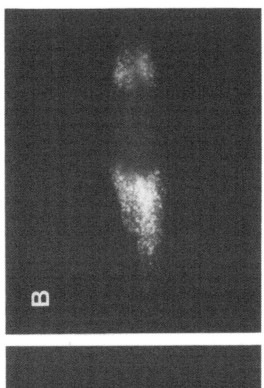

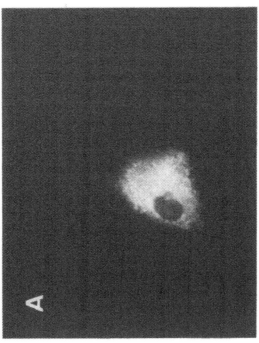

FIG. 9. Intracellular distribution of various organotin compounds in fixed cell. Human skin fibroblasts were fixed with 0.5% glutaraldehyde for 10 min at room temperature, washed, and incubated for 10 min at 37°C with 50 μM Bu$_2$SnCl$_2$ (A) or 100 μM Me$_2$SnCl$_2$ (B). The cells were then washed, incubated with 100 μg/ml morin for 5 min at 37°C, washed again, and photographed.

is maintained by the release of prostaglandins without the development of the first phase which is maintained by histamine and sero-
tonin. This inhibitory effect is almost to the same extent as that
of hydrocortisone. Inhibitory effects of triorganotins are much
less than that of dibutyltin. Neither dibutyltin, triphenyltin, nor
hydrocortisone inhibits the development of the first phase of edema.

3. CELLULAR AND BIOCHEMICAL ASPECTS OF THE PROPERTIES OF ALKYLTINS

3.1. Intracellular Distribution of Organotins

The toxicity of alkyltin derivatives depends on their solubility in
biological fluids and the extent of their incorporation into the
cells, i.e., their intracellular distribution, which are provided
by the number and length of their organic ligands.

Visualization of the intracellular distribution of organotin
compounds reveals that dibutyltin, triethyltin, tributyltin, and
triphenyltin, but not monomethyltin, monobutyltin, and dimethyltin,
are selectively distributed to the region of Golgi apparatus and
endoplasmic reticulum (ER) (Fig. 9) [63,64]. These results show
that the differences in the intracellular distribution of organotin
compounds are primarily due to the extent of their liposolubility or
their affinity to the intracellular lipids and lipophilic proteins.

3.2. Effects on Structure and Function of Golgi Apparatus and Endoplasmic Reticulum

Dibutyltin destroys the structure of the Golgi apparatus (Fig. 10B)
and suppresses the organelle functions such as the lipid metabolism
(Fig. 11). In particular, the metabolism of ceramide was significantly affected. These results show that the suppression of Golgi
functions by organotins may be due to the destruction of the Golgi
apparatus structure [63,65].

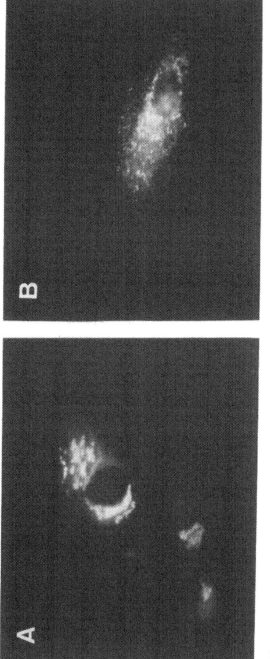

FIG. 10. Effects of organotin compound on the morphology of the Golgi apparatus in living cells. Human skin fibroblasts (SF-TY) were incubated in the absence (A) or presence (B) of 1 μM Bu$_2$SnCl$_2$ for 2 hr, and then were stained with C$_6$-NBD-ceramide-BSA and viewed in the fluorescence microscope.

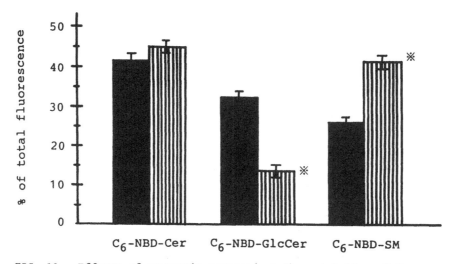

FIG. 11. Effects of organotin compound on the metabolism of C_6-NBD-ceramide in Golgi apparatus. CHO cells were incubated without (control, ■) and with 10 μM Bu_2SnCl_2 (▥) for 3 hr, and then incubated with C_6-NBD-Cer/BSA. Each value represents the mean ± SE of four independent experiments; those marked with asterisks differ significantly (Student's t test) from the corresponding control value (* $p < 0.001$). Cer: ceramide, GlcCer: glucosylceramide, SM: sphingomyelin.

Similarly, tributyltin and dibutyltin destroy the structure of the ER (Fig. 12B) [67] and inhibit the organelle functions such as inositol 1,4,5-triphosphate (IP_3)-induced intracellular Ca^{2+} mobilization by promoting Ca^{2+} release in a similar fashion to IP_3 (Fig. 13) [63,66]. These results show that the inhibition of Ca^{2+} mobilization by organotins may be due to the changes in the membrane structure and the destruction of ER structure. Monobutyltin and tetrabutyltin do not affect both the structure and the function of the ER (63,65-67].

From the results on the Golgi apparatus and the ER, the action of organotin compounds on each organelle appears to depend primarily on their intracellular distribution which are provided by their lipotropy, i.e., the number and length of their organic ligands. In addition, the above results suggest that certain organotin com-

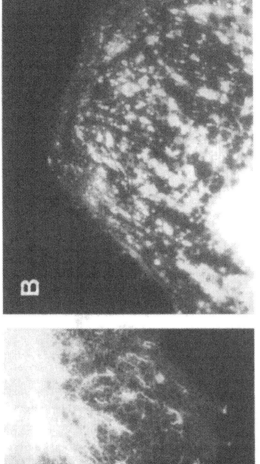

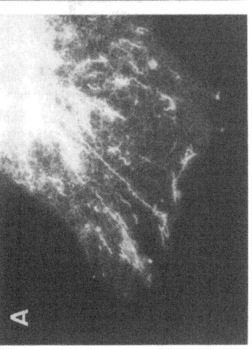

FIG. 12. Effects of organotin compounds on the morphology of the ER in living cell. African green monkey kidney epithelial cell lines (CV-1) were incubated in the absence (A) or presence (B) of 5 μM Bu₃SnCl for 10 min at 37°C. Further, the cells were fixed with 0.5% glutaraldehyde for 5 min at room temperature, washed, and incubated with 2.5 μg/ml DiOC₆ (3)(3.3'-dihexyloxacarbocyanine iodide) for 15 sec at room temperature. The cells were then washed and photographed.

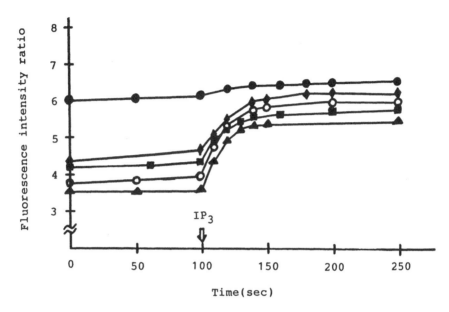

FIG. 13. Effects of organotin compounds on IP_3-induced intracellular
Ca^{2+} mobilization. Saponin-permeabilized RBL-2H3 cells (5 x 10^6 cells/
ml) were incubated with 5 x 10^{-6} M individual organotin compound for
10 min after the addition of 1.5 µM Fura-2. The kinetics of Ca^{2+}
release were resolved by determination of fluorescence intensity ratio
using dual excitation mode of 335 and 375 nm after the addition of
333 nM IP_3. Each point denotes the mean for five determinations:
control (o), $BuSnCl_3$ (▲), Bu_3SnCl (●), Bu_4Sn (■), TBTO (♦).

pounds may inhibit the phospholipid metabolism (mentioned later) by
causing the organelle's lesion in the Golgi apparatus and ER, e.g.,
by preventing the lipid transport from the Golgi apparatus to the
plasma membrane via the ER.

3.3. Effects on Mitochondrial Respiration

Trialkyltin compounds inhibit the oxidative phosphorylation that
would be conducted by the mitochondria. This inhibition is provoked
by binding of the alkyltins with high affinity to sites on the mitochondria or its membrane [68-71], and is coupled with the inhibition
of a step in the energy-transferring chain between electron transport

and ATP formation [70]. Further, trialkyltin compounds mediate an exchange of halide for hydroxyl ions across the mitochondrial membrane [72] and bind to a component of the ATP synthase complex, leading to a direct inhibition of ATP production [73] and cause gross mitochondrial swelling [71]. These results show that trialkyltin compounds act also as effective inhibitors of mitochondrial ATP synthesis. The order of effectiveness in the inhibition of oxidative phosphorylation by trialkyltin compounds is triethyltin > tributyltin > tripropyltin > triphenyltin > trimethyltin [74]. At higher concentrations, dibutyltin also inhibits the oxidative phosphorylation processes in the mitochondria [75,76].

Most of dialkyltin compounds inhibit the mitochondrial respiration by preventing the oxidation of α-keto acids which would be oxidized by the pyruvate and α-ketoglutarate dehydrogenase [77,78]. Dihexyltin is an active homolog, but the dioctyltin derivatives appear to be inactive.

Triethyltin also inhibits selectively the glucose oxidation [79]. The incorporation of ^{14}C from glucose into glutamate, glutamine, α-aminobutyrate, and aspartate is significantly decreased. The incorporation of ^{14}C from acetate into these amino acids is unaffected. These results show that the main action of triethyltin may be to decrease the rate at which pyruvate is oxidized.

3.4. Inhibition of Phospholipid Metabolism

Biochemical studies of alkyltin compounds on the cellular functions using the lymphocyte transformation reveal that dibutyltin inhibits DNA synthesis (Figs. 4 and 6), RNA synthesis, and RNA polymerase and phospholipid metabolism (Fig. 14) [41,80-83]. Ca^{2+} entry into the cells is not affected [41].

Particularly, at the earliest stage of lymphocyte transformation, dibutyltin inhibits the breakdown of inositol phospholipids, that is, phosphatidylinositol (PI) turnover, which appears to be one

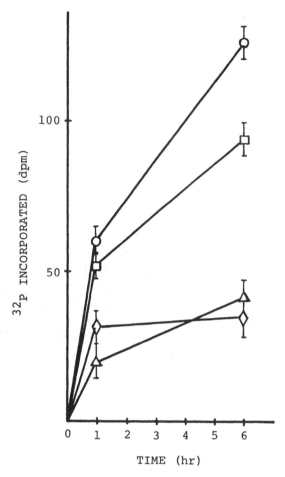

FIG. 14. Effect of organotin compounds on phospholipid synthesis of rat thymocytes. Phospholipid synthesis was measured by the incorporation of ^{32}P into the lipid fraction of the cultured cells (1.5×10^6 cells/ml) after stimulation with ConA (5 µg/ml) in the absence (o) and presence of 10^{-7} M n-Bu$_2$SnCl$_2$ ($\triangle$), 10^{-7} M MeSnCl$_3$ ($\square$), or 10^{-7} M Ph$_3$SnCl ($\diamond$). Each point is corrected for radioactivity incorporated without ConA at each incubation time. Vertical bars denote SE of the mean for five determinations.

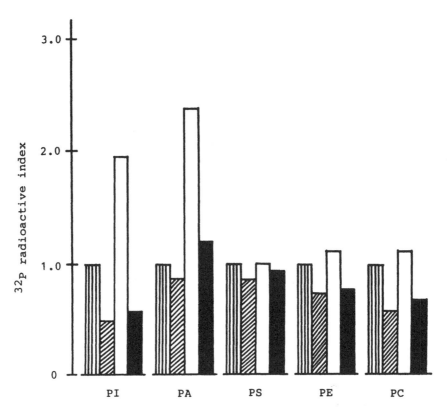

FIG. 15. Effect of Bu_2SnCl_2 on phospholipid metabolism. Rat thymocytes (3 x 10^6 cells/ml) prelabeled with $[^{32}P]$-phosphoric acid (10 μCi/ml) were treated without and with ConA (5 μg/ml) in the absence and presence of Bu_2SnCl_2 (5 x 10^{-7} M) for 5 min. The mean radioactivity of each phospholipid component separated from the control culture (▥) was taken as 1.0 and was compared to that of the corresponding phospholipid component from the experimental cultures treated with Bu_2SnCl_2 (▨), ConA (□), and ConA plus Bu_2SnCl_2 (■).

of the most important events in the lymphocyte transformation [80-83]. As shown in Fig. 15, the presence of dibutyltin inhibits the remarkable acceleration of PI and phosphatidic acid (PA) at the early stage of lymphocyte transformation.

On the other hand, the degradation of phospholipids is also inhibited by dibutyltin [41]. The presence of dibutyltin signifi-

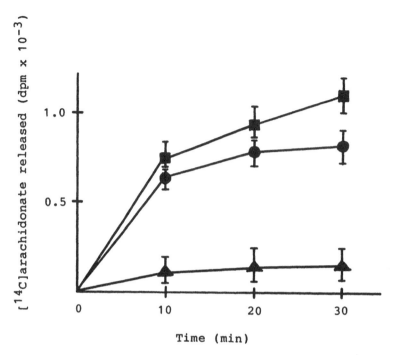

FIG. 16. Time course of release of arachidonic acid. [1-^{14}C]Arachidonate release was measured after stimulation with 5 μg of ConA in the absence (●) and presence of 10^{-6} M Bu_2SnCl_2 (▲) or 10^{-6} M $MeSnCl_3$ (■). Each point is corrected for radioactivity released without ConA at each incubation time. Vertical bars denote SE of the mean for five determinations.

cantly inhibits the acceleration of arachidonate release from phospholipids by phospholipase A_2 or C (Fig. 16). However, the direct inhibitory effects of dibutyltin on phospholipase A_2, C, and D are not found, although the substrate is hydrolyzed to three main products depending on the species of phospholipase (Table 3) [83,84]. These results suggest that the dibutyltin also may inhibit the activation system of phospholipase A_2 or C.

In fact, in the activation system of phospholipase A_2 or C, dibutyltin inhibits a transient increase of the phosphorylation of the lipocortin which is a phospholipase inhibitory protein and regulates the phospholipase activation (Fig. 17) [83,84]. This

TABLE 3

In Vitro Effect of Organotins on Various Phospholipase Activities

Enzyme source	Spot no.	Phosphatidylcholine hydrolyzed (% of total counts)			
		Organotin added (1 x 10^{-7} M)			
		Control	Bu_2SnCl_2	Bu_3SnCl	Ph_3SnCl
Phospholipase A_2	1	25.0 ± 0.7	27.3 ± 0.2	31.3 ± 0.8	30.1 ± 0.7
(porcine pancreas)	2	5.5 ± 0.2	7.3 ± 0.3	8.2 ± 0.9	9.0 ± 1.2
	3	4.8 ± 0.1	5.9 ± 0.5	7.1 ± 1.1	6.5 ± 0.7
Phospholipase C	1	7.5 ± 1.3	8.2 ± 1.0	5.3 ± 0.5	6.8 ± 0.3
(*B. cereus*)	2	0.7 ± 0.2	0.7 ± 0.1	0.6 ± 0.1	1.0 ± 0.2
	3	70.0 ± 4.2	63.0 ± 2.8	75.0 ± 5.9	63.0 ± 3.1
Phospholipase D	1	15.8 ± 0.9	17.4 ± 1.2	17.6 ± 1.2	16.8 ± 0.7
(cabbage)	2	1.0 ± 0.1	1.3 ± 0.1	1.4 ± 0.1	0.9 ± 0.1
	3	1.5 ± 0.3	2.3 ± 0.2	2.1 ± 0.2	1.4 ± 0.1

Note: The spot numbers indicate the hydrolyzed components of substrate, L-α-dipalmitoyl[cholinemethyl-14C-]phosphatidylcholine. The total recovered radioactivity from one sample was taken as 100% and the relative radioactivity in each spot was determined from this value. Each value is the mean ± standard error of five determinations

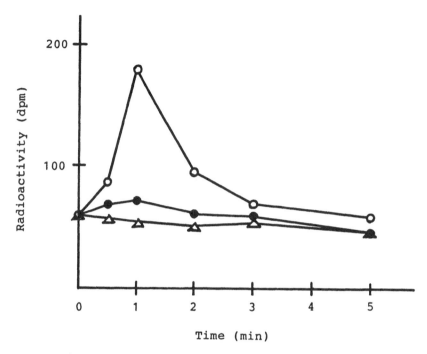

FIG. 17. Time course of phosphorylation of lipocortin. Neutrophils were preincubated with ^{32}P and then stimulated by 10^{-8} M fMet-Leu-Phe in the presence and absence of 10^{-7} M Bu_2SnCl_2. The ^{32}P-labeled lipocortin was immunoprecipitated with a serum from a patient with systemic lupus erythematosus. Each point represents the mean of three determinations. ($\triangle$) Control, (o) fMet-Leu-Phe, ($\bullet$) Bu_2SnCl_2 + fMet-Leu-Phe.

lipocortin is an extrinsic membrane protein, and it has a strong affinity toward charging phospholipids such as phosphatidylinositol and phosphatidylserine at the inside of a membrane. Therefore, this inhibition of the lipocortin phosphorylation by dibutytin may be the membrane-mediated inhibition which may be associated with the changes in physical properties and structure of the surrounding phospholipid membrane, e.g., the membrane order and the possible phase separation. Or this inhibition may be associated with the inhibition of ATP synthesis.

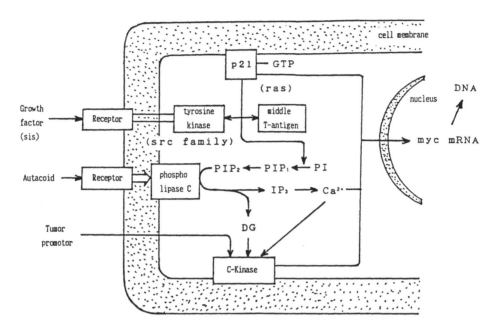

FIG. 18. Cell growth and transformation via oncogene products.

3.5. Inhibition of Membrane Signal Transduction

For the present, the mechanism of cell proliferation and transformation is not fully understood. However, some hypotheses are proposed
for a main pathway leading to DNA synthesis and ultimate mitotic
division (Fig. 18).

Either way, a signal transduction leading to DNA synthesis
appears to be initiated by stimulating the phospholipase activation
system and provoking PI turnover and arachidonate release (Fig. 19).

Studies on the mechanism for the inhibition of cell proliferation by dibutyltin reveal that the compound inhibits selectively PI
turnover (Fig. 15) [81,82], arachidonate release (Fig. 16) [41],
phosphorylation of lipocortin which may regulate the phospholipase
activation (Fig. 17) [83,84], and finally DNA synthesis (Figs. 4 and
6). Moreover, a parallelism is found between dose-dependent inhibition of acceleration of PI turnover or arachidonate release and that

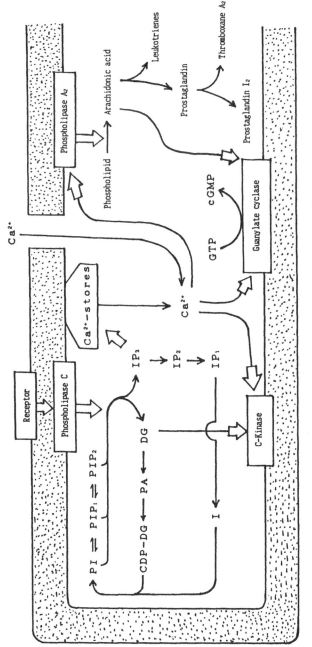

FIG, 19, Signal transduction via phospholipid turnover and Ca mobilization.

of DNA synthesis by dibutyltin. Ca^{2+} entry into the cells, which may be one of the main pathways leading to DNA synthesis, is not inhibited at all by dibutyltin [41].

Taking these results into consideration with the fact that dibutyltin does not directly inhibit phospholipase C, phospholipase A_2, or protein kinase C, the suppression of cell proliferation by dibutyltin appears to reflect the membrane-mediated inhibition of the signal transduction leading to DNA synthesis such as the phospholipase activation system.

3.6. Relationship Between Biological Activity and Physical Properties of Phospholipid Membrane

As shown in Fig. 20, the membrane fluidity of each phospholipid vesicle measured by the fluorescence polarization shows a significant decrease by the presence of dibutyltin [81,82]. Particularly, the dibutyltin shows a far stronger "ordering" effect on phosphatidyl-inositol 4-monophosphate (PIP_1) and phosphatidylinositol 4,5-diphosphate (PIP_2) vesicle membranes than on other phospholipid vesicle membranes. These effects are dose-dependent. This finding appears to provide clues for the inhibitory mechanism of dibutyltin on PI turnover, as mentioned above. The reason is that PIP_2 is the immediate target in provoking the breakdown of inositol phospholipid. PIP_2 is the substrate which is hydrolyzed to diacylglycerol and inositol phosphate or inositol polyphosphate by phospholipase C (Fig. 19). The significant decrease in the fluidity of PIP_2 vesicle membrane may be directly associated with the inhibitory effect of dibutyltin on the acceleration of PI turnover at the early stage of lymphocyte transformation. If the dibutyltin-induced inhibition of PI turnover is caused by the significant decrease in the fluidity of inositol phospholipid membranes, it may be an important clue and an effective tool for the elucidation of the mechanisms of inositol phospholipid breakdown and phospholipase C activation, because the alteration of membrane fluidity and lateral clustering of substrate may be directly linked to the activation of phospholipase C.

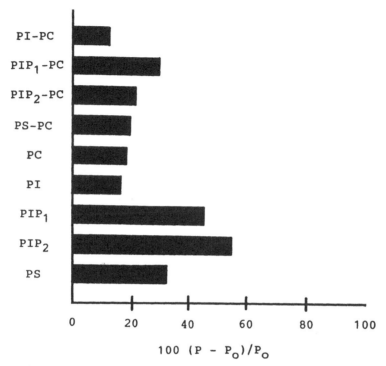

FIG. 20. Effect of Bu_2SnCl_2 on the membrane order of various phospholipid vesicles. Each phospholipid vesicle containing DPH in 20 mM Tris-HCl buffer (pH 7.5, containing 100 mM NaCl) was incubated without and with Bu_2SnCl_2 (10^{-4} M) for 5-7 min at 25°C and the membrane order was measured by fluorescence polarization. Each value was expressed as a percentage change in the degree of fluorescence polarization, $100(P - P_0)/P_0$. All the phospholipid vesicles contained 18 µg/ml total phospholipid. In case of the mixture vesicles, the ratio of each phospholipid to PC is 1/2.

ABBREVIATIONS, SYMBOLS, AND FORMULAS

ATP	adenosine 5'-triphosphate
BSA	bovine serum albumin
Bu	butyl
$BuSnCl_3$	monobutyltin trichloride
Bu_2SnCl_2	dibutyltin dichloride
Bu_3SnCl	tributyltin chloride
Bu_4Sn	tetrabutyltin

BW	body weight
Cer	ceramide
CHO	Chinese hamster ovary
ConA	concanavalin A
DAG	diacylglycerol
DMBA	7,12-dimethylbenzanthracene
DNA	deoxyribonucleic acid
DPH	1,6-diphenyl-1,3,5-hexatriene
ER	endoplasmic reticulum
fMet-Leu-Phe	N-formyl-methionylleucylphenylalanine
GlcCer	glucosyl ceramide
^{3}H	tritiated
IP_3	inositol 1,4,5-triphosphate
MCA	20-methylcholanthrene
Me	methyl
$MeSnCl_3$	monomethyltin trichloride
Me_2SnCl_2	dimethyltin dichloride
N	animal number
NBD	4-nitrobenzo-2-oxa-1,3-diazole
Oc	octyl
Oc_2SnCl_2	dioctyltin dichloride
p	significant difference
P	fluorescence polarization
PA	L-α-phosphatidic acid
PC	L-α-phosphatidylcholine
PE	L-α-phosphatidylethanolamine
Ph	phenyl
Ph_3SnCl	triphenyltin chloride
PHA	phytohemagglutinin
PI	L-α-phosphatidylinositol
PIP_1	L-α-phosphatidylinositol 4-monophosphate
PIP_2	L-α-phosphatidylinositol 4,5-diphosphate
Pr	propyl

Pr_2SnCl_2	dipropyltin dichloride
PS	L-α-phosphatidylserine
PVC	polyvinylchloride
RBL	rat basophilic leukemia
RNA	ribonucleic acid
SE	standard error
SM	sphingomyelin
TBTO	bistributyltin oxide
T/C	the ratio of survival time of treated animals to that of control animals
TPA	12-O-tetradecanoylphorbol-13-acetate

REFERENCES

1. Y. Arakawa, O. Wada, and T. H. Yu, *Toxicol. Appl. Pharmacol.*, *60*, 1 (1981).

2. H. B. Stoner, J. M. Barnes, and J. I. Duff, *Br. J. Pharmacol.*, *10*, 16 (1955).

3. J. M. Barnes and P. N. Magee, *J. Pathol. Bacteriol.*, 75, 267 (1958).

4. W. Seinen and M. I. Willems, *Toxicol. Appl. Pharmacol.*, *35*, 63 (1976).

5. Y. Arakawa, N. Yamazaki, T. H. Yu, and M. Nagahashi, *J. Toxicol. Sci.*, *5*, 258 (1980).

6. N. Yamazaki, Y. Arakawa, M. Nagahashi, and T. H. Yu, *Jpn. J. Hyg.*, *35*(1), 206 (1980).

7. Y. Arakawa, N. Yamazaki, T. H. Yu, and M. Nagahashi, *Jpn. J. Hyg.*, *35*(1), 207 (1980).

8. Y. Arakawa, T. H. Yu, M. Hihara, N. Yamazaki, and S. Fujita, *J. Toxicol. Sci.*, *6*, 237 (1981).

9. Y. Arakawa, *J. Pharm. Dyn.*, *6*, s-23 (1983).

10. T. W. Bouldin, N. D. Goines, C. R. Bagnell, and M. R. Krigman, *Am. J. Pathol.*, *104*, 237 (1981).

11. P. Mushak, M. R. Krigman, and R. B. Mailman, *Neurobehav. Toxicol. Teratol.*, *4*, 209 (1982).

12. N. J. Snoeij, A. A. J. Van Iersel, A. H. Penninks, and W. Seinen, *Toxicol. Appl. Pharmacol.*, *81*, 274 (1985).

13. K. Suzuki, *Exp. Neurol.*, *31*, 207 (1971).

14. P. N. Magee, H. B. Stoner, and J. M. Barnes, *J. Pathol. Bacteriol.*, *73*, 107 (1957).

15. L. W. Reiter, G. B. Haevner, K. F. Dean, and P. H. Ruppert, *Neurobehav. Toxicol. Teratol.*, *3*, 285 (1981).

16. R. S. Dyer, W. E. Howell, and L. W. Reiter, *Neurotoxicology, 2,* 609 (1981).

17. G. J. Harry and H. A. Tilson, *Neurotoxicology, 2,* 283 (1981).

18. R. E. Squibb, N. G. Carmichael, and H. A. Tilson, *Toxicol. Appl. Pharmacol.*, *55*, 188 (1980).

19. J. M. Barnes and H. B. Stoner, *Br. J. Industr. Med.*, *15,* 15 (1958).

20. A. W. Brown, W. N. Aldridge, B. W. Street, and R. D. Verschoyle, *Am. J. Pathol.*, *97*, 59 (1979).

21. L. W. Chang, T. M. Tiemeyer, G. R. Wenger, D. E. McMillan, and K. R. Reuhl, *Environ. Res.*, *29*, 435 (1982).

22. L. W. Chang, T. M. Tiemeyer, G. R. Wenger, D. E. McMillan, and K. R. Reuhl, *Environ. Res.*, *29*, 445 (1982).

23. L. W. Chang, T. M. Tiemeyer, G. R. Wenger, and D. E. McMillan, *Environ. Res.*, *30*, 399 (1983).

24. L. W. Chang, G. R. Wenger, and D. E. McMillan, *Environ. Res.*, *34*, 123 (1984).

25. L. W. Chang and R. S. Dyer, *Neurobehav. Toxicol. Teratol.*, *5* 673 (1983).

26. T. W. Bouldin, N. D. Goines, and M. R. Krigman, *J. Neuropathol. Exp. Neurol.*, *43*, 162 (1984).

27. A. W. Brown, R. D. Verschoyle, B. W. Street, W. N. Aldridge, and H. Grindley, *J. Appl. Toxicol.*, *4*, 12 (1984).

28. K. R. Reuhl, S. G. Gilbert, B. A. Mackenzie, J. E. Mallett, and D. C. Rice, *Toxicol. Appl. Pharmacol.*, *79*, 436 (1985).

29. E. Fortemps, G. Amand, A. Bomboir, R. Lauwerys, and E. C. Laterre, *Int. Arch. Occup. Environ. Health, 41,* 1 (1978).

30. W. D. Ross, E. A. Emmett, J. Steiner, and R. Tureen, *Am. J. Psychiatry, 138,* 1092 (1981).

31. J. R. Elsea and D. E. Paynter, *Arch. Industr. Health, 18,* 214 (1958).

32. Z. Pelikan and E. Cerny, *Arch. Toxikol.*, *27*, 79 (1970).

33. E. I. Krajnc, P. W. Wester, J. G. Loeber, F. X. R. Van Leeuwen, J. G. Vos, H. A. M. G. Vaessen, and C. A. Van der Heijden, *Toxicol. Appl. Pharmacol.*, *75*, 363 (1984).

34. R. D. Kimbrough, *Environ. Health Persp., 14,* 51 (1976).

35. M. C. Middleton and I. Pratt, *J. Invest. Dermatol., 71,* 305 (1978).

36. H. B. Stoner, *Br. J. Industr. Med., 23,* 222 (1966).

37. C. L. Winek, M. J. Marks, S. P. Shanor, and E. R. Davis, *Clin. Toxicol., 13,* 281 (1978).

38. Y. Arakawa, *Trace Nutrients Res., 6,* 43 (1989).

39. Y. Arakawa, *Sugiyama Chem. Industr. Lab. Ann. Rep.,* 151 (1990).

40. Y. Arakawa, T. Abe, T. H. Yu, and O. Wada, *Jpn. J. Hyg., 40*(1), 540 (1985).

41. Y. Arakawa and O. Wada, in *Tin and Malignant Cell Growth* (J. J. Zuckerman, ed.), CRC Press, Boca Raton, Florida, 1988, pp. 83-106.

42. W. Seinen, J. G. Vos, R. Van Krieken, A. H. Penninks, R. Brands, and H. Hooykaas, *Toxicol. Appl. Pharmacol., 42,* 213 (1977).

43. W. Seinen, in *Immunologic Considerations in Toxicology* (R. P. Sharma, ed.), CRC Press, Boca Raton, Florida, 1981, Vol. 1, pp. 103-119.

44. Y. Arakawa and O. Wada, *Proc. of the 2nd Int. Symp. on Tin upon Malignant Cell Growth,* Scranton, PA, 1985, pp. 4-5.

45. Y. Arakawa, T. Abe, T. H. Yu, and O. Wada, *J. Pharm. Dyn., 10,* s-2 (1987).

46. Y. Arakawa, T. Abe, T. H. Yu, and O. Wada, *Proc. of the 3rd Symp. on Roles of Metal in Biological Reactions, Biology and Medicine,* Nagoya, Japan, 1986.

47. Y. Arakawa, *Proc. of the 3rd Int. Symp. on Tin upon Malignant Cell Growth,* Padua, Italy, 1986, pp. 24-25.

48. Y. Arakawa, *Proc. of the 4th Int. Symp. on Tin upon Malignant Cell Growth,* Scranton, PA, 1988, p. 13.

49. Y. Arakawa, *Proc. of the NATO Advanced Research Workshop on Tin upon Malignant Cell Growth,* Brussels, 1989, p. 1.

50. Y. Arakawa, *Proc. of the 2nd Meeting of Int. Society for Trace Element Research in Humans* (ISTERH), Tokyo, 1989.

51. Y. Arakawa, *J. Trace Elements in Exp. Med., 2,* 114 (1989).

52. Y. Arakawa, *Proc. of the 1st A.N.A.I.C. Int. Chem. Conf. on Silicon and Tin,* Kuala Lumpur, Malaysia, 1989, p. 31.

53. Y. Arakawa, *Biomed. Res. Trace Elements, 1,* 135 (1990).

54. Y. Arakawa, *J. Pharm. Dyn., 14,* s-131 (1991).

55. Y. Arakawa, in *Chemistry and Technology of Silicon and Tin* (V. G. Kumar Das, ed.), Oxford University Press, Oxford, 1992, pp. 319-333.

56. Y. Arakawa, *Jpn. J. Hyg., 44*(1), 437 (1989).

57. Y. Arakawa, *Jpn. J. Hyg., 44*(1), 438 (1989).

58. Y. Arakawa and O. Wada, *Biochem. Biophys. Res. Commun., 123,* 543 (1984).

59. Y. Arakawa, O. Wada, T. H. Yu, and S. Fujita, *Jpn. J. Hyg.*, *39*(1), 394 (1984).

60. Y. Arakawa, *Jpn. J. Hyg.*, *40*(1), 541 (1985).

61. Y. Arakawa and O. Wada, *Biochem. Biophys. Res. Commun.*, *125*, 59 (1984).

62. Y. Arakawa, *Proc. of the 5th Int. Conf. on Organometal. Coord. Chem. of Germanium, Tin and Lead*, Padua, Italy, 1986, p. 3.

63. Y. Arakawa, T. Iizuka, and C. Matsumoto, *Biomed. Res. Trace Elements*, *2*(3), 321 (1991).

64. Y. Arakawa, *Proc. of the 6th Int. Conf. on Organometal. Coord. Chem. of Germanium, Tin and Lead*, Brussels, 1989.

65. T. Iizuka and Y. Arakawa, *Jpn. J. Hyg.*, *46*(1), 302 (1991).

66. T. Iizuka, C. Matsumoto, and Y. Arakawa, *Jpn. J. Hyg.*, *46*(1), 303 (1991).

67. T. Iizuka and Y. Arakawa, *Jpn. J. Hyg.*, *47*(1), 249 (1992).

68. W. N. Aldridge and B. W. Street, *Biochem. J.*, *124*, 221 (1971).

69. W. N. Aldridge and B. W. Street, *Biochem. J.*, *118*, 171 (1970).

70. W. N. Aldridge, *Biochem. J.*, *69*, 367 (1958).

71. W. N. Aldridge and B. W. Street, *Biochem. J.*, *91*, 287 (1964).

72. M. J. Selwyn, A. P. Dawson, M. Stockdale, and N. Gains, *Eur. J. Biochem.*, *14*, 120 (1970).

73. M. J. Selwyn, *Adv. Chem. Ser.*, *157*, 204 (1976).

74. M. Stockdale, A. P. Dawson, and M. J. Selwyn, *Eur. J. Biochem.*, *15*, 342 (1970).

75. K. Cain, R. L. Hyams, and D. E. Griffiths, *FEBS Lett.*, *82*, 23 (1977).

76. A. H. Penninks, P. M. Verschuren, and W. Seinen, *Toxicol. Appl. Pharmacol.*, *70*, 115 (1983).

77. W. N. Aldridge, *Adv. Chem. Ser.*, *157*, 186 (1976).

78. A. H. Penninks and W. Seinen, *Toxicol. Appl. Pharmacol.*, *56*, 221 (1980).

79. J. E. Cremer, *Biochem. J.*, *119*, 95 (1970).

80. Y. Arakawa, S. Fujita, T. H. Yu, and N. Yamazaki, *Jpn. J. Hyg.*, *37*(1), 174 (1982).

81. Y. Arakawa, *Main Group Metal Chem.*, *12*, 37 (1989).

82. Y. Arakawa, *Jpn. J. Hyg.*, *44*(1), 439 (1989).

83. Y. Arakawa, O. Wada, and T. H. Yu, *Jpn. J. Hyg.*, *39*(1), 393 (1984).

84. Y. Arakawa, *Main Group Metal Chem.*, *12*, 31 (1989).

5
Biological Properties of Alkyl Derivatives of Lead

Yukio Yamamura and Fumio Arai

Department of Public Health
St. Marianna University, School of Medicine
2-16-1 Sugao
Miyamae-ku
Kawasaki 216, Japan

1. INTRODUCTION

Tetraethyllead (Et_4Pb) and tetramethyllead (Me_4Pb) are the tetra-alkyllead (R_4Pb) compounds currently in use as antiknock agents for gasoline. The antiknock action of Et_4Pb was discovered by Midgley and Boyd [1] in 1921, and its industrial production was begun in the USA in 1924. It was already known at that time that Et_4Pb is a very toxic compound. At the Et_4Pb production plants, 138 workers suffered from severe Et_4Pb poisoning, including 13 deaths, during the first several months of production, and its production was then suspended for more than 1 year. Since then, with its use as an additive limited to gasolines for internal combustion engines only, and with the allowances set at 3 ml per gallon of automobile gasoline and at 4.6 ml per gallon of aeronautical gasoline, its production and sales have been permitted. In 1959, Me_4Pb was developed as another antiknock additive, and mixtures of Et_4Pb and Me_4Pb have since then been used as the antiknock agents of the mixed lead alkyls. The antiknock agents usually are composed of about 60% R_4Pb, and about 17% ethylene dibromide and about 18% ethylene dichloride as scavengers for the lead contained in the exhaust, respectively.

There are many reports on R_4Pb poisoning resulting from the production and accidental ingestion of antiknock agents, cleaning of leaded gasoline storage tanks, accidents in transit, and erroneous use of leaded gasoline in degreasing or dry cleaning. In Japan, a total of 61 deaths due to R_4Pb-related causes have been reported since 1938 [2].

In the advanced countries, air pollution with lead has become a serious problem along with the increasing consumption of leaded gasoline. From the technological point of view, on the other hand, the improvement of gasoline engines along with progress in the petroleum-refining process has been reducing the necessity for the addition of alkyllead antiknock agents to gasoline. Today, leaded gasoline is no longer used as the fuel for common automobiles in Japan and the USA. In the European Community member countries, there is a move to reducing the R_4Pb addition allowance to less than 0.15 g/liter, whereas no restrictions are enforced over the use of leaded gasoline in the eastern European countries, and in the Third World countries in South America, the Near East, Africa, and Asia. Under such a current situation, the output at the R_4Pb production plants is still on the increase on the worldwide basis.

2. CHEMISTRY

The alkylleads are the organolead compounds with several alkyl groups, such as methyl, ethyl, butyl, propyl, hexyl, etc. (C_nH_{2n+1}) attached directly to one lead atom. Of the alkylleads, Me_4Pb, trimethylethyllead (Me_3EtPb), dimethyldiethyllead (Me_2Et_2Pb), methyltriethyllead ($MeEt_3Pb$), and Et_4Pb are currently in use as antiknock agents.

R_4Pb is fat-soluble, but its degradation products, i.e., trialkyllead (R_3Pb^+) and dialkyllead (R_2Pb^{2+}) compounds, are water-soluble.

2.1. Tetraethyllead

$Pb(C_2H_5)_4$. Molecular weight: 323.4. Melting point: -136.8°C. Boiling point: about 200°C. Density: 1.653 at 20°C. Vapor pressure: 0.25 mm Hg at 20°C. Has an unpleasant, sweet odor, and is nauseating on smelling.

2.2. Tetramethyllead

$Pb(CH_3)_4$. Molecular weight: 267.3. Melting point: $-27.5°C$.
Boiling point: $110°C$. Density: 1.995 at $20°C$. Vapor pressure:
24 mm Hg at $20°C$. Has a peppermint-like odor.

3. ANALYSIS

R_4Pb is determined by atomic absorption spectrometry (AAS) [3,4],
gas chromatography (GC) [5,6], or GC-AAS [7]. R_3Pb^+ is determined
by colorimetry [8,9], AAS [4,10], hydride generation AAS [11],
hydride generation-nondispersive atomic fluorescence spectrometry
(NDAFS) [12], GC [6,13], GC-AAS [14], or high-performance liquid
chromatography (HPLC-AAS) [15]. R_2Pb^{2+} is determined by colorimetry
[8,16-18], AAS [4,19], hydride generation-AAS [11], hydride generation-
NDAFS [12], GC [13], GC-AAS [14,20], or HPLC-AAS [15]. Inorganic lead
(Pb^{2+}) is determined by at first producing precipitable lead by Bolanowska's method [21] and then determining the precipitable lead by AAS.

3.1. Determination of Ambient Alkylleads

The Lead-in-Air Analyzer Kit is an apparatus for trapping and determining ambient R_4Pb rapidly, which is commercially available from
Ethyl Corporation (Houston, Texas, USA, Tank Cleaning Manual). This
kit is used by the following procedure: an iodine scrubber is
inserted into the tip of an air-trapping pump, and an exactly 1-ft^3
sample of air is trapped through this scrubber. The R_4Pb so trapped
is removed from the scrubber with acidic potassium iodine solution
and then determined by the colorimetric dithizone method [22]. The
determination of ambient R_4Pb in the leaded gasoline storage tanks
by the use of this analyzer should never be performed before the
tanks are cleaned. In the tanks in which leaded gasoline has been
stored, considerable amounts of R_4Pb and its degradation compounds

remain. For this reason, no workers should be permitted to enter
such tanks which have not been completely cleaned without wearing
respiratory and skin protection devices. The analyzer is a device
used to ensure whether the insides of such tanks are completely
clean as a result of cleaning work.

There is another method by which to trap R_4Pb in a solution
of iodine monochloride in hydrochloric acid by the use of a midget
impinger. After trapping sufficient ambient air, the air sample is
brought back to the laboratory, and this iodine-trapping solution
is reduced to iodine with sodium sulfite and hydroxylamine hydro-
chloride. Lead ion is extracted as an ammonium pyrrolidine dithio-
carbamate (APDC)-lead complex into methyl isobutyl ketone (MIBK) by
the use of APDC, and determined by AAS [23-25]. Another determination
method is that of ambient alkyllead being trapped by a small vapor-
absorbing tube filled with porous polymer beads; the absorbing tube
is heated, and Et_4Pb and Me_4Pb are isolated and determined by AAS
[26]. In addition, five alkylleads, e.g., Me_4Pb, Me_3EtPb, Me_2Et_2Pb,
$MeEt_3Pb$, and Et_4Pb, may be identified respectively by sampling ambient
air into a glass column packed with a GC absorbant, and determining
the so trapped R_4Pb by GC [5,6] or GC-AAS [27].

3.2. Determination of Alkylleads in Biological Samples

Urine samples are transferred into polypropylene bottles which have
been washed with dilute nitric acid and redistilled water beforehand,
and kept below 4°C during transportation. When samples are to be
stored, they are frozen at -20°C to prevent organoleads from degrad-
ing. Blood samples are taken into heparinized, evacuated blood-
collecting tubes and stored at low temperature.

The methods for determination of R_3Pb^+, R_2Pb^{2+}, and Pb^{2+} in
the urine and blood are described below. After the addition of NaCl
to a urine or blood sample, R_3Pb^+ is extracted into organic solvents,
e.g., benzene, MIBK, etc., and then determined by colorimetry [21],
AAS [4], GC [6], or hydride generation AAS [11]. For the determina-

tion of R_2Pb^{2+}, a sample is adjusted to alkalinity; R_2Pb^{2+} in the sample is extracted into chloroform or MIBK by the use of a solution of glyoxalbis(2-hydroxyanil) in methanol [17] and determined by colorimetry [17,28], AAS [4,19], or hydride generation AAS [11]. Pb^{2+} is at first converted to precipitable lead by Bolanowska's method [21], and the precipitable lead is decomposed by heating in nitric acid, extracted with sodium diethyldithiocarbamate (SDDC)- MIBK, and determined by AAS [28]. Et_4Pb is at first extracted into organic solvents, e.g., MIBK and benzene, and then determined by AAS or GC.

4. METABOLISM

R_4Pb compounds when absorbed into the body are not excreted in unchanged form. Kehoe [29] suggested in experiments in rabbits that the latent period to evolution of the characteristic mental manifestations after Et_4Pb administration may indicate that the manifestations potentially derive from the water-soluble organolead compounds resulting from the conversion of fat-soluble Et_4Pb in the body. Mortensen [30] described that Et_4Pb, a volatile lead, would not be detected in the brain, liver, kidneys, or blood several hours after exposure. At that time, there was no analytical method available for the analysis of triethyllead (Et_3Pb^+). Cremer [31] later detected Et_3Pb^+ in the liver, kidneys, and blood of rats treated with Et_4Pb. Bolanowska et al. [32] found Et_3Pb^+ in the liver, kidneys, brain, blood, and urine of human victims of Et_4Pb poisoning. Furthermore, other investigators showed that lead is excreted in three chemical forms—Et_3Pb^+, diethyllead (Et_2Pb^{2+}) and Pb^{2+}—in the urine of Et_4Pb-exposed humans [33,34].

4.1. Absorption

Et_4Pb is a liquid volatile at room temperature. Although it has a characteristic sweet odor, it is less irritant, so that it is liable

to be absorbed into the body through the airway. Because it is fat-
soluble, it is readily absorbed through the skin. Because Me_4Pb is
more volatile than Et_4Pb, it is less absorbed through the skin. In
the experiments of exposure to low concentrations of ^{203}Pb-labeled
Et_4Pb and Me_4Pb in human volunteers, both compounds proved to be
absorbed into the body in amounts equivalent to 50-40% of the respec-
tive inhaled amounts [35]. There are reported deaths from drinking
Et_4Pb [19,32]. The cleaning work in leaded gasoline storage tanks
is associated with the risk of exposure to R_3Pb^+ dust in addition to
the risk of exposure to R_4Pb. In this instance, because Et_3Pb^+ is
very irritative, the irritation of the upper airway may be manifested
as, say, "sneezing."

4.2. Distribution

Because R_4Pb is fat-soluble, it is liable to be accumulated in the
liver, kidneys, brain, muscle, and adipose tissue. The brain is a
critical organ to R_4Pb poisoning, and lead is detected in such high
concentrations as about 10 mg/kg wet weight in the brain of people
who died of Et_4Pb poisoning. In a report on three victims who died
30 hr, 10 days, and 20 days, respectively, after drinking Et_4Pb,
with the lead in the brain measured as total lead and Et_3Pb^+, it was
shown that Et_3Pb^+ made up 100% of total lead in the brain, about 50%
in the liver, and about 90% in the kidneys of the man who had died
30 hr after drinking Et_4Pb; and that in the man who had died 20 days
after drinking Et_4Pb, total lead was decreased in the liver and
kidneys, but the ratio of Et_3Pb^+ was increased. The results suggested
that Et_3Pb^+ might be less liable to be excreted [32]. In the experiments in rabbits for determining total lead and Et_3Pb^+ in the organs
and tissues of Et_4Pb- and Pb^{2+} compound-treated animals, total lead
and Et_3Pb^+ levels at 24 hr and those 30 days after the treatment
were compared. It was revealed that the lead decreasing rate up to
30 days after the treatment was higher in all the organs of Et_4Pb-
treated rabbits than in those of Pb^{2+} compound-treated ones. The

level in the bone was higher in the Pb^{2+}-treated group and tended to increase with time after the treatment, whereas it did not increase in the Et_4Pb-treated group [36]. In the experiments in rhesus monkeys (*Macaca mulatta*) it was shown, by comparing the accumulation of lead in organs following administration of Et_4Pb and Me_4Pb, that the lead level in the brain was higher in the Et_4Pb-treated group than in the Me_4Pb-treated one [37].

4.3. Degradation of Tetraalkyllead Compounds

Et_4Pb, upon exposure to direct sunlight, is rapidly degraded to produce Et_3Pb^+, which is further degraded slowly to produce Et_2Pb^{2+}, which is further degraded to produce Pb^{2+}. Et_4Pb, if heated vigorously at once in the laboratory, is degraded to metal lead and hydrocarbon gas [8].

Et_4Pb is degraded to Et_3Pb^+ in the body by the action of cytochrome P-450-dependent monooxygenases occurring in the smooth endoplasmic reticulum in liver cells [38,39]. This reaction can also take place in the kidneys and brain in addition to the liver [40]. Figure 1 shows the ratio of Et_3Pb^+, Et_2Pb^{2+}, and Pb^{2+} in the liver, kidneys, blood, urine, bile, and cecal feces of rabbits 24 hr after intravenous injection of Et_4Pb. The liver, kidneys, and blood are rich in Et_3Pb^+; the urine and bile in Et_2Pb^{2+}; and the cecal feces in Pb^{2+} [41].

4.4. Excretion

In the experiments in rabbits by application of Et_4Pb to the skin, it has been reported that 80% of total lead excretion occurred in the feces and 20% in the urine [42]. It may be surmised that the lead present in the feces derives from the Et_2Pb^{2+} eliminated in the bile to Pb^{2+} being degraded in the intestinal tract. In the experiments in rabbits by intravenous injection of Et_4Pb [28], lead was

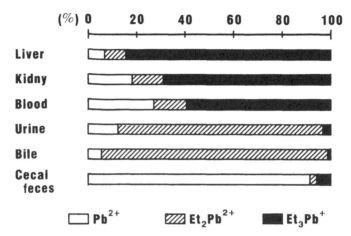

FIG. 1. Percentages of triethyllead (Et_3Pb^+), diethyllead (Et_2Pb^{2+}), and inorganic lead (Pb^{2+}) to total lead in the liver, kidney, blood, urine, bile, and cecal feces of three rabbits 24 hr after injection of tetraethyllead (Et_4Pb) at a dose of 12 mg/kg body weight.

excreted in the urine and feces in amounts equivalent to about 4% and about 56%, respectively, of the injected dose during the first 7 days after the administration of Et_4Pb; in other words, lead was excreted in the feces in amounts about 10 times as much as in the urine. In their next experiments [43], rabbits were injected intravenously with Et_4Pb; a group of rabbits were measured for lead in the cecal contents, and another for lead in the bile 24 hours afterward, respectively. In the former group, the lead occurred in the cecal contents in amounts equivalent to about 12% of the administered dose, and in the latter the lead occurred in the bile in amounts equivalent to about 8% of the administered dose. The finding that the results in these two experiments were mostly consistent may support the thinking that the Et_2Pb^{2+} eliminated in the bile is degraded into Pb^{2+} in the intestinal tract.

Chiesura [33] isolated and determined the urinary lead in Et_4Pb poisoning patients at an Et_4Pb production plant by Bolanowska's method as unprecipitable lead (organolead), benzene-extractable lead (Et_3Pb^+), and precipitable lead (Pb^{2+}). As a result, the urinary total lead

consisted of about 90-70% organolead (with Et_3Pb^+ making up about
10% of total lead), and the same author surmised that 80-60% of
the difference between unprecipitable lead and benzene-extractable
lead would be Et_2Pb^{2+}. Yamamura et al. [34] isolated and determined
by hydride generation AAS [11] the organolead in the urine of an
Et_4Pb poisoning patient from 21 days after acute exposure and on,
and achieved the results shown in Fig. 2. The urinary lead 21 days
after the acute exposure was made up of 2% Et_3Pb^+, about 50% Et_2Pb^{2+},
and about 48% Pb^{2+}. This finding revealed that Et_2Pb^{2+} would be the
major chemical species of lead excreted in the urine following massive
exposure to Et_4Pb. Figure 3 shows the lead species excreted in the
urine following a single injection of Et_4Pb in rabbits. On the day
following the injection, the urinary total lead was composed of about
69% Et_2Pb^{2+}, about 27% Pb^{2+}, and about 4% Et_3Pb^+ [28]. In a similar

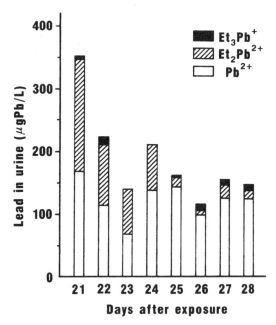

FIG. 2. Concentrations of Et_3Pb^+, Et_2Pb^{2+}, and Pb^{2+} in the urine
of a patient at 21-28 days after exposure to Et_4Pb.

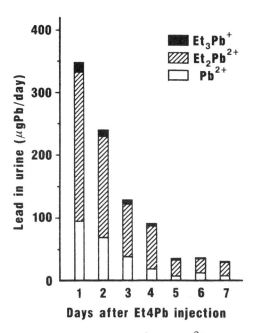

FIG. 3. Mean excretions of Et_3Pb^+, Et_2Pb^{2+}, and Pb^{2+} in the urine
of three rabbits after intravenous injection of Et_4Pb at a dose of
12 mg/kg body weight.

study by a single injection of Et_4Pb, it was reported that the lead
excreted in the urine on the day following the injection consisted
of about 65% Et_2Pb^{2+} [44]. In three patients with acute Et_4Pb
intoxication, Turlakiewicz et al. [19] assayed the urine for Et_2Pb^{2+}
and reported that the urinary total lead consisted of 70-30% Et_2Pb^{2+}
in the patients. They further reported in another study that the
total lead concentration in the urine of petrol company employees
exposed to Et_4Pb was 73.8 µg/liter, and the Et_2Pb^{2+} concentration,
6.6 µg/liter on the average, accounting for about 9% of the former.
In this study, no Et_2Pb^{2+} was detected in the urine of control subjects [45]. Isolation of workers from Et_4Pb exposure at another,
similar working site was followed by the decrease in both urinary
total lead and Et_2Pb^{2+} [44].

In a Me_4Pb poisoning patient, lead is excreted in large
amounts in the urine as in Et_4Pb poisoning patients [46]. It has
been reported that in the urine of organolead poisoning patients at
an offset printing plant using leaded gasoline, no Et_4Pb, Me_4Pb, or
Et_3Pb^+ but only trimethyllead (Me_3Pb^+) was detected [47].

Rabbits were given a single injection of Me_4Pb, and the chemical
species of lead excreted in the urine were determined. Dimethyllead
(Me_2Pb^{2+}) accounted for 67% of total lead excreted in the urine on the
day following the injection, but decreased to 8% 7 days after the
injection, whereas Me_3Pb^+ accounted for 14% of total lead on the day
following the injection, and increased to 74% 7 days after the injection. The results suggested that the exposure to Me_4Pb might be
followed by the urinary excretion of Me_3Pb^+ over considerably long
periods of time even after withdrawal from the exposure [48].

It is known in human cases of Et_4Pb poisoning that lead is
excreted in large amounts in stools [49]. In the experiments in
rabbits by injection of Et_4Pb, the lead excreted in the feces 2 days
after the injection was made up of about 6% Et_3Pb^+, about 9% Et_2Pb^{2+},
and about 85% Pb^{2+}. The reason for the feces being rich in Pb^{2+} is
that the lead excreted in the bile as Et_2Pb^{2+} is dealkylated in the
intestinal tract [28].

In the Me_4Pb poisoning patient, lead also is excreted in large
amounts in stools [46]. In the experiments in rabbits by injection
of Me_4Pb, the lead in the feces was composed of 100% Pb^{2+} 2 days
after injection [48].

5. EXPOSURE AND INTOXICATION

5.1. Exposure

Et_4Pb and Me_4Pb differ greatly in toxicity. The reported cases of
Et_4Pb poisoning include a number of lethal poisoning cases, whereas
to our knowledge there are no published data on lethal cases of
Me_4Pb poisoning.

Exposure to Et_4Pb in an ambient concentration equivalent to 100 mg of lead per m^3 for an hour results in an illness. The absorption of a high concentration of highly fat-soluble Et_4Pb through the skin alone is causative of lethal poisoning. The occupation with the greatest frequency of poisoning is the cleaning of leaded gasoline storage tanks. The ambient Et_4Pb concentrations in the tanks often exceed 60 mg of lead per m^3 [50]. Because R_4Pb has a higher density than gasoline, it precipitates at the bottom of the tanks. The scales on the walls of the tanks are rich in R_3Pb^+. For this reason, scraping out the precipitate and scrubbing the scales are associated with the exposure to large amounts of alkylleads.

The incidence of organolead poisoning has been reported in adolescents and children of the native American tribes in the USA and Canada due to leaded gasoline sniffing [51,52].

5.2. Effects, Dose-Response Relationship

5.2.1. Lead in Blood

Because the blood lead has a short half-life in Et_4Pb poisoning, there is no distinct dose-response relationship between the blood lead levels and the manifestations as in inorganic lead poisoning. In patients who died of acute lead poisoning, attention has been called to the higher accumulation of lead in the brain than in the blood, e.g., 565-960 µg/100 g of brain vs. 73-255 µg/100 g of blood [53]; about 900 µg/100 g vs. 171 µg/100 g [54]; about 851 µg/100 g vs. 84 µg/100 g [55]. The determination of blood Et_3Pb^+ concentrations is a useful means for the differential diagnosis of Et_4Pb poisoning from inorganic lead poisoning. In our experience, a trace of Et_3Pb^+ was detected in the blood even 7 months after onset of acute Et_4Pb poisoning [56].

5.2.2. Lead in Urine

Exposure to R_4Pb is associated with the marked increase in urinary lead. It has been reported that the lead content in the bladder

urine of patients who had died of Et_4Pb poisoning was 405-8371 µg/
liter [53], 1200 µg/liter [54], and 811 µg/liter [55]. Generally
in the patients who die or in severe poisoning patients, however,
there is a tendency to a decrease in urinary lead excretion due to
renal hypofunction. Cassells and Dodds [57] reported that the
urinary lead excretion was 150 µg/liter or more in moderate poisoning patients and 300 µg/liter or more in severe poisoning patients.
Schlang [58] reported such a high urinary lead content as 13,500
µg/liter in patients in the delirious state. Chiesura [59] reported
in a study of 38 patients with Et_4Pb poisoning that the mean urinary
lead concentration was 477 µg/liter in severe patients, 450 µg/liter
in moderate patients, and 376 µg/liter in mild patients. However,
no data on severe patients at the excitatory stage or in a delirious
state was available. The long-term observation of the urine of con-
valescent patients revealed that the decrease in urinary lead concentration was paralleled by the remission of manifestations.

5.3. Clinical Signs and Symptoms

Kehoe [60] reported that the clinical manifestations in patients who
died of Et_4Pb poisoning were characterized by the following.

  1. *Insomnia*. Insomnia ranges from frequent awakening at
     night to sleeplessness for several days. Severe patients
     cannot sleep due to severe twitching of the whole body,
     frequent body movements, and crying out. The patient,
     too excited, complains that he is not sleepy and that as
     he is attacked by grotesque dreams and hallucinations,
     regardless of whether he is awake or asleep, it is more
     tiring to try to sleep than to stay awake.
  2. *Nausea, anorexia, and vomiting*. These symptoms are prom-
     inent in the early morning. Nausea and vomiting are associ-
     ated with the odor of Et_4Pb and other disgusting odors.
  3. *Vertigo and headache*. These symptoms are not frequent.
     However, dull headache is frequent. Vertigo may occasion-
     ally be prominent.

4. *Muscular weakness*. Muscular weakness is the typical manifestation. It is prominent especially after the evolvement of symptoms, such as anorexia. This manifestation persists for considerably long periods of time.

5. *Pallor*. A very impressive symptom. This symptom presumably results from peripheral circulatory failure due to fall in blood pressure.

6. *Subnormal blood pressure*. When a normotensive man suffers from Et_4Pb poisoning, the maximum blood pressure may fall to 80 mm Hg.

7. *Subnormal temperature*. Body temperature falls extremely low in the early morning.

8. *Loss of weight*. Weight loss is a common manifestation. The patient may lose several kilograms of body weight in a couple of weeks.

9. *Tremor*. Rough tremors may evolve systemically. This is a considerably characteristic manifestation. Tendon reflexes are markedly exaggerated. Severe patients may suffer from systematic convulsions, with body temperature rising to more than 40°C, and eventually die. Unlike inorganic lead poisoning, even severe patients are not associated with peripheral neuropathy.

10. *Blood chemistry*. There is no characteristic finding in acute patients. In the patients with long-term exposure over several months, the percentage of stipple cells is increased among red blood cells. Compared with inorganic lead poisoning, it is associated with fewer changes in the heme synthetase system but only with the decreased activity of δ-aminolevulinic acid dehydratase (ALA-D, EC 4.2.1.24) of erythrocytes [61]. This activity is mostly in negative correlation with blood lead concentrations.

11. *Electroencephalography (EEG)*. It was thought until recently that neurotoxicity in Et_4Pb poisoning is a reversible change not associated with structural changes, and recovers without sequelae. Today, however, it is known to cause morphological

changes in neurons [62]. Chiesura [59] reported in the
analysis of EEGs of 37 workers at an R_4Pb production plant
that the EEG was normal in 35%, of low voltage in 22%, and
pathological in 43% of them, and that the pathological
findings were present in 5 of 6 severe patients, 2 of 4
moderate patients, and 9 of 11 mild patients. In one of
our patients, mild nonspecific abnormal findings were
present about 2 months after cessation of exposure [56].

5.4. Clinical Types

Slight exposure to Et_4Pb is associated with clinical manifestations
by anorexia, somnipathy, and slight falls in blood pressure and body
temperature. Sleep may be interrupted by nightmares. Such symptoms
may disappear completely in a couple of months, but hypotension may
persist for a couple of weeks.

More severe exposure is associated with further aggravation of
the above-mentioned manifestations, especially with marked weight
loss, insomnia, and anorexia. In such a degree of poisoning, malaise,
weakness, hypotension, and low body temperature persist for considerably long periods, but disappear in 6-10 weeks. It is seldom associated with sequelae. Further severe, acute exposure may be associated
with such mental manifestations as delirious, manic, confused, or
schizophrenic state. In a delirious state, the patient is excited
with slight stimuli, and falls into a frenzied state. This state is
similar to other cases of toxic delirium. The patient is frightened
by hallucinations of grotesque animals, and often may jump out of the
ward window in an attempt to escape from them. The patient with such
mental manifestations generally runs a poor prognosis. Tremors become
more severe, associated with systemic convulsions, extreme sweating,
and rise in body temperature. The shorter the interval from exposure
to onset of mental manifestations, the higher is the fatality rate.

5.5. Diagnosis

The steps to diagnosis are as follows:

1. Investigation of clinical symptoms and signs, and history taking of R_4Pb exposure.

2. Measurement of urinary total lead: The detection of R_2Pb^{2+} in the urine is definitively diagnostic of R_4Pb poisoning.

3. Measurement of blood R_3Pb^+: Hayakawa's method [6] is used. Detection of R_3Pb^+ in the blood facilitates the differentiation from other diseases.

4. Measurement of red blood cell ALA-D activity: The activity is reduced with R_4Pb exposure, but this parameter alone is not capable of differentiating the poisoning from inorganic lead poisoning.

5. Increased urinary δ-aminolevulinic acid and coproporphyrin excretion: In chronic exposure, the excretions of these substances show abnormal changes. This disordered heme metabolism appears to result from the action of Pb^{2+} derived from the degradation of R_4Pb. These abnormalities are absent in acute exposure patients.

5.6. Treatment

Ethylenediaminetetraacetate (EDTA)-$CaNa_2$ or British anti-Lewisite (BAL), which accelerate the excretion of Pb^{2+}, are not effective in accelerating the excretion of Et_3Pb^+ and Et_2Pb^{2+} because they do not form chelates with the latter compounds [31]. They are effective, however, in accelerating the excretion of Pb^{2+} as a metabolite [41].

1. *Forced diuresis*: Attempts are made at diuresis by massive dose intravenous drip of lactated Ringer's solution and by administration of furosemide on an as-needed basis while exercising caution on the potential risk of electrolyte imbalance.

2. *Purgatives*: Magnesium sulfate may be administered in an
   attempt to accelerate the excretion of intestinal contents.
3. *Direct blood perfusion*: Because R_3Pb^+ is eliminated by a
   hydron adsorbing cylinder, it is effective in the treatment
   of acute poisoning.

A patient who is excited should be sedated with a potent tran-
quillizer. Treatment for complications, e.g., pneumonia, is also
necessary.

6. BIOLOGICAL MONITORING

Because large amounts of lead are excreted in the exposure to R_4Pb,
compared with the exposure to Pb^{2+}, the urinary lead excretion is
used as a biological monitoring means in patients exposed to R_4Pb
[44,45]. Kehoe reported that urinary lead concentrations of 150
µg/liter and more were the criterion for the management of employees
at Et_4Pb production plants as indicative of dangerous exposure [63].
Linch et al. [64] reported that the mean lead level in the urine of
workers at an Et_4Pb production plant in the USA was 63 µg/liter, and
that in the urine of workers at the Me_4Pb production plant, 71 µg/
liter; the level in neither group of workers exceeded 150 µg/liter.
Cope et al. [65] reported that the mean 6-week urinary lead concentration at another R_4Pb production plant in the USA was 34-98 µg/
liter, and that at this plant there had occurred no accidents at
this exposure level in the preceding 50 years. At the R_4Pb production plant, the urinary total lead concentration had been managed
not to exceed 150 µg/liter. On exposure to high concentrations of
Et_4Pb, the total lead excreted in the urine in the early stage is
composed of 30-70% Et_2Pb^{2+} and only 2-3% Et_3Pb^+. In a report on the
workers in occupational exposure to Et_4Pb, there is a positive correlation between the ambient Et_4Pb concentrations at the working site
and the urinary total lead and Et_2Pb^{2+} excretions (correlation coefficient, r = 0.84 and 0.70) [45]. In mixed exposure to inorganic lead
and R_4Pb as in the workers engaged in the production of R_4Pb, the

measurement of urinary Et_3Pb^+ is a useful indicator of whether there was an exposure to Et_4Pb. In the exposure to Me_4Pb, on the other hand, Me_3Pb^+ is more readily detectable than Me_2Pb^{2+}, and the measurement of urinary Me_3Pb^+ is useful [47]. Because of mixed exposure to organic and inorganic leads at R_4Pb production plants, it appears that the urinary lead allowance levels are set at such considerably high levels. When it is taken into account that the lead level in the urine of normal men is as low as below 10 μg/liter, a higher level than the level in unexposed workers should be taken as indicative of potential exposure, and such workers should be closely monitored. Because there is no distinct relationship between blood lead concentrations and manifestations of lead exposure, the measurement of blood lead concentrations is not as frequently used in biological monitoring as the measurement of urinary lead [66]. Because Et_3Pb^+ is detected in the blood for considerably long periods of time in Et_4Pb exposure, however, the measurement of blood Et_3Pb^+ levels is a useful monitoring means as an indicator of R_4Pb exposure [56]. The measurement of lead in the hair is not adequate as an indicator of biological monitoring because the lead cannot be differentiated from the exogenous lead contamination.

7. CONCLUSIONS

The R_4Pb compounds comprise a number of analogous compounds varying in the kind and number of alkyl groups, but Et_4Pb and Me_4Pb, which are used as antiknock agents for gasoline, are the R_4Pb compounds which are industrially produced in the greatest quantities. These compounds are extremely toxic, and because there is no adequate therapy for poisoning with the compounds even today, the progression of their toxic manifestations is associated with a very high fatality rate. When R_4Pb is absorbed by the body, its alkyl group is degraded to produce alkylleads, e.g., R_3Pb^+, R_2Pb^{2+}, and Pb^{2+}. Of these metabolites, R_3Pb^+ compounds are the most neurotoxic. Pb^{2+} is retained in the body for long periods of time; its poisoning runs a chronic course, characteristically accompanied by the onset of anemia and peripheral neuropathy. On the other hand, alkyllead poisoning is characterized by extremely acute onset and mental manifestations. Studies of the in vivo behavior of alkylleads in recent years have revealed the differences between inorganic lead poisoning and alkyllead poisoning to a considerable extent.

The measurement of R_4Pb degradation products in biological samples is useful in the biological monitoring of workers with occupational exposure to R_4Pb compounds. Especially, the measurement of urinary Et_3Pb^+ and Et_2Pb^{2+} in addition to the measurement of total lead in the urine of subjects exposed to Et_4Pb is of greater value than the measurement of urinary total lead only as employed in the past with respect to elucidation of whether there has been an exposure to Et_4Pb. Me_4Pb is considerably less toxic than Et_4Pb. In Me_4Pb exposure, the measurement of urinary Me_3Pb^+ excretion is useful in finding whether there has been an exposure to Me_4Pb.

ABBREVIATIONS AND FORMULAS

AAS	atomic absorption spectrometry
ALA-D	δ-aminolevulinic acid dehydratase
APDC	ammonium pyrrolidine dithiocarbamate
BAL	British anti-Lewisite
CNS	central nervous system
EEG	electroencephalography
EDTA-CaNa$_2$	ethylenediaminetetraacetic acid calcium salt
Et_4Pb	tetraethyllead
Et_3Pb^+	triethyllead
Et_2Pb^{2+}	diethyllead
GC	gas chromatography
HPLC	high performance liquid chromatography
Me_3EtPb	trimethylethyllead
Me_2Et_2Pb	dimethyldiethyllead
$MeEt_3Pb$	methyltriethyllead

Me_4Pb	tetramethyllead
Me_3Pb^+	trimethyllead
Me_2Pb^{2+}	dimethyllead
MIBK	methyl isobutyl ketone
NDAFS	non-dispersive atomic-fluorescence spectrometry
Pb^{2+}	inorganic lead
r	correlation coefficient
R_4Pb	tetraalkyllead
R_3Pb^+	trialkyllead
R_2Pb^{2+}	dialkyllead
SDDC	sodium diethyldithiocarbamate

REFERENCES

1. T. Midgley, Jr. and T. A. Boyd, *Ind. Eng. Chem.*, *14*, 894 (1922).

2. K. Akatsuka, *Jpn. Industr. Health*, *15*, 3 (1973).

3. G. R. Sirota and J. F. Uthe, *Anal. Chem.*, *49*, 823 (1977).

4. F. Arai, *Industr. Health*, *24*, 139 (1986).

5. H. J. Dawson, Jr., *Anal. Chem.*, *35*, 542 (1963).

6. K. Hayakawa, *Jpn. J. Hyg.*, *26*, 377 (1971).

7. Y. K. Chau, P. T. S. Wong, G. A. Bengert, and J. L. Dunn, *Anal. Chem.*, *56*, 271 (1984).

8. S. R. Henderson and L. J. Snyder, *Anal. Chem.*, *33*, 1172 (1961).

9. S. Imura, K. Fukutaka, H. Aoki, and T. Sakai, *Jpn. Analyst, 20*, 704 (1971).

10. T. Nielsen, K. A. Jensen, and P. Grandjean, *Nature, 274*, 602 (1978).

11. H. Yamauchi, F. Arai, and Y. Yamamura, *Industr. Health, 19*, 115 (1981).

12. A. D'Ulivo, R. Fuoco, and P. Papoff, *Talanta, 33*, 401 (1986).

13. D. S. Forsyth and W. D. Marshall, *Anal. Chem.*, *55*, 2132 (1983).

14. Y. K. Chau, P. T. S. Wong, and O. Kramar, *Anal. Chim. Acta, 146*, 211 (1983).

15. J. S. Blais and W. D. Marshall, *J. Anal. Atom. Spectrom.*, *4*, 641 (1989).

16. G. Pilloni and G. Plazzogna, *Anal. Chim. Acta.*, *35*, 325 (1966).

17. S. Imura, K. Fukutaka, and T. Kawaguchi, *Jpn. Analyst, 18,* 1008
     (1969).

18. F. G. Noden, in *Lead in the Marine Environment,* Proc. Int.
     Experts. Disc., 1980, pp. 83-91.

19. Z. Turlakiewicz, M. Jakubowski, and J. Chmielnicka, *Br. J.
     Industr. Med., 42,* 63 (1985).

20. P. T. S. Wong, Y. K. Chau, J. Yaromich, P. Hodson, and M.
     Whittle, *Appl. Organomet. Chem., 3,* 59 (1989).

21. W. Bolanowska, *Br. J. Industr. Med., 25,* 203 (1968).

22. L. J. Snyder and S. R. Henderson, *Anal. Chem., 33,* 1175 (1961).

23. L. J. Purdue, R. E. Enrione, R. J. Thompson, and B. A. Bonfield,
     Anal. Chem., 45, 527 (1973).

24. S. Hancock and A. Slater, *Analyst, 100,* 422 (1975).

25. Y. Yamamura, H. Yamauchi, M. Yoshida, F. Hirayama, and E.
     Ohkura, *Jpn. J. Industr. Health, 18,* 480 (1976).

26. D. T. Coker, *Ann. Occup. Hyg., 21,* 33 (1978).

27. Y. K. Chau, P. T. S. Wong, and H. Saitoh, *J. Chromatogr. Sci.,
     14,* 162 (1976).

28. F. Arai, Y. Yamamura, and M. Yoshida, *Jpn. J. Industr. Health,
     23,* 496 (1981).

29. R. A. Kehoe, *J. Lab. Clin. Med., 12,* 554 (1927).

30. R. A. Mortensen, *J. Industr. Hyg. Toxicol., 24,* 285 (1942).

31. J. E. Cremer, *Br. J. Industr. Med., 16,* 191 (1959).

32. W. Bolanowska, J. Piotrowski, and H. Garczynsky, *Arch. Toxicol.,
     22,* 278 (1967).

33. P. Chiesura, *Med. Lavoro., 61,* 437 (1970).

34. Y. Yamamura, F. Arai, and H. Yamauchi, *Industr. Health, 19,* 125
     (1981).

35. M. J. Heard, A. C. Wells, D. Newton, and A. C. Chamberlain, in
     *Human Uptake and Metabolism of Tetraethyl and Tetramethyl Lead
     Vapor Labelled with* ^{203}Pb, CEP Consultants Ltd., Edinburgh, 1979,
     pp. 103-108.

36. Y. Yamamura, F. Arai, M. Yoshida, and E. Shimada, *St. Marianna
     Univ. Med. J., 7,* 10 (1979).

37. R. Heywood, R. W. James, R. J. Sortwell, D. E. Prentice, and
     P. S. I. Barry, *Toxicol. Lett., 2,* 187 (1978).

38. J. E. Cremer, *Occup. Health Rev., 17,* 14 (1965).

39. A. A. Jensen, in *Biological Effects of Organolead Compounds*
     (P. Grandjean, ed.), CRC Press, Boca Raton, Florida, 1984,
     pp. 97-115.

40. W. Bolanowska and J. M. Wiśniewska-knypl, *Biochem. Pharmacol.*, *20*, 2108 (1971).

41. Y. Yamamura, F. Arai, and H. Yamauchi, in *Toxicovigilance industrielle*, Actes du Xe Congrés du Medichem, Paris, 1982, pp. 57-64.

42. R. A. Kehoe and F. Thamann, *Am. J. Hyg.*, *13*, 478 (1931).

43. F. Arai, Y. Yamamura, H. Yamauchi, and M. Yoshida, *Jpn. J. Industr. Health*, *25*, 175 (1983).

44. Z. Kozarzewska and J. Chmielnicka, *Br. J. Industr. Med.*, *44*, 417 (1987).

45. Z. Turlakiewicz and J. Chmielnicka, *Br. J. Industr. Med.*, *42*, 682 (1985).

46. J. Gething, *Br. J. Industr. Med.*, *32*, 329 (1975).

47. K. Hayakawa, *Jpn. J. Hyg.*, *26*, 526 (1972).

48. F. Arai and Y. Yamamura, *Industr. Health*, *28*, 63 (1990).

49. W. F. Machle, *J. Am. Med. Assoc.*, *105*, 578 (1935).

50. L. W. Sanders, *Arch. Environ. Health*, *8*, 270 (1964).

51. R. O. Robinson, *J. Am. Med. Assoc.*, *240*, 1373 (1978).

52. S. S. Seshia, K. R. Rajani, R. L. Boeckx, and P. N. Chow, *Dev. Med. Child Neurol.*, *20*, 323 (1978).

53. S. Yamaga, K. Saruta, K. Ohmori, H. Ueda, R. Ohshima, and Y. Nitoh, *Jpn. J. Hyg.*, *14*, 852 (1959).

54. G. Salvi, U. Ambanelli, and M. Gherardi, *Folia Medica*, *43*, 421 (1960).

55. T. Nakano, M. Imaeda, T. Takeuchi, and T. Inose, *Adv. Neurol. Sci.*, *22*, 376 (1978).

56. Y. Yamamura, J. Takakura, F. Hirayama, H. Yamauchi, and M. Yoshida, *Jpn. J. Industr. Health*, *17*, 223 (1975).

57. D. A. K. Cassells and E. C. Dodds, *Br. Med. J.*, *2*, 681 (1946).

58. H. A. Schlang, *Aerospace Med.*, *32*, 333 (1961).

59. P. Chiesura, *Lav. Um.*, *22*, 249 (1970).

60. R. A. Kehoe, *J. Am. Med. Assoc.*, *85*, 108 (1925).

61. J. A. Millar, G. G. Thompson, and A. Goldberg, *Br. J. Industr. Med.*, *29*, 317 (1972).

62. A. A. Seawright, A. W. Brown, J. C. Ng, and J. Hrdlicka, in *Biological Effects of Organolead Compounds* (P. Grandjean, ed.), CRC Press, Boca Raton, Florida, 1984, pp. 177-206.

63. R. Kehoe, in *ILO Encyclopaedia of Occupational Health and Safety*, Vol. 2 (L. Parmeggiani, ed.), ILO, Geneva, 1983, pp. 1197-1199.

64. A. L. Linch, E. G. Wiest, and M. D. Carter, *Am. Industr. Hyg. Assoc. J., 31,* 170 (1970).

65. R. F. Cope, B. P. Pancamo, W. E. Rinehart, and G. L. Ter Haar, *Am. Industr. Hyg. Assoc. J., 40,* 372 (1979).

66. L. Alessio, A. Dell'Orto, and A. Forni, in *Biological Indicators for the Assessment of Human Exposure to Industrial Chemicals* (L. Alessio, A. Berlin, M. Boni, and R. Roi, eds.), Commission of European Communities, Luxembourg, 1987, pp. 3-15.

6

Metabolism of Alkyl Arsenic and Antimony Compounds

Marie Vahter

Institute of Environmental Medicine
Karolinska Institutet
P.O. Box 60208
S-104 01 Stockholm, Sweden

and

Erminio Marafante

Commission of the European Communities
Joint Research Centre
Environment Institute
I-210 20 Ispra, Italy

1. INTRODUCTION

1.1. Use of Alkyl Arsenic Compounds

A number of alkyl arsenic compounds are industrially produced, most
of them from arsenic trioxide. Methylarsonic acid (MAA, methane
arsonate) and dimethylarsinic acid (DMAA, cacodylic acid), or their
sodium salts, are used mainly in herbicides, insecticides, silvi-
cides, and cotton desiccants [1]. DMAA is also frequently used in
biochemical analysis because of its buffering capacity at physiological pH.

Sodium cacodylate was once used in medication for the same
purposes as drugs containing inorganic arsenic, i.e., skin diseases
and leukemia [2]. Ferric cacodylate, strychnine cacodylate, and
sodium cacodylate were also popular ingredients of proprietary
injections advocated for their stimulant effects in debilitated
conditions.

Trimethylarsine (TMA) is used as a dopant in the production of
gallium arsenide semiconductors [3]. The chlorinated alkyl arsenic
compound Lewisite, $Cl_2AsCH=CHCl$, has been extensively used as war
gas.

1.2. Environmental Cycling of Alkyl Arsenic Compounds

Alkyl arsenic compounds are also produced naturally in the environment. During the second half of the nineteenth century and the
beginning of this century, a number of intoxications were associated
with wallpaper or paint used indoors containing arsenic [4,5]. In

1901 Gosio reported that a garlic-smelling, volatile, methylated arsenic compound was released from molds growing in the presence of inorganic arsenic, and shortly thereafter Challenger and coworkers [6] identified the gas as trimethylarsine. Since then, several fungi and bacteria have been shown to be able to methylate inorganic trivalent arsenic [As(III)] to MAA, DMAA, and trimethylarsine oxide (TMAO), via the corresponding arsines [7-9]. Under anaerobic conditions, the methylation proceeds to dimethylarsine (DMA) only [10].

Reduction of MAA, DMAA, and TMAO to the corresponding arsines, as well as demethylation of MAA to inorganic arsenic, may take place in soil, water, or sediment [11,12]. Probably the demethylation is mediated by microorganisms. Methylated arsenicals may also be formed in plants [13,14].

Arsenate, which is the major form of inorganic arsenic in surface waters, is readily taken up by algae [15,16]. After being absorbed, arsenate is reduced to As(III) and converted to MAA and DMAA via the corresponding arsines. Both MAA and DMAA have been found in the edible seaweed *Hizikia fusiforme* [17]. The methylated arsenicals may be further transformed to arsenic-containing ribofuranosides and arsenic-containing phospholipids (PL) [18,19]. These are transferred to the food web of marine animals and transformed into arsenobetaine, the arsenic analog of betaine, in which the N atom is substituted by As [19,20]. Arsenobetaine is the main form of arsenic in fish and crustacea. Interestingly, arsenobetaine is retained in the fish muscle, probably by the same mechanism as the naturally occurring osmolyte betaine [21]. Arsenocholine, the arsenic analog of choline, has been found in shrimps [22-25]. Dimethyloxarsylethanol is formed from anaerobic decomposition of the brown kelp *Ecklonia radiata* and may be a precursor of arsenobetaine in the marine fauna [26].

Low levels of TMAO have been found in four species of fish from the Baltic Sea [27], and as much as 40% of the arsenic in estuary catfish was reported to be in the form of TMAO [28]. TMAO was also detected in estuary catfish and school whiting following

TABLE 1

Some Commonly Occurring Alkyl Arsenic Compounds

Methylarsonic acid	$CH_3AsO(OH)_2$
Methylarsine	CH_3AsH_2
Dimethylarsinic acid	$(CH_3)_2AsO(OH)$
Dimethylarsine	$(CH_3)_2AsH$
Trimethylarsine oxide	$(CH_3)_3As=O$
Trimethylarsine	$(CH_3)_3As$
Tetramethylarsonium ion	$(CH_3)_4As^+$
Arsenobetaine	$(CH_3)_3As^+CH_2COO^-$
Arsenocholine	$(CH_3)_3As^+CH_2CH_2OHX^-$

oral administration of sodium arsenate, probably as a result of bacterial action in the gut [28]. Garlic-like off-flavor of deep-sea
prawns has been associated with naturally occurring TMA [29]. TMAO
and TMA are formed in fish during postmortem storage, contributing
to the garlic-like off-flavor [27,29]. TMAO may also be formed from
arsenobetaine by biodegradation in bottom sediments [30]. Tetramethylarsonium salts have been found in the clam *Meretrix lusoria*
and in some marine animals [31,32].

Table 1 shows the chemical formulas of some commonly occurring
alkyl arsenic compounds. For further discussion of the environmental
cycling of alkyl arsenic compounds, reference is made to Hood [1],
Andreae [33], Cullen and Reimer [20], and Phillips [34].

1.3. Exposure to Alkyl Arsenic Compounds

The total daily intake of arsenic by humans is greatly influenced by
the amount of seafood in the diet, but it is usually less than 0.2
mg/day [35,36]. People living in the coastal areas of Norway, where
the consumption of fish products is fairly high, have considerably

higher concentrations of arsenic in blood (7-18 µg/liter) than
people living in the inland regions (0.8-3.1 µg/liter), where the
consumption of fish is much lower [37]. Ingestion of a single meal
of seafood may increase the urinary concentration of arsenic from
10-50 µg/liter to several hundred µg/liter [38,39].

As mentioned above, arsenobetaine is the main form of arsenic
in seafood [20,36]. Normally, inorganic arsenic constitutes only
a few percent of the total arsenic content [36]. Duplicate diets
collected by four Japanese subjects were found to contain 6%
inorganic arsenic, 4% MAA, 27% DMAA, and 48% trimethylarsenic compounds, probably mainly arsenobetaine [40]. It should be noted
that certain types of edible seaweed, which are common in the
Japanese diet, may contain significant levels of MAA and DMAA [17].
Possibly the fraction of MAA and DMAA is less in food from many
other countries.

There are great variations in the concentration of arsenic
between fish species and locations. Marine fish normally have higher
arsenic concentrations than freshwater fish [27,41]. In general,
teleost fish contain 1-10 mg As/kg, but individuals of bottom-feeding
species may contain more than 100 mg As/kg [20,36]. Elasmobranchs
appear to have somewhat higher levels, 5-30 mg As/kg, than teleosts.
High arsenic concentrations have also been reported for crustaceans,
with 1-50 mg As/kg as normal values.

Normally, the uptake of airborne arsenic is less than 1 µg/day.
The concentration of arsenic in outdoor air may range from less than
0.1 ng/m^3 to a few ng/m^3 [35]. Much higher levels have been recorded
near point sources. In one study, approximately 15% of the arsenic
in outdoor air was found to be in the form of di- and trimethylarsenic
compounds [42]. Higher concentrations of methylated arsenic compounds
were found in greenhouse air. Monomethylated arsenic was detected in
air in areas where iron methane arsonate had been spread over rice
fields [43].

Occupational exposure to alkyl arsenic compounds may occur
during production and use of arsenic-containing pesticides [1,44-46]
as well as in the electronics industry [3].

2. CHEMICAL AND TOXICOLOGICAL PROPERTIES
OF ALKYL ARSENIC COMPOUNDS

MAA is a water-soluble strong dibasic acid with pK_{a1} 3.6 and pK_{a2} 8.2. DMAA is a water-soluble amphoteric electrolyte with pK_a 6.2 [1,47]. Both MAA and DMAA are easily reduced to the corresponding arsines by zinc/hydrochloric acid or sodium borohydride, a reaction utilized in the analytical method described by Braman and Foreback [48]. The methylated arsines are colorless liquids with boiling points +2, +36, and +50°C for mono-, di-, and trimethylarsine, respectively [49]. All the methylated arsines are readily oxidized by air to the corresponding oxide or acid.

Reactions of DMAA with organic thiols such as 2-mercaptoethanol, cysteine, glutathione, and dithiothreitol have been described [50,51]. However, most alkyl arsenic compounds have a lower affinity for tissue constituents than inorganic arsenic, especially the trivalent form [52,53]. The mechanism of action of alkyl arsenic compounds was extensively studied by Paul Ehrlich and Carl Voegtlin in the beginning of this century in the search for efficient arsenical chemotherapeutic agents. This research was continued during the 1930s and 1940s by R. A. Peters and coworkers, who studied antidotes for the war gas Lewisite. An excellent review of this early work on the mechanisms of action of arsenic was published by Holmstedt and Liljestrand [54]. Ehrlich's idea that the action of the organic arsenic compounds involved interaction with sulfydryl groups was definitely proven by Voegtlin and his collaborators [55]. They showed that arsenic in the form RAs=O, where R is an aliphatic or aromatic radical, was a specific poison for glutathione and related SH compounds. Peters and coworkers [56] reported that the high toxicity of Lewisite and trivalent arsenicals was due to their ability to combine with SH groups in tissue proteins to form stable arsenical rings:

$$= C-S \diagdown$$
$$\qquad\qquad As\ R$$
$$= C-S \diagup$$

In general, the acute toxicity of alkyl arsenic compounds is
considerably lower than that of inorganic arsenic. Kaise et al.
[57] reported LD_{50} values in mice of 1.8 and 1.2 g/kg for MAA and
DMAA, respectively, compared to an LD_{50} of about 30 mg/kg for
arsenic trioxide in mice.

TMA has been found to have an even lower acute toxicity. The
oral LD_{50} of TMA in mice was about 8 g/kg [3]. A single oral dose
of 750 mg/kg TMA caused mild, transient hemolysis in hamsters. Even
at that high dose level the effect was very mild compared to the
hemolysis caused by arsine [58]. The LD_{50} for TMAO in mice was
found to be as high as 10.6 g/kg [57].

The oral, intraperitoneal, and intravenous LD_{50} values for
tetramethylarsonium iodide in mice are 890, 175, and 82 mg/kg,
respectively [32]. Sublethal doses caused transient ataxia and
tremor. An LD_{50} of more than 10 g/kg has been reported for arsenobetaine [59].

It is well documented that inhalation of inorganic arsenic
may cause lung cancer, while ingestion of inorganic arsenic may
give rise to skin lesions, including skin cancer [35,60]. One
possible mechanism for the carcinogenicity of inorganic arsenic is
its interference with the DNA repair mechanisms [61,62]. There are
no reports on the carcinogenic potential of alkyl arsenic compounds.
Very high oral doses of DMAA (1500 mg/kg) have been shown to give
rise to DNA damage in mouse and rat lung [63-65]. The authors suggested the DNA damage to be caused by active oxygen and dimethyl-
arsenic peroxyl radical, produced in the metabolism of DMAA. DMA,
but not TMA, was found to be mutagenic in *E. coli* [66].

In vitro studies on the initiating and promoting effects of
arsenobetaine have been negative even at concentrations as high as
10 mg/ml [67]. No cytotoxicity or morphological transformation was
seen in the mouse embryo cell line BALB/3T3 using 500 μM arsenobetaine [68].

3. METHYLATION OF INORGANIC ARSENIC

Most mammals are able to methylate inorganic arsenic to MAA and
DMAA (for review, see Vahter and Marafante [69]). The marmoset
monkey is the only species known not to methylate inorganic
arsenic [70,71].

The methylation of inorganic arsenic can be considered a
detoxification mechanism, since the methylated metabolites have a
lower affinity for tissue constituents and lower toxicity than
inorganic arsenic [53,72]. The methylation takes place mainly in
the liver. It is enzymatically mediated and involves the transfer
of carbonium ions from S-adenosylmethionine to arsenic in its trivalent form [73-75]. Glutathione (GSH) is involved in the reduction
of arsenic prior to addition of the carbonium ions [76-78]. The
methylation of inorganic arsenic is decreased by inhibition of the
methyl transfer reactions, by low intake of methyl groups (methionine,
choline, or proteins), or by depletion of tissue GSH. The inhibition
of arsenic methylation results in increased tissue retention of
arsenic [73,78-80].

In human subjects exposed to "normal" environmental levels of
inorganic arsenic, the urinary excretion consists of 10-20% inorganic
arsenic, 10-15% MAA, and 60-80% DMAA [38,72,81,82]. The methylation
efficiency decreases with increasing dose level [83], and in cases of
acute arsenic intoxication it may take a couple of days before the
methylated metabolites can be detected in the urine [84,85]. However,
Swedish smelter workers with 16-328 µg As/g creatinine in the urine
had the same rate of methylation as the general population, i.e.,
10-20% inorganic arsenic, 10-15% MAA, and 60-80% DMAA [86].

4. METABOLISM OF METHYLARSONIC ACID, DIMETHYLARSINIC ACID, TRIMETHYLARSINE OXIDE, AND TETRAMETHYLARSONIUM SALTS

4.1. Uptake

Most methylated arsenicals seem to be readily absorbed in the gastrointestinal tract. In human subjects ingesting 500 µg arsenic in the

form of MAA or DMAA, about 75% of the dose was excreted in the urine within 4 days, indicating a high degree of absorption [72]. The rate of absorption seems to be lower in some experimental animals. In hamsters, 56% of an oral dose of MAA was recovered in feces compared to less than 1% of a corresponding dose given intraperitoneally [87]. Studies on mice and rats have indicated that 70-80% of DMAA is absorbed in the gastrointestinal tract [88-90]. A somewhat lower rate of absorption of DMAA was seen in hamsters [91]. DMAA is also rapidly absorbed from the lungs following intratracheal instillation in rats [88].

Following oral administration of TMAO to hamsters, less than 1% of the dose was recovered in feces [92], indicating a very efficient absorption from the gastrointestinal tract. Also, tetramethyl-arsonium was readily absorbed (more than 70%) from the gastrointestinal tract of mice [32].

4.2. Biotransformation

In the above-mentioned study, where human volunteers ingested MAA (500 μg As/person), about 10% of the dose was further methylated and excreted in the urine as DMAA and about 70% as unchanged MAA [72]. Similar results were obtained in hamsters administered 5 mg/kg body weight of MAA [87]. The methylation of MAA decreased with increasing dose level, and at a dose of 250 mg/kg less than 1% of the dose was methylated and excreted in the urine as DMAA.

DMAA has generally been considered the end point of the methylation of arsenic in mammals. However, following administration of DMAA to mice, hamsters, and a human subject, about 5% of the administered dose was excreted as TMAO in the urine [91].

Older literature mentions that administration of sodium dimethylarsinate preparations to human subjects gave rise to a strong garlic-like odor of the breath, perspiration, and urine due to the release of cacodylic oxide (dimethylarsine oxide) [2,93]. However, these observations have not been confirmed in modern studies. Following

oral administration of a single, very high dose of sodium dimethylarsinic acid to mice (1500 mg/kg body weight, corresponding to 704
mg As/kg), 0.003% of the dose was recovered as a dimethylated compound, assumed to be DMA, in the expired air within 24 hr [64].
There was also a garlic-like odor from the animals. It may be of
interest to compare this metabolic pathway with that of selenium.
Exhalation of dimethylselenide has been observed following exposure
to toxic levels of selenium [94]. However, at lower exposure
levels, selenium is further methylated to trimethylselenium ion,
which is excreted in the urine.

Following oral administration of TMA (10 mg As/kg body weight)
in olive oil to hamsters, about 76% of the dose was oxidized in the
body and excreted as TMAO in the urine within 24 hr [3]. At a dose
of 50 mg/kg body weight given intraperitoneally to hamsters, TMA
was detected in the expired air 5 min after administration, together
with a garlic-like odor. Following oral administration of TMAO
(10 mg As/kg body weight) to hamsters, about 87% of the dose was
excreted in the urine in unchanged form within 24 hr [92]. Tetramethylarsonium iodide administered to mice was excreted unchanged
in the urine [32].

Older literature mentions that the effects of DMAA preparations were due to the conversion to inorganic arsenic in the body
[2,93]. However, according to modern literature, it seems unlikely
that the methylated arsenicals are demethylated in the body [3,72,
88,91,92].

4.3. Tissue Distribution

There are major differences in the metabolism of arsenic compounds
among animal species. Specific binding of DMAA to rat hemoglobin,
for example, results in a significant accumulation of arsenic in
the blood following exposure to both inorganic arsenic and DMAA
[88,95]. Such an accumulation in the red blood cells is not seen
in other species.

In human subjects exposed to normal environmental levels of arsenic, hair and nails show the highest tissue concentrations of arsenic, while fairly high tissue concentrations are found in skin and lungs [96]. Studies on experimental animals exposed to inorganic arsenic show that the tissues with the longest retention of arsenic are skin, hair, squamous epithelium of the upper gastrointestinal tract (oral cavity, tongue, and esophagus), stomach wall, intestinal walls, epididymis, thyroid, skeleton, and lens [70,97,98].

Oral administration of MAA to hamsters showed the highest concentrations in kidney, spleen, and lungs 6-24 hr after dosing [87]. No MAA was found in hair.

Six hours after a single i.v. administration of [^{74}As]-DMAA (0.4 mg As/kg) to mice the highest concentrations were found in lungs, kidneys, and intestine [90]. By 24 hr after dosing there was very little ^{74}As left in the body. Whole-body autoradiographic studies revealed that the ^{74}As was retained particularly in the thyroid, the peripheral part of the lens, and the intestinal walls. DMAA was partly incorporated in the skin and fur, but was rapidly eliminated. In human hair, inorganic arsenic accounted for 73% and DMAA for 27% of the total arsenic present [99]. Accumulation of arsenic in the hair of cattle exposed to MAA or DMAA has been reported [100]. Daily feeding of DMAA in doses corresponding to 5.4 mg As/kg body weight gave rise to arsenic concentrations of about 3 mg As/kg in hair after 10 days and about 20 mg As/kg after 48 days. A summary of the biotransformation and tissue distribution of DMAA is given in Fig. 1.

Mice administered a single dose of sodium dimethylarsinate (600 mg/kg) or disodium methane arsonate (800 mg/kg) on day 12 of gestation showed rapid uptake in the fetuses [1]. After 12 hr the concentrations in the fetal tissues had declined from about 30 to 2 µg/g.

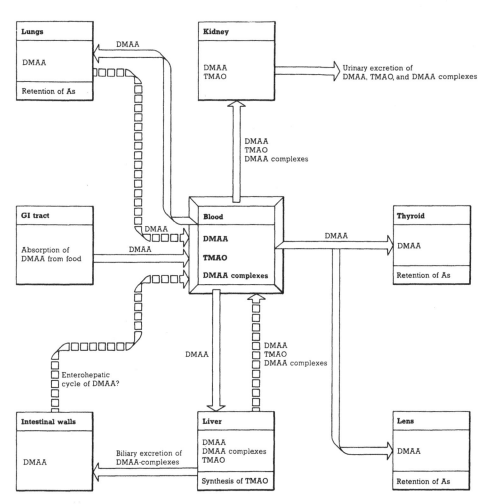

FIG. 1. Scheme of the biotransformation and distribution of
dimethylarsinic acid in mammals.

4.4. Excretion and Half-time

The major route of excretion of most arsenic compounds is via the
kidneys, but the rate of excretion varies with the chemical form.
In human subjects ingesting 500 μg arsenic in the form of MAA or
DMAA, about 75% of the dose was excreted in the urine within 4 days

[72]. Following ingestion of the same amount of arsenic in the form
of arsenite, only about 45% of the dose was excreted in the urine
within 4 days. In mice and hamsters, less than 1% of a single oral
dose of ^{74}As-labeled DMAA was retained in the body 48 hr after dosing
[90,91]. In the mice, about 85% of the dose was eliminated with a
half-time of 2.5 hr, about 14% with a half-time of 10 hr, and less
than 0.5% with a half-time of 20 days. In rats administered [^{74}As]-
DMA (0.4 mg As/kg body weight) about 45% of the dose was eliminated
with a half-time of 13 hr, while the remaining 55% was eliminated
with a half-time of as much as 50 days, reflecting a long-term retention in the red blood cells.

The rate of excretion of TMA has been studied in hamsters
given single peroral doses of TMA corresponding to 10 mg As/kg body
weight [3]. About 69% of the dose was excreted in the urine within
12 hr and 76% within 24 hr. The whole-body half-time was 3.7 hr.
The main form of arsenic in the urine was TMAO. When TMAO was administered orally to hamsters, 66% of the dose (10 mg As/kg body weight)
was excreted in the urine within 12 hr and 87% within 24 hr [92].

Following peroral administration of tetramethylarsonium salt
(5-400 mg/kg body weight) to mice, about 55% of the dose was recovered
in the urine within 6 hr and about 70% within 24 hr after administration [32].

5.   METABOLISM OF ARSENOBETAINE AND ARSENOCHOLINE

5.1.   Uptake

Arsenobetaine, the major form of arsenic in seafood, is efficiently
taken up in the gastrointestinal tract. In human subjects who had
ingested flounder containing about 10 mg of arsenic in the form of
arsenobetaine, less than 1% of the dose was recovered in feces
within 8 days, indicating an almost complete absorption [101]. A
similar high degree of absorption has been observed in mice and rats
following oral administration of [^{75}As]-arsenobetaine or [^{73}As]-
arsenocholine [102,103].

5.2. Biotransformation

Arsenobetaine is excreted in the urine without being biotransformed [102,104]. Following exposure to [73]As-labeled arsenobetaine to mice and rabbits, no labeled arsenic compounds other than arsenobetaine were found in urine or tissues [102].

Arsenocholine is to a great extent oxidized to arsenobetaine in mammals, probably by a mechanism similar to that of choline [103, 105]. In mice, rats, and rabbits, 55-70% of the administered arseno-choline was oxidized and excreted in the urine as arsenobetaine within 3 days [103]. Unmetabolized arsenocholine was found in the urine on the first day only (about 10% of the dose). Incubation of arsenocholine with rat liver mitochondria showed the formation of arsenobetaine via arsenobetaine aldehyde [106]. Trimethylarsine oxide was formed via a side reaction from arsenobetaine aldehyde.

The mammalian metabolism of various arsenosugars, the main arsenic compounds in certain brown kelp, is not yet known.

5.3. Tissue Distribution

Data on the tissue distribution and retention of the organic arsenic compounds ingested with seafood are limited. In mice and rabbits the tissues with the longest retention of [[73]As]-arsenobetaine were cartilage, epididymis, testes, semen ducts, thymus, and muscles [102]. Siewicki [95] administered a flounder diet containing 29 mg As/kg to rats for 42 days. In the treated rats, the arsenic levels in liver and spleen were 7 and 2 times higher, respectively, than in the control rats. In contrast to inorganic arsenic, arsenobetaine is not incorporated into hair.

Administration of [[73]As]-arsenocholine to mice, rats, and rab-bits caused higher tissue concentrations and longer tissue retention of [73]As than the administration of [[73]As]-arsenobetaine [103], probably due to incorporation of arsenocholine in phospholipids in the same way as choline [105]. Following administration of [[73]As]-arsenocholine,

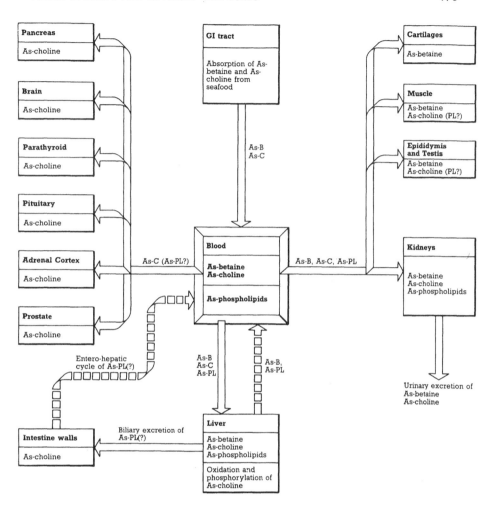

FIG, 2, Scheme of the biotransformation and distribution of arsenobetaine and arsenocholine in mammals.

[73]As was accumulated in muscles and several parenchymatous and endo-
crine organs, i.e., epididymis, semen ducts, testes, prostate, para-
thyroid, pancreas, adrenal cortex, liver, lungs, salivary glands,
and thymus [103].

The tissue distribution and biotransformation of arsenobetaine
and arsenocholine are summarized in Fig. 2.

5.4. Excretion

The biological half-time of organic arsenic compounds ingested with
fish and crustacea is shorter than that of inorganic arsenic. In
human subjects eating fish with a high arsenic content, on average
about 70% of the ingested arsenic (total range 52-85%, N = 32) was
excreted in the urine within 3 days [41,101,107,108]. About 76%
(range 61-86%) of the arsenic (10 mg As) ingested with witch flounder
was excreted in the urine within 8 days [101]. Less than 1% was recovered in feces. In a study where six volunteers ingested ^{74}As-
labeled arsenobetaine with a fish meal, less than 10% of the ^{74}As
was retained in the whole body after 8 days, as measured by an
array of six NaI(Tl) crystals [109]. After 3 weeks, less than 1%
of the dose remained in the subjects.

A similar rate of excretion of arsenic in the urine has been
reported for rabbits given a single oral dose of ^{73}As-labeled arsenobetaine [102]. About 75% of the dose was excreted in the urine within
3 days. The urinary excretion of [^{73}As]-arsenobetaine in mice and rats
was found to be essentially complete within 3 days [102].

Following administration of [^{73}As]-arsenocholine, almost 80% of
the dose was excreted in the urine in mice and rats and 66% in rabbits
within 3 days [103].

6. ALKYL ANTIMONY COMPOUNDS

Compared to arsenic, little information is available on the occurrence and behavior of alkylated antimony compounds. The only organoantimony species detected in the natural environment are monomethyl-
stibinic acid, $CH_3SbO(OH)_2$, and dimethylstibinic acid, $(CH_3)_2SbO(OH)$,
which have been found in river waters and seawaters at levels ranging
from 1 to 12 ng/liter [110]. Although the biological alkylation of
antimony by marine algae has been postulated by several authors, the
source of these methylated antimony species is not known.

The pioneer work carried out by Challenger in the 1930s and
1940s on the methylation of heavy elements included the demonstra-
tion of the production of volatile antimony compounds, following
the addition of $KSbO_3$ to cultures of the mold *Penicillium notatum*
[111]. However, the chemical identity of these compounds has
never been clarified, and a possible reduction to stibine as a
confounding factor cannot be excluded [112]. Modern analytical
methods to detect antimony species are based on gas chromatographic
separation and atomic absorption spectrophotometric determination
of stibine derivatives, generated by sodium borohydride reduction
[110].

Biological alkylation of antimony by marine algae, similar to
that of arsenic, has been suggested by Benson and Cooney [113].
They reported on the formation of stibinolipids, similar to arseno-
lipids, in cultures of a marine diatom, *Thalassiosira nana,* exposed
to $^{125}SbCl_3$. Based on the behavior of ^{125}Sb following hydrolysis
and paper chromatography of the methanol extract from these cells,
the algal metabolism of antimony species appears to parallel that
of arsenate.

Interestingly, an analog of choline in which nitrogen is replaced by antimony has been synthesized and neurochemically characterized in rat cortical synaptosomes as a marker for neuronal
cholinergic pools [114]. The behavior of stibinocholine was similar
to that of choline. It proved to be a substrate for several cholinergic enzymes. Sb-choline was transported by a Na^+-dependent, hemi-
cholium-3-sensitive process, acetylated, and subsequently released
by depolarization in a calcium-dependent manner as the normal choline.

Antimony-containing drugs are widely used in the treatment of
leishmaniasis. Pentavalent antimony, Sb(V), complexed to carbo-
hydrates to form stibogluconate (Pentostam), is administered in
solution, or encapsuled in niosomes or liposomes for stabilization of the release of antimony in order to reduce the dose required for therapy [115]. The pharmacokinetics of antimony in
patients treated with stibogluconate is described by a biphasic

excretion mode, in which a rapid initial elimination of Sb(V) (half-time 2 hr) is followed by a slow elimination (half-time 76 hr), possibly related to conversion of Sb(V) to Sb(III) [116].

There is no indication of an alkylation of inorganic antimony in mammals. Unlike arsenic, inorganic trivalent antimony is not methylated in vivo. Metabolic studies of the behavior of Sb(III) and Sb(V) in laboratory animals indicate that all antimony is eliminated in inorganic forms [117].

Trivalent antimony compounds are generally more toxic than the pentavalent ones [118,119].

ABBREVIATIONS

DMA   dimethylarsine
DMAA   dimethylarsinic acid (cacodylic acid)
GSH   glutathione
MAA   methylarsonic acid (methane arsonate)
PL   phospholipids
TMA   trimethylarsine
TMAO   trimethylarsine oxide

REFERENCES

1. R. D. Hood, *Cacodylic Acid: Agricultural Uses, Biological Effects, and Environmental Fate,* Veterans Administration Central Office, Washington, D.C., 1985.

2. W. Martindale, *The Extra Pharmacopoeia,* 27th ed. (A. Wade and J. E. F. Reynolds, eds.), Pharmaceutical Press, London, 1977, p. 1811.

3. H. Yamauchi, T. Kaise, K. Takahashi, and Y. Yamamura, *Fund. Appl. Toxicol., 14,* 399 (1990).

4. B. Gosio, *Arch. Ital. Biol., 35,* 210 (1901).

5. The Swedish Arsenic Commission (1919), *To Discover, Prevent and Counteract the Risk of Chronic Arsenic Poisoning*, Report of a Swedish governmental expert group, Håkan Ohlssons boktryckeri, Lund, 1919. In Swedish.

6. F. Challenger, C. Higginbottom, and L. Ellis, *J. Chem. Soc. Trans.*, 95 (1933).

7. E. A. Woolson, *Weed Sci.*, 25, 412 (1977).

8. B. C. McBride, H. Merilees, W. R. Cullen, and W. Pickett, in *Organometals and Organometalloids* (F. E. Brinckman and J. M. Bellama, eds.), American Chemical Society, Washington, D.C. (ACS Symp. Ser. 82), 1978, pp. 94-115.

9. J. M. Wood, L. J. Cheh, L. J. Dizikes, W. P. Ridley, S. Rakow, and J. R. Lakowicz, *Fed. Proc.*, 37(1), 16 (1978).

10. B. C. McBride and R. S. Wolfe, *Biochemistry*, 10, 4312 (1971).

11. A. W. Pickett, B. C. McBride, W. R. Cullen, and H. Manji, *Can. J. Microbiol.*, 27, 773 (1981).

12. M. S. Mohan, R. A. Zingaro, P. Micks, and P. J. Clark, *Int. J. Environ. Anal. Chem.*, 11, 175 (1982).

13. P. Nissen and A. A. Benson, *Physiol. Plant.*, 54, 446 (1982).

14. R. A. Pyles and E. A. Woolson, *J. Agric. Food Chem.*, 30, 866 (1982).

15. M. O. Andreae and D. Klumpp, *Environ. Sci. Technol.*, 13, 738 (1979).

16. J. G. Sanders, *Chemosphere*, 3, 135 (1979).

17. S. Tagawa, *Bull. Jpn. Soc. Sci. Fisheries*, 46, 1257 (1980).

18. J. S. Edmonds and K. A. Francesconi, *Nature*, 289(5798), 602 (1981).

19. J. S. Edmonds and K. A. Francesconi, *Experientia*, 43, 553 (1987).

20. W. R. Cullen and K. J. Reimer, *Chem. Rev.*, 89, 713 (1989).

21. P. H. Yancey, M. E. Clark, S. C. Hand, R. D. Bowlus, and G. N. Somero, *Science*, 217, 1214 (1982).

22. H. Norin, R. Ryhage, A. Christakopoulos, and M. Sandström, *Chemosphere*, 12(3), 299 (1983).

23. J. F. Lawrence, P. Michalik, G. Tam, and H. B. S. Conacher, *J. Agric. Food Chem.*, 34, 315 (1986).

24. A. Christakopoulos, B. Hamasur, H. Norin, and I. Nordgren, *Biomed. Environ. Mass Spectrom.*, 15, 67 (1988).

25. B. P.-Y. Lau, P. Michalik, C. J. Porter, and S. Krolik, *Biomed. Environ. Mass Spectrom.*, 14, 723 (1987).

26. J. S. Edmonds, K. A. Francesconi, and J. A. Hansen, *Experientia*, *38*, 643 (1982).

27. H. Norin, M. Vahter, A. Christakopoulos, and M. Sandström, *Chemosphere*, *14*, 325 (1985).

28. J. S. Edmonds and K. A. Francesconi, *Sci. Tot. Environ.*, *64*, 317 (1987).

29. F. B. Whitfield, *Water Sci. Technol.*, *20*(8/9), 63 (1988).

30. T. Kaise, K. Hanaoka, and S. Tagawa, *Chemosphere*, *16*, 2551 (1987).

31. K. Shiomi, Y. Kakehashi, H. Yamanaka, and T. Kikuchi, *Appl. Organomet. Chem.*, *1*, 177 (1987).

32. K. Shiomi, Y. Horiguchi, and T. Kaise, *Appl. Organomet. Chem.*, *2*, 385 (1988).

33. M. O. Andreae, in *Organometallic Compounds in the Environment* (P. J. Craig, ed.), Longman Group, Burnt Mill, Harlow, UK, 1986, pp. 198-228.

34. D. J. H. Phillips, *Aquatic Toxicol.*, *16*, 151 (1990).

35. WHO, *Environmental Health Criteria 18*, World Health Organization, Geneva, 1981.

36. WHO, in *IMO/FAO/UNESCO/WMO/WHO/IAEA/UN/UNEP Joint Group of Experts on the Scientific Aspects of Marine Pollution—GESAMP: Review of Potentially Harmful Substances: Arsenic, Mercury and Selenium*, World Health Organization, Geneva, 1986, pp. 17-73.

37. V. Blekastad, J. Jonsen, E. Steinnes, and K. Helgeland, *Acta Med. Scand.*, *216*, 25 (1984).

38. E. A. Crecelius, *Environ. Health Persp.*, *19*, 147 (1977).

39. H. Norin and M. Vahter, *Scand. J. Work Environ. Health*, *7*, 38 (1981).

40. T. Mohri, A. Hisanaga, and N. Ishinishi, *Food Chem. Toxicol.*, *28*(7), 521 (1990).

41. G. Westöö and M. Rydälv, *Vår Föda*, *24*, 21 (1972) (in Swedish with English summary).

42. D. L. Johnson and R. S. Braman, *Chemosphere*, *6*, 333 (1975).

43. H. Mukai and Y. Ambe, *Atm. Environ.*, *21*(1), 185 (1987).

44. R. F. Tarrant and J. Allard, *Arch. Environ. Health*, *24*, 277 (1972).

45. S. L. Wagner and P. Weswig, *Arch. Environ. Health*, *28*, 77 (1974).

46. A. A. Abdelghani, A. C. Anderson, M. Jaghabir, and F. Mather, *Arch. Environ. Health*, *4*(3), 163 (1986).

47. EPA, *Initial Scientific Review of Cacodylic Acid*, U.S. Environ-
     mental Protection Agency, Washington, D.C., 1975.

48. R. S. Braman and C. C. Foreback, *Science, 182,* 1247 (1973).

49. P. J. Durrant and B. Durrant, *Introduction to Advanced Inorganic
     Chemistry*, Longmans, London, 1962.

50. K. B. Jacobsen, B. Murphy, and B. D. Sarma, *FEBS Lett., 22,* 80
     (1972).

51. C. H. Banks, J. R. Daniel, and R. A. Zingaro, *J. Med. Chem.,
     22,* 572 (1979).

52. K. S. Squibb and B. A. Fowler, in *Biological and Environmental
     Effects of Arsenic*. Topics in Environmental Health, Vol. 6
     (B. A. Fowler, ed.), Elsevier, Amsterdam, 1983, pp. 233-269.

53. M. Vahter and E. Marafante, *Chem. Biol. Interact., 47,* 29
     (1983).

54. B. Holmstedt and G. Liljestrand, *Readings in Pharmacology*,
     Raven Press, New York, 1981.

55. C. Voegtlin, H. A. Dyer, and C. S. Leonard, *U.S. Public Health
     Rep., 38,* 1882 (1924).

56. R. A. Peters, L. A. Stocken, and R. H. S. Thompson, *Nature,
     156,* 616 (1945).

57. T. Kaise, H. Yamauchi, Y. Horiguchi, T. Tani, S. Watanabe,
     T. Hirayama, and S. Fukui, *Appl. Organomet. Chem., 3,* 273
     (1989).

58. H. L. Hong, B. A. Fowler, and G. A. Boorman, *Toxicol. Appl.
     Pharmacol., 97,* 1173 (1989).

59. T. Kaise, S. Watanabe, and K. Ito, *Chemosphere, 14,* 1327
     (1985).

60. *IARC Monographs on the Evaluation of the Carcinogenic Risk
     of Chemicals to Humans*. Some Metals and Metallic Compounds,
     Vol. 23, International Agency for Research on Cancer, Lyon,
     1980, pp. 39-141.

61. T. Okui and Y. Fujiwara, *Mut. Res., 172,* 69 (1986).

62. J.-H. Li and T. G. Rossman, *Mol. Toxicol., 2,* 1 (1989).

63. K. Yamanaka, A. Hasegawa, R. Sawamura, and S. Okada, *Biol.
     Trace Element Res., 21,* 413 (1989).

64. K. Yamanaka, A. Hasegawa, R. Sawamura, and S. Okada, *Biochem.
     Biophys. Res. Commun., 165*(1), 43 (1989).

65. K. Yamanaka, A. Hasegawa, R. Sawamura, and S. Okada, *Toxicol.
     Appl. Pharmacol., 108,* 205 (1991).

66. K. Yamanaka, H. Ohba, A. Hasegawa, R. Sawamura, and S. Okada,
     Chem. Pharm. Bull., 37(10), 2753 (1989).

67. W. M. F. Jongen, J. M. Cardinaals, and P. M. J. Bos, *Food Chem. Toxicol.*, *23*(7), 669 (1985).

68. E. Sabbioni, M. Fischbach, G. Pozzi, R. Pietra, M. Gallorini, and J. L. Piette, *Carcinogenesis*, *12*(7), 1287 (1991).

69. M. Vahter and E. Marafante, in *The Biological Alkylation of Heavy Elements* (P. J. Craig and F. Glockling, eds.), Royal Society of Chemistry, London, 1988, pp. 105-119.

70. M. Vahter, E. Marafante, A. Lindgren, and L. Dencker, *Arch. Toxicol.*, *51*, 65 (1982).

71. M. Vahter and E. Marafante, *Arch. Toxicol.*, *57*, 119 (1985).

72. J. P. Buchet, R. Lauwerys, and H. Roels, *Int. Arch. Occup. Environ. Health*, *48*, 71 (1981).

73. E. Marafante and M. Vahter, *Chem. Biol. Interact.*, *50*, 49 (1984).

74. J. P. Buchet and R. Lauwerys, *Arch. Toxicol.*, *57*, 125 (1985).

75. B. Georis, A. Cardenas, J. P. Buchet, and R. Lauwerys, *Toxicology*, *63*, 73 (1990).

76. F. Bertolero, G. Pozzi, E. Sabbioni, and U. Saffiotti, *Carcinogenesis*, *8*(6), 803 (1987).

77. J. P. Buchet and R. Lauwerys, *Biochem. Pharmacol.*, *37*(16), 3149 (1988).

78. M. Hirata, A. Hisanaga, A. Tanaka, and N. Ishinishi, *Appl. Organomet. Chem.*, *2*, 315 (1988).

79. E. Marafante, M. Vahter, and J. Envall, *Chem. Biol. Interact.*, *56*, 225 (1985).

80. M. Vahter and E. Marafante, *Toxicol. Lett.*, *37*, 41 (1987).

81. G. K. H. Tam, S. M. Charbonneau, F. Bryce, C. Pomroy, and E. Sandi, *Toxicol. Appl. Pharmacol.*, *50*, 319 (1979).

82. M. Vahter, *Acta Pharmacol. Toxicol.*, *59*(7), 31 (1986).

83. M. Vahter, *Environ. Res.*, *25*, 286 (1981).

84. V. Foa, A. Colombi, M. Maroni, M. Buratti, and G. Calzaferri, *Sci. Tot. Environ.*, *34*, 241 (1984).

85. M. A. Lovell and J. G. Farmer, *Human Toxicol.*, *4*, 203 (1985).

86. M. Vahter, L. Friberg, B. Rahnster, Å. Nygren, and P. Nolinder, *Int. Arch. Occup. Environ. Health*, *57*, 79 (1986).

87. H. Yamauchi, N. Yamato, and Y. Yamamura, *Bull. Environ. Contam. Toxicol.*, *40*, 280 (1988).

88. J. T. Stevens, L. L. Halle, J. D. Farmer, L. C. DiPasquale, N. Chernoff, and W. F. Durham, *Environ. Health Persp.*, *19*, 151 (1977).

89. T. C. Siewicki, and J. S. Sydlowski, *Nutr. Rep. Int.*, *24*(1), 121 (1981).

90. M. Vahter, E. Marafante, and L. Dencker, *Arch. Environ. Contam. Toxicol.*, *13*, 259 (1984).

91. E. Marafante, M. Vahter, H. Norin, J. Envall, M. Sandström, A. Christakopoulos, and R. Ryhage, *J. Appl. Toxicol.*, *7*(2), 111 (1987).

92. H. Yamauchi, K. Takahashi, and Y. Yamamura, *Toxicol. Environ. Chem.*, *22*, 69 (1989).

93. L. S. Goodman and A. Gilman, *The Pharmalogical Basis of Therapeutics*, 2nd Ed., Macmillan, New York, 1955, pp. 948-969.

94. WHO, *Environmental Health Criteria 58*, Selenium, World Health Organization, Geneva, 1987.

95. T. C. Siewicki, *J. Nutr.*, *111*, 602 (1981).

96. K. Liebscher and H. Smith, *Arch. Environ. Health*, *17*, 881 (1968).

97. O. DuPont, I. Ariel, and S. L. Warren, *Am. J. Syph. Gon. Vener. Dis.*, *26*, 96 (1941).

98. A. Lindgren, M. Vahter, and L. Dencker, *Acta Pharmacol. Toxicol.*, *51*, 253 (1982).

99. N. Yamato, *Bull. Environ. Contam. Toxicol.*, *40*, 633 (1988).

100. J. O. Dickinson, *Am. J. Vet. Res.*, *33*, 1889 (1972).

101. G. K. H. Tam, S. M. Charbonneau, F. Bryce, and E. Sandi, *Bull. Environ. Contam. Toxicol.*, *28*, 669 (1982).

102. M. Vahter, E. Marafante, and L. Dencker, *Sci. Tot. Environ.*, *30*, 19 (1983).

103. E. Marafante, M. Vahter, and L. Dencker, *Sci. Tot. Environ.*, *34*, 223 (1984).

104. J. R. Cannon, J. S. Edmonds, K. A. Francesconi, and J. B. Langsford, in *Proc. of the International Conference on Management and Control of Heavy Metals in the Environment* (London, September 1979), CEP Consultants Ltd., Edinburgh, 1979.

105. P. J. Mann, H. E. Woodward, and J. W. Quastel, *Biochem. J.*, *32*, 1024 (1938).

106. A. Christakopoulos, H. Norin, M. Sandström, H. Thor, P. Moldeus, and R. Ryhage, *J. Appl. Toxicol.*, *8*(2), 119 (1988).

107. H. C. Freeman, J. F. Uthe, R. B. Fleming, P. H. Odense, R. G. Ackman, G. Landry, and C. Musial, *Bull. Environ. Contam. Toxicol.*, *22*, 224 (1979).

108. J. B. Luten and G. Riekwel-Booy, in *Trace Element: Analytical Chemistry in Medicine and Biology*, Vol. 2 (P. Brätter and P. Schramel, eds.), de Gruyter, Berlin, 1983, pp. 277-286.

109. R. M. Brown, D. Newton, C. J. Pickford, and J. C. Sherlock, *Human Exp. Toxicol.*, *9,* 41 (1990).

110. M. O. Andreae, J. F. Asmodé, P. Foster, and L. Van't dack, *Anal. Chem.*, *53,* 1766 (1981).

111. P. Barnard, Ph.D. thesis, University of Leeds, UK, 1947.

112. P. J. Craig, in *Organometallic Compounds in the Environment* (P. J. Craig, ed.), Longman Group, Harlow, UK, 1986, pp. 345-364.

113. A. A. Benson and R. V. Cooney, in *The Biological Alkylation of Heavy Metals* (P. J. Craig and F. Glockling, eds.), Royal Chemistry Society, London, 1988, pp. 135-137.

114. E. M. Meyer, R. J. Barnett, and J. R. Cooper, *J. Neurochem.*, *39,* 321 (1982).

115. C. A. Hunter, T. F. Dolan, G. H. Coombs, and A. J. Baillie, *J. Pharm. Pharmacol.*, *40,* 161 (1988).

116. J. D. Chulay, L. Fleckenstein, and D. H. Smith, *Trans. Roy. Soc. Trop. Med. Hyg.*, *82,* 69 (1988).

117. A. Gellhorn, N. A. Tubikova, and H. B. Van Dyke, *J. Pharmacol. Exp. Ther.*, *87,* 169 (1946).

118. C.-G. Elinder and L. Friberg, in *Handbook on the Toxicology of Metals,* Vol. 2 (L. Friberg, G. F. Nordberg, and V. V. Vouk, eds.), Elsevier, Amsterdam, 1986, pp. 26-42.

119. T. Norseth and I. Martinsen, in *Biological Monitoring of Toxic Metals* (T. W. Clarkson, L. Friberg, G. F. Nordberg, and P. R. Sager, eds.), Plenum Press, New York, 1988, pp. 337-367.

7

Biological Alkylation of Selenium and Tellurium

Ulrich Karlson

Department of Marine Ecology and Microbiology
National Environmental Research Institute
P.O. Box 358, Frederiksborgvej 399
DK-4000 Roskilde, Denmark

and

William T. Frankenberger, Jr.

Department of Soil and Environmental Sciences
University of California
Riverside, California 92521, USA

1. OCCURRENCE OF ALKYLATED SELENIUM AND TELLURIUM IN THE ENVIRONMENT

For about a century, alkylselenides and alkyltellurides have been recognized as biologically produced malodorous compounds. Although one may find small amounts of these compounds produced by anthropogenic sources in some industrial work spaces and in a few research laboratories, the bulk of environmental alkylselenides and alkyltellurides are thought to be produced by biological processes. While the main biologically produced species are methylated forms, longer alkyl chains may also exist naturally, although in most cases when the presence of alkylselenides was investigated in natural environments only methylated forms were detected. These compounds are highly volatile and exist primarily in the gaseous form. The vapor pressures at 25°C of $(CH_3)_2Se$ and $(CH_3)_2Se_2$ measured at 30.46

and 0.38 kPa, respectively [1]. It is important to note that $(CH_3)_2Se$ is moderately water-soluble. Its solubility was determined at 24.4 g/liter [1]. In contrast, $(CH_3)_2Se_2$ is not stable in water.

In addition to the volatile dialkyl species, e.g., $(CH_3)_2Se$, $(CH_3)_2Se_2$, $(CH_3)_2SeO_2$, $(CH_3)_2Te$, $(CH_3)_2Te_2$, the trimethylated species $(CH_3)_3Se^+$ and possibly $(CH_3)_3Te^+$ are important naturally occurring alkylated compounds. These compounds are nonvolatile and water-soluble. There is no distinct odor from these trimethylated species.

Published data on concentrations of alkylated Se and Te compounds in environmental compartments are scarce, most likely because they are not frequently monitored.

1.1. Air

Measurable levels of alkylselenides occur in continental air spaces, but alkylselenides have not been detected in marine air samples [2]. This may be due to the fact that ocean air has not yet been analyzed for these Se species. Vapor phase Se in coastal regions is reported to be 0.01-0.6 ng/m^3, amounting to 13-48% of the atmospheric concentration [3]. Vapor phase Se was operationally defined in this study as that portion of Se passing through either a 0.2- or a 0.4-μm pore size membrane filter and being adsorbed on activated charcoal. The Se fraction presumably consists of vapor phase SeO_2, $(CH_3)_2Se$, $(CH_3)_2Se_2$, or elemental Se. In a continental location, outdoor air was found to contain ~2.10 ng/m^3 of Se in the vapor phase, based on adsorption sampling with gold-coated beads and charcoal [4]. Further analysis of the charcoal filters revealed that organoselenium compounds constituted about 15% of the total vapor phase Se. The air in various rural and municipal locations tested contained $(CH_3)_2Se$ and $(CH_3)_2Se_2$ at concentrations ranging from nondetectable to 2.40 and 0.63 ng Se/m^3, respectively [5]. A third methylated compound, tentatively identified as $(CH_3)_2SeO_2$, was present in a few samples. Diethylselenide, $(C_2H_5)_2Se$, was detected only in laboratory air, where the compound was being handled as a standard [6].

The only report on environmental Te is for a continental location, where outdoor air was found to contain ca. 0.24 ng/m^3 of Te in
the vapor phase, based on adsorption sampling with gold-coated beads
and charcoal. Although the filters were not analyzed for organic Te,
the author suggested that the predominant species were Te^0, TeO$_3^{2-}$ and
TeO$_4^{2-}$ [4]. There are no environmental data specifically for alkylated
tellurides.

1.2. Freshwater Bodies

A number of selenide compounds have been found to exist in freshwater
samples. Organoselenium compounds were detected in samples from
various freshwater environments, but only in association with high
dissolved organic C content [7]. The only species identified in this
study, $(CH_3)_3Se^+$, measured at a concentration of 13 ng/liter, accounting for 6% of the total Se and 8% of the dissolved organic Se. The
other organic Se species were thought to exist in the form of peptides,
although this was not confirmed analytically. In another study, a
lake water sample was found to contain 19 ng Se/m^3, and a groundwater
sample 9 ng Se/m^3 in the form of $(CH_3)_3Se^+$ [8]. In surface waters
from high-Se areas, one nonvolatile Se species, $(CH_3)_2Se^+$-R (possibly
Se-methylselenomethionine), and two volatile species, $(CH_3)_2Se$ and
$(CH_3)_2Se_2$, were detected [9].

1.3. Oceans

Marine water studies of selenide compounds are limited, although
detection of some selenide compounds has been reported in the last
decade. In Atlantic Ocean estuary water, 0.5 μg/m^3 of Se in the
$(CH_3)_2Se$ form was determined [10]. Organic selenides constituted
~80% of the total dissolved Se in surface waters from the North and
South Pacific Oceans, with concentrations ranging from 15 to 55 ng
Se/liter [11]. However, organic selenides were thought to consist

mainly of selenoamino acids in peptides, although this was not confirmed analytically. In another study, $(CH_3)_2Se$ was detected in
seawater in trace quantities [12].

2.   SOURCES OF ENVIRONMENTAL ALKYLSELENIDES

2.1.   Formation by Soil Microorganisms

Conversion of selenite and selenate to $(CH_3)_2Se$ by pure cultures of
Scopulariopsis brevicaulis growing on bread crumbs was reported as
early as 1934 [13]. The same organism did not produce $(C_2H_5)_2Se$.
Evolution of volatile Se from soil supplemented with $[^{75}Se]$-selenite
was first shown in 1958; however, the species of volatile Se was not
determined [14]. Autoclaving was shown to eliminate Se evolution
from soil [15], whereas supplying organic C and adequate moisture
favored the process [14-16], which provided indirect evidence that
microorganisms were involved. Conversion of Se to volatile malodorous
compounds by cultures of bacteria, actinomycetes, and fungi isolated
from soil had been reported in an early publication [17] that received
little attention. Eventually, several identified fungal soil isolates
were shown to convert selenite and selenate to $(CH_3)_2Se$ [18], and thus
the link between soil organisms and the chemical identity of their
product was established.

In experiments involving the incubation of environmental soil
samples, without an attempt to isolate the organisms responsible,
indigenous soil Se, as well as supplemented elemental Se, selenite,
selenate, $(CH_3)_3Se^+$, Se-methionine, and Se-cystine were all found to
be converted to $(CH_3)_2Se$ through soil microbial activity [19-21].
Selenomethionine was also converted to $(CH_3)_2Se_2$ [20]. A third volatile Se species, $(CH_3)_2SeO_2$, was reported to evolve from soil samples
due to microbial activity, in addition to $(CH_3)_2Se$ and $(CH_3)_2Se_2$ [22].
Evolution rates from experimentally undisturbed soils were 273, 143,
and 63 ng Se/kg soil/day for $(CH_3)_2Se$, $(CH_3)_2Se_2$, and $(CH_3)_2SeO_2$,
respectively, and were attributed to microbial activity [23].

Sewage sludge is another environmental source of alkylated Se. A strain of *Penicillium,* isolated from raw sewage, was reported to convert selenite, selenate, and inorganic selenide to $(CH_3)_2Se$ [24]. Three volatile Se species, $(CH_3)_2Se$, $(CH_3)_2Se_2$, and $(CH_3)_2SeO_2$, were found to evolve from sewage sludge samples due to microbial activity [22]. At the outlet of a sealed sewage digestion tower, high air concentrations of $(CH_3)_2SeO_2$ (18.8 ng Se/m^3) and $(CH_3)_2Se$ (2.40 ng Se/m^3) were determined, while measurable amounts of $(CH_3)_2Se_2$ were not found [23].

Until 1974, the only isolated and identified Se-alkylating organisms were fungi (Table 1), suggesting that fungi were the predominant group performing this process. However, the selective media typically used for isolation of Se alkylators contained very high amounts of Se in the selenite form. Selenite concentrations greater than 220 mg Se/liter inhibit a number of bacteria, and growth of most bacteria is inhibited at 1200 mg/liter [25]. However, one exception to this rule was found in a Se-requiring *Bacillus* sp. (Table 1), which was isolated from a Se container [26]. A few bacterial genera have now been added to the list of microorganisms capable of alkylating Se, some of which were isolated using low-Se media.

Working with soil samples Koval'skii [17] reported the isolation of Se-volatilizing actinomycetes and bacteria, in addition to fungi, although he did not identify his isolates. Nevertheless, Abu-Erreish et al. [15] concluded, based on observations of reduced Se evolution from water-logged soils, that soil fungi were the predominant group of organisms performing Se alkylation. One might argue, however, that such observations indicate the predominance of aerobic microorganisms (including both fungi and obligate aerobic bacteria) as responsible for this process. In contrast, Doran and Alexander [20] reported slow conversion of Se-methionine to $(CH_3)_2Se$ and $(CH_3)_2Se_2$ under anaerobiosis, indicating the possible contribution of anaerobic bacteria to Se volatilization in the environment. In aerobic soil systems, addition of the bactericidal antibiotic chloramphenicol resulted in a 50% reduction of Se volatilization

TABLE 1

Microorganisms Capable of Forming Alkylselenides

Microorganism	Source of isolate	Isolation medium (mg Se/liter)	Ref.
Scopulariopsis brevicaulis	Air and cheese	0	13,27
Schizophyllum commune	Decaying wood	0	28
Aspergillus niger	Bread[a]	0	29
Penicillium sp.	Sewage	300	24
Candida humicola	Sewage[a]	0	30
Cephalosporium spp., *Fusarium* spp., *Scopulariopsis* sp., *Penicillium* sp.	Soil	300	18
Aeromonas sp., *Flavobacterium* sp., *Pseudomonas* sp.	Lake sediment	5	31
Corynebacterium sp.	Soil	0	32
Candida humicola	Decaying seed[b]	6222	33,34
Penicillium chrysogenum	Decaying seed[b]	6222	34
Mortierella spp.	(Culture collections)	(unknown)[c]	35
Pseudomonas fluorescens	(unknown)	(unknown)	36
Acremonium falciforme, Penicillium citrinum, Ulocladium tuberculatum	Soil	300	37
Alternaria alternata	Pond water	10-200	38
Pseudomonas fluorescens K27	Dewatered pond sediment	790	39,40
Bacillus sp.	Se container	1500-9000	41

[a]From Ref. 85.
[b]Seed of *Neptunia amplexicaulis*, containing 10 mg Se/g naturally.
[c]Concentrations of ≥30 mg Se/liter were inhibitory.

from a loam soil spiked with selenite [33]. The same antibiotic
caused a 33% increase of $(CH_3)_2Se$ evolution from a naturally
seleniferous pond sediment, and streptomycin even caused a 3.7-fold
increase in $(CH_3)_2Se$ formation (see Fig. 5 in Sec. 4.8). It appears
that the capability to alkylate Se is not limited to one microbial
group and that the relative contribution to Se volatilization by
the different microbial groups varies among ecosystems. A compila-
tion of the quantities of Se evolved from soils due to microbial
activity was published by Haygarth et al. [42]. Measured volatilization rates range from 0.05% to several hundred percent of the
original soil Se content if calculations are projected on a per-
year basis, depending on experimental design and environmental
conditions.

2.2. Formation by Higher Plants

Formation of alkylselenides occurs in many plant species, including
salt grass (*Distichlis spicata*) [43], milk vetch (*Astragalus racemo-
sus*) [44-46], alfalfa (*Medicago sativa*) [44,47], barley (*Hordeum
vulgare*) [48], sunflower (*Helianthus annuus*) [49], spinach (*Spinacia
oleraria*) [49], brown top (*Agrostis tenuis*) [48], cabbage (*Brassica
oleracea*) [50], and four other cruciferous species [49]. The plant
leaves appear to be the organs of alkylselenide synthesis, whereafter
the volatile products are released into the surrounding air [50].
Production of alkylselenides occurred when plants were grown on
soils or in nutrient solutions containing selenite or selenate [48,
51]. Volatilization of Se occurred when the plants were exposed to
solution concentrations as low as 7 µg/liter, concentrations which
can be found naturally in many soil solutions [48], indicating that
this process may be widely distributed on a global scale. The
alkylated Se products identified to date are $(CH_3)_2Se$ and $(CH_3)_2Se_2$
[46]. At least two additional species have been observed but not
yet identified [45,46]. For milk vetch, production of $(C_2H_5)_2Se$ was

excluded experimentally [46]. With barley growing in nutrient solution (7 μg selenite-Se/liter), between 20% and 25% of the Se taken
up by the whole plant and between 35% and 40% of the Se translocated
from root to shoot were recovered in the volatile form [48]. Of the
Se added to several *Astragalus* species in nutrient solution (1-4 mg
selenate Se/liter), between 8% and 13% were lost in the form of
$(CH_3)_2Se$ and $(CH_3)_2Se_2$ over a period of 4 weeks [52]. Under field
conditions, a 1 m^2 salt grass plot produced up to 180 μg volatile Se
per day from seleniferous dewatered pond sediment (1-3 mg Se/kg).

2.3. Formation in Freshwater Bodies

Chau et al. [31] reported production of $(CH_3)_2Se$, $(CH_3)_2Se_2$, and
$(CH_3)_2SeO_2$ in lake sediments. It occurred upon addition of selenite
or selenate in every sample of 12 lake sediments tested and was
attributed to aerobic and anaerobic microorganisms in the sediments.
Volatilization of Se has also been detected from the water column of
seleniferous evaporation ponds. Evolution of alkylselenides in the
aquatic environment is attributed to bacteria, cyanobacteria, and
fungi [38,53-56]. Addition of bactericidal but not fungicidal antibiotics inhibited $(CH_3)_2Se$ formation [54]. Amendments with organic
C, in particular with proteins, stimulated Se removal from the water
column [55,56].

In surface waters and oxic groundwaters $(CH_3)_2Se$ and $(CH_3)_2Se_2$
were detected, but no $(CH_3)_2SeO_2$ was found [9]. This study also provided some preliminary evidence for the release of $(CH_3)_2Se^+$-R from
detrital plant material and its subsequent abiotic conversion to
$(CH_3)_2Se$. The identity of $(CH_3)_2Se^+$-R was speculated to be Se-
methylselenomethionine. This may indicate that some of the volatile
alkylselenides found in freshwater are not microbially produced but
are instead derived nonenzymatically from cell constituents of algae
or higher plants decaying in those water bodies. However, the evidence is still incomplete at this point, and this hypothesis remains
unconfirmed.

Three freshwater green algae, *Ankistrodesmus* sp., *Chlorella vulgaris*, and *Selenastrum* sp. were found to form $(CH_3)_3Se^+$ and $(CH_3)_2Se$ in pure cultures [57]. Most of the $(CH_3)_3Se^+$ formed was retained in the algal cells, but approximately 1% of it was released into the culture medium. Total $(CH_3)_3Se^+$ formation was up to 0.3% of the Se added (50 mg/liter), and the amount of $(CH_3)_2Se$ formed ranged from 0.001% to 0.2% in 4 days. These findings provide an explanation for the detection of $(CH_3)_3Se^+$ in freshwater bodies by other researchers.

2.4. Formation in Marine Environments

Information on production of alkylselenides in marine environments is both incomplete and hypothetical. A marine bacterium, *Pseudomonas marina*, was found to release an unidentified organic Se compound into the culture medium when growing in the presence of selenite but not selenate [58]. The product may have been $(CH_3)_2Se$, since the presence of $(CH_3)_2Se$ was not tested for in this study. Preliminary observations indicate the formation of $(CH_3)_2Se$ and $(CH_3)_2Se_2$ from selenite and selenate during anaerobic incubation of the marine red alga *Ceramium rubrum* in nonsterile seawater [59]. Other preliminary evidence has also been obtained in the laboratory for the formation of $(CH_3)_2Se$ by marine algae and a pathway has been proposed [60].

Model calculations of Se fluxes in a coastal salt marsh indicate that sedimentary Se is continuously removed from the marsh through volatilization [61]. In the open ocean, aerosol Se concentrations were found to be high in areas of high primary productivity and low selenite concentrations in the water. Since these areas correlate with high non-sea-salt sulfate concentrations, Mosher et al. [62], drawing an analogy to the known production of dimethylsulfide and its atmospheric conversion to sulfate, suggest that the source of aerosol Se in the marine atmosphere is biologically produced organoselenium compounds which have undergone gas-to-particle conversion. Considering the few recent analytical data for $(CH_3)_2Se$ in seawater (Sec. 1.3), one may speculate that these biologically produced Se compounds are alkylselenides. However, even if this could be confirmed analytically, the source of alkylselenide production and the possible involvement of microorganisms in this process remain unclear.

2.5. Formation by Mammals

As early as 1893, volatile Se was suggested to occur in the breath of Se-poisoned animals [63]. This notion has been confirmed in much detail. In rats, up to 10% of the injected selenate [64] and over 25% of the injected selenite [65] were exhaled in 24 hours, when the Se dose approached acute toxic levels. As the dose is lowered, the portion exhaled decreases [66,67], and at subacute chronic doses less than 2% of the ingested Se was exhaled [68]. In experiments with mice, adding Se-cystine, Se-methionine, and selenite to the feed to make a total intake of 0.35 mg/kg body weight, 0.1%, 0.24%, and <0.02% were excreted through the breath, respectively. Exhaled metabolites were identified as $(CH_3)_2Se$ and $(CH_3)_2Se_2$ [69]. One other unidentified volatile species was produced from Se-methionine, but was determined not to be $(C_2H_5)_2Se$ or $(CH_3)_2SeO_2$ [70].

The highest activity of conversion from selenite to $(CH_3)_2Se$ has been found in the liver and kidneys, followed by the lungs, leg muscles, spleen, and heart [71-73]. In the rat, exhalation of $(CH_3)_2Se$ was stimulated by methylmercury [74,75], inhibited by arsenic [76], inhibited by mercuric or cadmium salts, and not significantly affected by zinc salts [77]. The main route for elimination of volatile Se from the body seems to be exhalation.

Excretion of $(CH_3)_3Se^+$ in the urine following dietary supplementation with various forms of Se has been reported for rats [78,79] as well as for humans [80], and has been shown to be a major excretory product from selenate, Se-methionine, Se-cystine, Se-methylseleno-

cysteine, dimethylselenocysteineselenonium, Se-methylselenomethionine-
selenonium, and seleniferous wheat [79,81].

3. MECHANISMS OF SELENIUM ALKYLATION

3.1. Microorganisms

In microorganisms, the capability to form alkylselenides is associated with resistance to high Se levels [40,82]. Transformations of
cellular Se to volatile species is a conversion to less toxic forms
(see Sec. 6) and initiates its permanent removal from the cell. An
alternative resistance mechanism is the oxidation or reduction of
soluble Se species to the elemental form, which is insoluble in water.
In pure microbial cultures, accumulation of elemental Se is recog-
nizable through the appearance of brick red colonies. Pure cultures
of the Se-methylating fungi *Acremonium falciforme, Penicillium citrinum, Ulocladium tuberculatum* [37], and *Alternaria alternata* [38]
showed signs of simultaneous elemental Se accumulation when growing on
selenite medium, but not on selenate. Several species of *Mortierella*
produced $(CH_3)_2Se$ from selenite and selenate, and formed red colonies
at ≥ 100 mg/liter selenite-Se [35]. *Scopulariopsis brevicaulis* was
found to volatilize selenate and selenite, but also reduced selenite
to elemental Se, resulting in the production of less methylated Se
[13]. In one study, bacterial soil isolates forming red colonies on
selenite medium were unable to produce volatile Se [18]. Formation
of $(CH_3)_2Se$ did not correlate with selenate toxicity in 11 Se methylating fungal soil isolates; hence it was concluded that Se methylation
was not responsible for fungal resistance to selenate [83]. Out of
approximately 200 selenite-resistant (elemental Se-forming) bacterial
isolates, only three were resistant to selenate [40]. However, almost
all of them formed $(CH_3)_2Se$ when grown on selenite medium. From these
findings one may conclude that Se alkylation is a secondary Se resistance mechanism for some microorganisms, whereas the primary Se resistance determinant for most microorganisms is Se reduction.

Several pathways for microbial Se alkylation have been proposed. Challenger [84] suggested that $(CH_3)_2Se$ formation begins with methylation of selenite to form methaneselenonic acid, followed by an ionization and reduction step to form methaneseleninic acid, a second methylation step to form $(CH_3)_2SeO_2$, and a last reduction step to produce $(CH_3)_2Se$. Reamer and Zoller [22] expanded this pathway to include $(CH_3)_2Se_2$: Methaneseleninic acid is alternatively reduced to methaneselenol and/or methaneselenenic acid which would rapidly produce $(CH_3)_2Se_2$. This alternative pathway could be regulated by Se concentration (active at high substrate Se concentrations), or could depend on the microbial species (genotype). The intermediates of Challenger's pathway, except for $(CH_3)_2SeO_2$, have not been isolated in culture solutions. Based on methylation studies with bacterial cell extracts, Doran [85] proposed an alternative pathway from selenite to selenide in two reduction steps and from selenide to $(CH_3)_2Se$ in two methylation steps via methaneselenol. Although Doran does not discuss it, his pathway would allow for $(CH_3)_2Se_2$ formation from methaneselenol, which is analogous to the pathway proposed by Reamer and Zoller [22]. However, it does not explain the presence of $(CH_3)_2SeO_2$ in microbial cultures and environmental samples. None of the proposed pathways elucidates the mechanism by which the initial reduction of selenate to selenite proceeds.

Evidence supporting Doran's hypothesis comes from reports of traces of methaneselenol in fungal cultures [39,86], and from the discovery of a selenide methyltransferase in *Tetrahymena thermophila,* which reportedly generates methaneselenol from selenide [87]. Doran's pathway is similar to the pathway proposed for mammalian tissues (Sec. 3.3) in that it requires methaneselenol as an intermediate. This would suggest that *S*-adenosylmethionine, and not methylcobalamin, acts as the methyl donor for Se methylation in microbial cells. Methylcobalamin has been suggested as the methyl donor in bacterial methylation of mercury, tin, arsenic, palladium, and platinum [88-91].

3.2. Higher Plants

The production of alkylselenides by plants occurs in seemingly healthy tissues, under normal physiological conditions, and does not require toxic levels of Se in the growth medium [48]. As far as is known, alkylselenides are a naturally occurring byproduct of normal biochemical reactions within plants. In this process, Se is metabolized along the same pathways as sulfur. Even in plants adapted to high soil Se levels, so-called Se accumulators, volatilization of Se appears to play a minor role in the plant's tolerance for seleniferous growth media [49]. Two pathways for Se methylation have been suggested. One is found in nonaccumulator species, such as alfalfa, sunflower, spinach, and cruciferous plants. In this pathway following initial assimilation and reduction of selenite or selenate, dimethylselenide is formed via Se-methionine and Se-methylseleno-methionine. The second pathway occurs in accumulator species, such as milk vetch and brown top. In this pathway $(CH_3)_2Se_2$ is produced via selenocystathionine, Se-methylselenocysteine and Se-methylseleno-cysteineselenoxide. Se-methylation in plants has been thoroughly reviewed by Lewis [49]. Neither proposed pathway elucidates the initial step in which selenate or selenite is reduced to form selenide.

A pathway similar to that described in nonaccumulator species has been proposed for the formation of $(CH_3)_2Se$ by marine algae [60].

3.3. Mammals

The exhalation of $(CH_3)_2Se$ by mammals is apparently a detoxification mechanism. Exhalation of $(CH_3)_2Se$, sometimes called "garlic breath," indicates acute Se poisoning. In Se poisoning, the formation rate of $(CH_3)_2Se$ exceeds its rate of methylation to $(CH_3)_3Se^+$, a reaction which occurs at physiological Se levels. Under normal physiological conditions, the major product of Se metabolism appears to be $(CH_3)_3Se^+$, which is excreted in the urine [81].

In the proposed metabolic pathway of selenite to $(CH_3)_3Se^+$, selenite is reduced in three steps. In the first step, selenite is reduced to H_2Se with the aid of glutathione, NADPH, and glutathione reductase. The intermediate product, H_2Se, is methylated to $(CH_3)_2Se$ via methylselenol involving S-adenosylmethionine and two methyltransferases [72,73,92,93]. In the final step, $(CH_3)_2Se$ is thought to be converted to $(CH_3)_3Se^+$ [93,94]. Experiments in which rats received 5 mg selenate-Se/kg in their diet, excretion of $(CH_3)_3Se^+$ increased under dietary supplementation with methionine and vitamin E [95]. It is not clear whether the entire pathway or only the last step was stimulated by the addition of these cofactors.

4. FACTORS AFFECTING MICROBIAL ALKYLATION OF SELENIUM

4.1. Temperature

As a microbial process, Se alkylation is strongly influenced by temperature. In pure culture studies with *Alternaria alternata*, the optimum temperature for $(CH_3)_2Se$ formation was 30°C [38]. The temperature optimum for Se alkylation in a pure culture of a *Bacillus* species was also reported at 30°C [41]. In studies involving the incubation of environmental samples, the maximum evolution of $(CH_3)_2Se$ from lake sediments was observed at 20°C [31], from seleniferous pond water at 35°C [55], from a loamy soil at 20°C [33], from various dewatered seleniferous sediments at 35°C [96], and from a sandy seleniferous soil at 35°C [97]. In the last two mentioned studies, the apparent Q_{10} between 5°C and 35°C was calculated at 2.0 and 2.6, respectively. However, in each of these studies, the maximum emission of $(CH_3)_2Se$ was observed at the highest temperature used in the experiment; hence the optimum temperature for environmental evolution of alkylselenides may not have been reached. In a separate study, Karlson and Frankenberger [98] determined volatilization of ^{75}Se added in the selenite form to Los Baños soil with soil temperature (5-55°C) as the variable. Irrespective of an organic carbon amend-

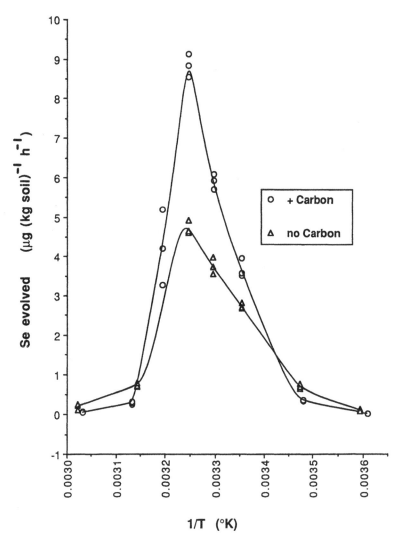

FIG. 1. Dependency of Se volatilization in Los Baños clay loam upon incubation temperature. Experimental temperatures were (from left to right): 55°C, 45°C, 40°C, 35°C, 30°C, 25°C, 15°C, and 5°C. Addition of ^{75}Se-labeled Na_2SeO_3, 1 mg Se/kg soil; pectin amendment, 0.2% carbon where applicable; soil moisture potential, -33kPa [98].

ment, there was a clear temperature optimum at 35°C (Fig. 1). Above
that temperature, there was a sharp decrease in Se evolution, presumably due to heat inactivation of the microbial process.

In field measurements, the diurnal peak of Se volatilization
coincided with the peak temperature, a close correlation between soil
temperature and gaseous Se evolution was determined, and the apparent
Q_{10} between 5°C and 15°C was calculated at 2.6 [99].

4.2. Moisture and Aeration

In soil environments, the water status has a dual effect on microbial
activities due to its influence on aeration. Air drying the soil
strongly reduced Se alkylation due to lack of water for microbial
growth [33,97], whereas water saturation caused anaerobiosis and
moderately reduced Se volatilization rates, possibly also due to
inhibited gas exchange from soil to the atmosphere [97]. The moisture optimum for Se evolution ranged between 18% and 70% of the
water-holding capacity, i.e., water contents that permit maximum
aerobic microbial activity [15,16,33,97]. Sequential wetting and
drying of soil appears to stimulate Se volatilization [16]. The
decomposition of organic matter in soil is also stimulated by wetting-
drying cycles, which leads to an increased availability of nutrients
for the soil microflora [100].

Anaerobiosis, experimentally created by replacing oxygen in the
sample air with N_2, reduced Se alkylation by a factor of 80 and 10 in
sewage sludge and soil, respectively [22]. Seleniferous pond water
samples exposed to air produced larger quantities of volatile Se than
corresponding samples exposed to N_2 [54]. Evolution of $(CH_3)_2Se$ was
reduced or inhibited under argon, depending on soil type [19] and
species of Se added [20]. However, formation of $(CH_3)_2Se_2$ from Se-
methionine was increased under the same conditions [20]. Reamer and
Zoller [22] speculated that different microorganisms are responsible

for the formation of each volatile Se species, and that shifts in
volatile species composition may be due to the differing tolerances
of the microorganisms to environmental stress. Thus the formation
of the different alkylselenides in the environment may be the result
of a rather complex microbial ecology.

4.3. pH and Redox Potential

The hydrogen ion activity is an important factor for microbial growth.
Optimum pH values range between 5.0 and 8.0; this variation depends
largely on whether the isolate is a fungus or a bacterium [101]. In
natural environments, pH and redox potential affect the solubility
and species distribution of Se [102-104]; hence its availability for
alkylation. In soils, the optimum pH values for Se volatilization
were found to be in the range most favorable for microbial growth,
and close to the native pH of the soil, indicating adaptation of the
soil microflora to the prevailing chemical conditions. In seleniferous sediments (pH 7.7), the optimum pH for $(CH_3)_2Se$ formation
was 8.0 [97]. The addition of lime to a peat soil increased Se
volatilization to some extent [105]. Liming a sandy soil increased
its pH from about 6 to 7, and volatilization of added selenite
increased by about 15% in 121 days [16]. It was not determined in
either study whether the pH effect resulted from microbial activity
or Se availability. In seleniferous sediments, bioalkylation of Se
only occurred under oxidized conditions (measured redox potential =
450 mV), which coincided with the highest solubility of inorganic
Se [106]. With a pure culture of A. alternata, isolated from seleniferous pond water (pH 7.4 to 8.8), the optimum for $(CH_3)_2Se$ formation from selenate occurred at pH 6.5 [38]. In an acidic aquatic
environment, Se methylation was found to be low at pH 3.5, and
increased with increasing pH.

4.4. Salinity

Some Se-alkylating organisms appear to tolerate highly saline conditions. Methylation of selenate by *Candida humicola* in a glucose
salts medium was not affected by 1000 mg/liter KH_2PO_4, NaH_2AsO_4, or
Na_2TeO_4, and 1000 mg/liter Na_2SO_4 reduced methylation by only 25%
[30]. Although species diversity is known to decrease as environmental salinity and alkalinity increase [107,108], considerable Se
volatilization has been demonstrated in saline soils [97] and aquatic
[55] environments with natural EC_e values occurring as high as 22
dS/m and 10-30 dS/m, respectively. Formation of $(CH_3)_2Se$ from
selenite by *Penicillium* sp. increased with increasing sulfate concentrations [24]. With *Fusarium* sp., $(CH_3)_2Se$ production from
selenite was decreased by sulfite and sulfate, presumably because
of competitive inhibition of Se uptake mechanisms [82]. The addition
of small quantities of Na_2SO_4 and $CaCl_2$ stimulated Se volatilization
from one nonsaline soil, while in three soils artificial salinization
up to 20 dS/m using Na_2SO_4, NaCl, and $CaCl_2$ resulted in an average
inhibition of 16.8, 18.3, and 24.3%, respectively [109]. Slightly
stronger inhibitory effects were observed from chloride anions than
from sulfate anions, and from sodium cations than from calcium
cations. In this study, the soils were not naturally saline, i.e.,
the soils did not contain an indigenous microbial population adapted
to high salinity. One might expect that in naturally saline environments the alkylating microflora has adapted to those conditions and
can tolerate salinity stress.

4.5. Selenium Species and Concentration

For Se alkylation in soils, optimum concentrations of inorganic Se
species have been reported [37]. Without an organic C amendment, the
optima, defined as the concentration of inorganic Se resulting in the

highest fraction of added Se volatilized, measured between 5 and
20 mg Se/kg for selenite, and at approximately 0.025 mg Se/kg for
selenate. Addition of organic C raised the optimum concentration
for selenate toward that for selenite. In terms of total Se volatilized, evolution rates from various soils without C amendment
ranged between 2 ng Se/kg/day and 300 μg Se/kg/day, depending on
both level and species of Se added: rates were significantly higher
(as much as one order of magnitude) for selenite than for selenate
(Fig. 2), and Se evolution increased with increasing inorganic Se
levels (0.025-1000 mg Se/kg soil), except that at the highest concentrations of selenate, the Se evolution rates were below maximum,
which could be due to selenate toxicity [37]. When monitoring the
water-soluble Se in soil during Se volatilization, the Se evolution
rate from the soil was found to decrease substantially, after the
initially added 100 μg Se/kg had been depleted to about 25 μg/kg
[33].

In addition to affecting volatilization rates, the Se concentration can also change the ratio of alkylated species that are
formed. The major volatile Se species evolved from sewage sludge
at concentrations of 1 and 10 mg Se/kg was $(CH_3)_2Se$, while at ≥100

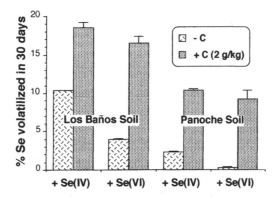

FIG. 2. Effect of pectin amendment (0.2% C) on Se volatilization
rates in two California soils, with Se added at 5 mg/kg as the
selenite [Se(IV)] and selenate [Se(VI)] species (after [37]).

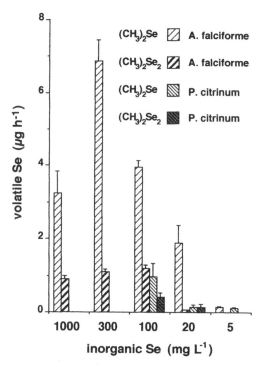

FIG. 3. Production of $(CH_3)_2Se$ and $(CH_3)_2Se_2$ by pure cultures [37] of *Acremonium falciforme* and *Penicillium citrinum*. Concentrations of selenite-Se as specified; incubation temperature, 22°C [98].

mg Se/kg the proportion of $(CH_3)_2Se$ decreased markedly and that of $(CH_3)_2Se_2$ and $(CH_3)_2SeO_2$ increased accordingly [22]. The major product from naturally seleniferous pond water and sediments is $(CH_3)_2Se$ [54-56,110]. However, following the addition of >500 µg/ liter of selenate-Se to pond water that had received an organic C amendment, a small quantity of $(CH_3)_2Se_2$ evolved in addition to $(CH_3)_2Se$ [111]. The same observation was made with seleniferous soil above 100 mg Se/kg [112]. With pure cultures of *Acremonium falciforme* and *Penicillium citrinum*, addition of selenite-Se at 5-1000 mg/liter resulted in production of $(CH_3)_2Se$ and $(CH_3)_2Se_2$, the latter, however, in significant quantities only at ≥100 mg Se/ liter (Fig. 3). Other possible alkylated forms were not determined in this study.

Selenium in certain organic forms, including $(CH_3)_3Se^+$, Se-methionine, selenocysteine, Se-cystine, selenoguanosine, selenoinosine, selenoethionine, selenopurine, and selenourea, is usually more rapidly transformed into volatile species than inorganic forms, as determined by incubation of soils and dewatered sediment samples [20,31,97,113]. However, a pure culture of *A. alternata,* isolated from seleniferous pond water, methylated inorganic Se at higher rates than selenoamino acids [38]. Most of the soluble Se in soil environments whose Se species distribution has not been disturbed by experimental Se addition or by recent contamination episodes occurs in the organic form according to some authors [114], whereas the main Se species in seleniferous pond water are selenite and selenate [115].

In laboratory experiments involving experimental addition of Se to soil and sewage sludge, elemental Se was only slowly methylated compared with other inorganic species [22]. Poor solubility of elemental Se is the probable reason. However, in microcosm experiments with dewatered naturally seleniferous sediments, changes in individual Se species distribution (water-soluble Se and elemental Se) were monitored and elemental Se was found to be rapidly converted to volatile forms, following prior oxidation to water-soluble forms [116]. In a 5-week period, 29% of the elemental Se was volatilized from the nutrient-amended soil system. Apparently the dynamics of Se transformations in natural environments are quite complex and can only be approximated by more intricate laboratory models.

According to Doran's hypothesis [85], the pathway of Se bio-alkylation requires reduction to the Se^{2-} oxidation state prior to transalkylation. It should therefore require less energy to volatilize selenite than selenate, and alkylation of selenite should proceed at higher rates. In natural soil environments, however, the situation is made more complex by the strong adsorption tendency of selenite, particularly at low pH [117-119]. Formation rates of volatile Se compounds in soil were higher with 50 mg/kg of Se added in the selenite than in the selenate form [20]. Karlson and Frankenberger [37] found that without organic C amendments, volatilization rates were one order

of magnitude higher after selenite addition, compared with selenate
addition. Amendments with organic C largely canceled out this difference, presumably because of energy availability for the additional
reduction step from selenate to selenite (Fig. 2). Without the C
amendment, this first reduction step may have been rate limiting. In
work with pure cultures of soil microorganisms [18], all 11 fungal
species studied were capable of methylating selenite, but only six
species volatilized selenate. All of the bacterial isolates were
unable to methylate inorganic Se, although they had the ability to
reduce it to elemental Se. On the other hand, seleniferous shale
[14], seleniferous pond water [56], pure cultures of A. alternata
[38] and of Scopulariopsis brevicaulis [13] volatilized selenate more
rapidly than selenite, with volatilization rates reported as occurring
independently of the availability of organic C.

4.6. Organic Carbon

Availability of organic C is a limiting factor for microbial Se
alkylation in natural environments. Organic C is required to provide
the alkyl groups for transalkylation and it is a source of energy for
reduction of selenate and selenite. Amendment of soil and water
samples with various natural and reagent grade C substrates resulted
in significant increases in Se evolution rates, up to 30-fold [14-16,
19,20,37,96,105,113,116,120,121]. The use of different C sources led
to volatilization increases of differing magnitudes, depending on the
chemical structure, water solubility, availability for microbial
decomposition, and carbon/nitrogen ratio of the C compound. Some of
this evidence suggests the predominant role of fungal microflora in
Se alkylation in certain habitats. This topic is further discussed
in a recent review article [122].

4.7. Heavy Metals

There are few reports on the effects of heavy metals in bioalkylation
of Se. The addition of 5 mmol/kg of molybdate, mercuric ion, chrom-
ate, and lead ion to soil strongly inhibited Se volatilization, while
the addition of arsenate, borate, and manganate had little or no
effect [120]. However, the addition of cobalt, zinc, and nickel ion
had a stimulatory effect, in particular at high (25 mmol/kg) concentrations. In an earlier study, the same observation had been made
with cobalt [123]. In field experiments with dewatered seleniferous
sediments, increased Se volatilization was noted using cobalt and
zinc as a soil amendment (50 mg/kg) [99]. Considering the functions
of Co, Zn, and Ni in various other enzyme systems, a possible cofactor
role of these elements in Se bioalkylation was postulated [120] but
remains unconfirmed. Indirect evidence that the apparent stimulation
may be based on a bacteriostatic effect was found by Karlson and
Timmis [124]. Pure cultures of fungal isolates from seleniferous
environments [37,38] did not respond to cobalt, zinc, or nickel amendments with increased $(CH_3)_2Se$ production. Instead, they showed a
remarkable tolerance to those metal ions, as measured by growth and
$(CH_3)_2Se$ formation (Fig. 4). Appreciable growth and activity was
noted with metal additions as high as 100 mg/liter. Most bacteria
would be inhibited at these concentrations.

Soil incubation studies with high concentrations of cobalt,
zinc, or nickel [120] had revealed a proliferation of filamentous
fungi by visual observation. A shift in microbial populations
through toxicity of zinc (72-288 mg/kg), leading to a decrease in
bacterial and an increase in fungal numbers, was observed in soils
[125]. Assuming that at least in some soil environments the predominant Se-alkylating microflora consists of fungi, one could interpret
the apparent stimulatory effect of cobalt, zinc, and nickel as an
indirect mechanism, by which the fungal population is favored at the
expense of the soil bacteria, allowing the fungi to grow to unusually
high densities and to alkylate higher percentages of the soil Se. An

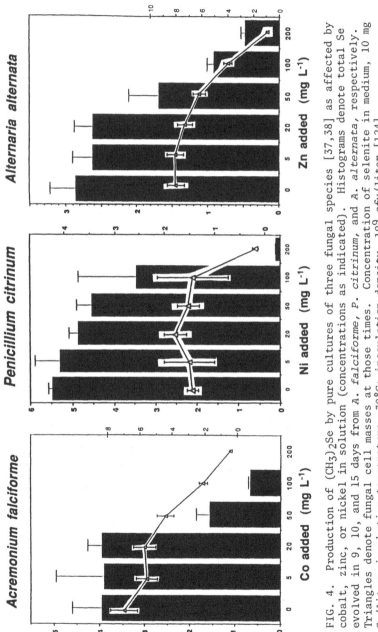

FIG. 4. Production of (CH₃)₂Se by pure cultures of three fungal species [37,38] as affected by cobalt, zinc, or nickel in solution (concentrations as indicated). Histograms denote total Se evolved in 9, 10, and 15 days from *A. falciforme*, *P. citrinum*, and *A. alternata*, respectively. Triangles denote fungal cell masses at those times. Concentration of selenite in medium, 10 mg Se/liter; incubation temperature, 30°C; inoculation density, 10⁹ cfu/liter [124].

alternative explanation may be that Se, cobalt, zinc, and nickel
resistance are genetically linked traits in soil microorganisms,
and application of cobalt, zinc, or nickel selects for Se alkylating
genotypes, regardless of whether they are bacteria or fungi.

4.8. Antibiotics

Due to their selective mode of action, antibiotics can be employed
for growth inhibition of certain groups of microorganisms in complex
habitats. With regard to Se alkylation, inhibition of the Se-
alkylating microflora, or of their competitors for organic substrates, may be used to study the ecology of microbial Se transformations, and to accelerate or inhibit Se volatilization rates
from the habitats of concern. The addition of 28 mg/kg of (bactericidal) chloramphenicol to a loam soil spiked with Se caused a
50% reduction in Se volatilization [33]. On the other hand, working
with a naturally seleniferous dewatered pond sediment, Frankenberger
[126] observed a more than threefold increase in $(CH_3)_2Se$ evolution
by the addition of streptomycin, a less than 50% increase with chloramphenicol, and a 5% decrease with cycloheximide (Fig. 5). Similar
results were reported for Japanese soils [121].

   In a different study, also using naturally seleniferous
dewatered pond sediment, the alkylating microflora was found to be
resistant to the bactericides penicillin G (40 mg/liter) and streptomycin (40 mg/liter) [116]. In seleniferous pond water, cycloheximide
(200 mg/liter) stimulated biomethylation and nystatin (200 mg/liter)
had no effect (both are fungicides), whereas the bactericides penicillin G (100 mg/liter), polymyxin B (100 mg/liter), and crystal
violet (10 mg/liter) were strongly inhibitory, and (bactericidal)
chlortetracyclin (100 mg/liter) and streptomycin (100 mg/liter) had
little effect [54].

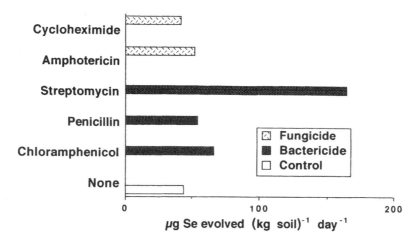

FIG. 5. Effect of antibiotics on $(CH_3)_2Se$ production by a dewatered
seleniferous sediment. Indigenous Se concentration, 60.7 mg/kg;
moisture potential, -33 kPa; incubation temperature, 23°C, antibiotic
addition, 100 mg/kg [126].

5. ENVIRONMENTAL FATE OF ALKYLSELENIDES

5.1. Adsorption by Soil Constituents

Urinary $(CH_3)_3Se^+$ deposited in soil samples is adsorbed to the soil
matrix or readily converted to volatile forms by soil microorganisms
[21]. At high concentrations of $(CH_3)_2Se$ (2 mg Se/m^3 of soil/air),
between 10% and 50% of the volatile Se can be adsorbed to organic and
inorganic soil particles in one week [127]. Microorganisms play a
small role in this process. Given the high volatility of methyl-
selenides [1], rapid dispersal by diffusion would be expected, and
only a small portion would actually be retained in the top soil.
However, volatile Se released from the deeper soil layers would
probably be trapped within the soil profile.

5.2. Demethylation by Microorganisms and Plants

Some soil microorganisms are capable of demethylating alkylselenides. Several species utilizing alkylselenides as a sole source of C were isolated from aerobic soils, including a *Pseudomonas* sp. [utilizing $(CH_3)_3Se^+$], two *Pseudomonas* spp. [utilizing $(CH_3)_2Se$] and strains of *Xanthomonas* and *Corynebacterium* [utilizing $(CH_3)_2Se_2$] [32]. The importance of this process in natural soil environments has not been elucidated, but the role of these microorganisms in the Se cycle is most likely negligible in view of the rapid dispersal of alkylselenides, at least from the top soil. In anoxic sediments, methylotrophic methanogenic bacteria seem to convert $(CH_3)_2Se$ to CO_2 and CH_4 [128]. The existence of this bacterial pathway might explain some of the inconsistent results in experiments involving soil Se volatilization under anaerobic conditions. In either case, the fate of the Se atom in these conversions has not been investigated.

Higher plants are capable of assimilating $(CH_3)_2Se$. Between 0.1% and 10% of the Se in the shoots of red clover plants (*Trifolium pratense*) were found to be derived from atmospheric $(CH_3)_2Se$, but under typical conditions up to 2% can be expected to be from that source [129,130]. The assimilated Se is converted to inorganic selenite and to the Se analogs of glutathione and methionine [131]. Transfer of volatile Se from soil to plants [127] and between plants of the same species [47] has been attributed to emission and subsequent absorption of alkylselenides.

5.3. Atmospheric Oxidation

Numerous investigations indicate that considerable portions of the atmospheric Se are produced by natural processes [132-134]. This may explain findings of higher atmospheric Se levels [135] and higher Se concentrations in rainwater [136] during seasons with higher biological activity. Over recent decades, atmospheric sulfur, but not atmospheric Se, has increased due to fossil fuel burning,

evident from analyses in Greenland [133] and in Antarctic [137] ice
cores. This suggests that Se from fossil fuel burning is not transported long distances in the atmosphere. Thus atmospheric Se in
rural areas and over the oceans probably originates from natural
sources.

Comparisons of methylated Se concentrations in soil samples
and in related air samples revealed a consistent relationship for
$(CH_3)_2Se$, $(CH_3)_2Se_2$, and $(CH_3)_2SeO_2$ [23]. This finding indicates
that local air values are controlled by gaseous emissions from the
ground. After release into the atmosphere, $(CH_3)_2Se$ reacts with OH
and NO_3 radicals and with O_3 within a few hours, yielding products
that are as yet unknown [138]. There is evidence, however, that
these oxidized products may be scavenged onto aerosols or sorbed
onto particulates which have a residence time of 7-9 days in the
atmosphere [139] and can travel considerable distances [140], before
returning to the Earth's surface as dry and/or wet deposition. For
soils, deposition can be the major source of Se [42,134]. The global
atmospheric flux of Se from natural sources is estimated at 9.3×10^9
g/year, in contrast to 6.3×10^9 g/year for anthropogenic emissions
[141]. Ninety percent of the natural emissions are thought to be
from biogenic processes, approximately one-third of them being continental and two-thirds marine [141]. On the continents, approximately
two-thirds of the biological emissions is thought to be produced by
microorganisms and one-third by higher plants [42]. Thus, the biological formation of alkylselenides is an important part of the
global Se cycle (see Chap. 1 of this volume).

Loss of Se from soils through volatilization may be one of
the factors contributing to Se deficiency in agricultural crops in
a region of China [34,142]. For Se-contaminated soils, Karlson and
Frankenberger suggest microbial volatilization as a means of bioremediation through stimulation of this natural process [120]. An
analogous suggestion has been made for the bioremediation of seleniferous pond water [55]. These methods are patented [143] and details
about their application were recently reviewed by Frankenberger and
Karlson [144].

TABLE 2

Toxicity of Some Selenium Species

Compound	Parameter	Value mg Se/kg	Value mg Se/liter	Ref.
$(CH_3)_2Se$	ipr-rat LD_{50}	2200		145
	scu-rat LD_{50}	2180		146
	ipr-mouse LD_{50}	1800		145
	air-rat LD_{50}		>8034	147
$(CH_3)_3SeCl$	ipr-rat LD_{50}	100		148
Na_2SeO_4	ipr-rat LD_{Lo}	9		149
	scu-rat LD_{Lo}	19		146
	ipr-rat LD_{75}	13		150
Na_2SeO_3	ipr-rat LD_{50}	3		149
	scu-rat LD_{50}	2		151
	ipr-rat LD_{75}	8		150
Se-methionine	ipr-rat LD_{Lo}	4250		152
	ipr-rat LD_{50}	11		153
Se-cystine	ipr-rat LD_{50}	9		153

6. TOXICOLOGY OF ALKYLSELENIDES

Among the different inorganic and organic Se species, $(CH_3)_2Se$ has
the lowest toxicity levels, as determined by intraperitoneal or sub-
cutaneous injection into mice or rats (Table 2). Arsenite was found
to increase the toxicity of subcutaneously injected $(CH_3)_2Se$ and
$(CH_3)_3Se^+$ in rats [154]. Pretreatment with selenite reduced the
toxicity of $(CH_3)_2Se$ intraperitoneally applied to rats [155]. The
decreased toxicity in pigs of intravenously applied $(CH_3)_2Se$ (2 mg
Se/kg) as compared with selenite is documented in detail [156,157].

The toxicity of inhaled $(CH_3)_2Se$ in rats was recently determined
[147]. Eighty-five rats were exposed to air concentrations of up

to 8034 mg/liter of $(CH_3)_2Se$ for 1 hr. All of the rats survived,
i.e., no LD values were determined. During the first 24 hr of
postinhalation, animals exhaled $(CH_3)_2Se$. Necropsis at 1 or 7 days
postinhalation revealed that all organs were either unaffected or
showed only a minor temporary response. The Se level in the blood
serum was unaffected, and lung levels were slightly elevated at one
day postinhalation. These findings reveal the remarkably low toxicity
of inhaled $(CH_3)_2Se$. Apparently most of the inhaled $(CH_3)_2Se$ is
exhaled again after inhalation, without significantly affecting body
tissues.

Some Se compounds have a positive physiological effect. For
example, dialkyldiselenides were found to prevent dietary necrotic
liver degeneration [158]. Protection was found with alkyl chain
lengths between 2 and 44 carbon atoms, but the most effective treatment was noted with chain lengths of 9-11 atoms. The same class of
compounds has been claimed to possess antitumor activity [159].

Like $(CH_3)_2Se$, $(CH_3)_3Se^+$ has a reduced toxicity compared with
inorganic Se species or Se-amino acids (Table 2). Methylation is one
mechanism of action which appears to render Se biologically inactive.
This is not surprising, since both $(CH_3)_2Se$ and $(CH_3)_3Se^+$ are metabolites involved in the excretory and detoxification functions in mammals.

7. FORMATION, FATE, AND TOXICITY OF ALKYLTELLURIDES

Due to its chemical similarities to Se, Te could be expected to have
similar biological effects as Se. However, a number of Se-resistant
bacterial strains isolated from a natural seleniferous environment
showed no resistance to Te [40]. Minimal inhibitory concentration
(MIC) values for tellurite and tellurate measured 1-2 orders of magnitude lower than for selenite and selenate, respectively.

The addition of inorganic Te to a Se-methylating culture of
Penicillium sp. led to reduced Se volatilization but resulted in
$(CH_3)_2Te$ production. However, $(CH_3)_2Te$ was not formed in media

devoid of Se [24]. These findings can be interpreted as abiotic transmethylation of Te from $(CH_3)_2Se$ produced by this *Penicillium* strain. In a different study, several *Penicillium* strains and a strain of *Scopulariopsis brevicaulis* were found to methylate Te in the absence of Se. The volatile metabolite was identified as $(CH_3)_2Te$, and $(C_2H_5)_2Te$ was not produced [160]. The same fungal strains also methylated Se. In a separate laboratory study, fungi and bacteria capable of methylating selenite were found to produce $(CH_3)_2Te$, $(CH_3)_2Te_2$, and an unidentified compound from tellurite. The reaction occurred in the absence of Se [39].

Garlic breath in industrial workers was attributed to previous exposure to Te [161]. Garlic breath has also been noted after administration of Te to rats [162] and dogs [63]. In these reports, $(CH_3)_2Te$ was implied to be the odorous ingredient in the breath, but was not positively identified.

More information about Te alkylation is not presently available; however, the above citations suggest that the biology of Te is similar to that of Se in some cases. There are no environmental data on alkyltellurides. Due to lack of more precise information, Te is speculated to follow the pathways of Se in biological alkylation and in its environmental fate [163-165].

The toxicity of $(C_2H_5)_2Te$ in air is significantly higher than that of $(CH_3)_2Se$ (Table 3). Ethylation of Te apparently does not render it less toxic as is the case with alkylation of Se. No toxicity data were available for other alkyltellurides. Diethyltelluride in workspace air has recently received attention in certain industries. Although few specifics of tellurium toxicity are known, arsenite was found to increase the toxicity of subcutaneously injected $(CH_3)_3Te^+$ in rats [154].

8. CONCLUSIONS

The present knowledge base for alkyltellurides is extremely limited. The absence of substantial reserach efforts on the biology of Te in

TABLE 3

Toxicity of Some Tellurium Species

Compound	Parameter	Value		Ref.
		mg/kg	mg/liter	
$(C_2H_5)_2Te$	Air-rat LD_{50}		0.055	166
	Air-guinea pig LD_{50}		0.045	166
	Air-mouse LD_{50}		0.154	166
Na_2TeO_4	ipr-rat LD_{75}	47		150
Na_2TeO_3	ipr-rat LD_{75}	4		150

general, and of alkyltellurides in particular, may be due to the fact that Te, unlike Se, is not known to be essential to vertebrates. There are also no reported incidences of environmental Te poisoning in animals, an effect that has been well documented in the case of Se.

Selenium has been an important element in widespread nutritional deficiencies in humans and in domestic and wild animals, and in episodes of animal poisoning. Regarding both its beneficial and its toxic properties, it continues to be of major concern in many parts of the world. However, in view of its biological importance, limited knowledge does exist regarding the biological transformations and conversions of Se to the vapor phase, particularly in marine environments. More research is needed on species distribution and the fate of volatile Se in the atmosphere. Although very limited data on the presence of alkylselenides exist, this is probably due more to sampling techniques than to the absence of those compounds in the environment. Another important area of research that requires expansion is the marine biology of Se.

ABBREVIATIONS AND DEFINITIONS

C	carbon
cfu	colony-forming units
EC_e	electrical conductivity of a saturated extract
ipr	intraperitoneal
LD_{50}	dose which is lethal to 50% of animals tested
LD_{Lo}	minimum lethal dose
MIC	minimal inhibitory concentration (lowest metal concentration that suppresses growth of microbial cultures [167])
N_2	nitrogen
NADPH	nicotinamide adenine dinucleotide phosphate, reduced form
Q_{10}	temperature coefficient (factor by which reaction rate increases by raising the temperature 10°C)
scu	subcutaneous
Se	selenium
Se^0	elemental selenium {Se(0)}
SeO_3^{2-}	selenite {Se(IV)}
SeO_4^{2-}	selenate {Se(VI)}
Se^{2-}	selenide {Se(-II)}
H_2Se	hydrogen selenide
$(CH_3)_2Se$	dimethylselenide
$(C_2H_5)_2Se$	diethylselenide
$(CH_3)_2Se_2$	dimethyldiselenide
$(CH_3)_2SeO_2$	dimethylselenone
$(CH_3)_3Se^+$	trimethylselenonium ion
$(CH_3)_2Se^+$-R	dimethylselenonium ion
CH_3SeO_3H	methaneselenonic acid
CH_3SeO_2H	methaneselenenic acid
CH_3SeOH	methaneseleninic acid
CH_3SeH	methaneselenol
Se-methionine	selenomethionine
Se-cystine	selenocystine

Te	tellurium
Te^0	elemental tellurium {Te(0)}
TeO_3^{2-}	tellurite {Te(IV)}
TeO_4^{2-}	tellurate {Te(VI)}
Te^{2-}	telluride {Te(-II)}
$(CH_3)_2Te$	dimethyltelluride
$(C_2H_5)_2Te$	diethyltelluride
$(CH_3)_2Te_2$	dimethylditelluride
$(CH_3)_3Te^+$	trimethyltelluronium ion

REFERENCES

1. U. Karlson, W. T. Frankenberger, Jr., and W. F. Spencer, unpublished.

2. B. W. Mosher, R. A. Duce, J. M. Prospero, and D. L. Savoie, *J. Geophys Res., D: Atmos., 92,* 13277 (1987).

3. B. W. Mosher and R. A. Duce, *J. Geophys. Res., (Sect.) C, 88* 6761 (1983).

4. S. Muangnoicharoen, *Diss. Abstr. Int. B, 51,* 176 (1990).

5. S. Jiang, H. Robberecht, and F. Adams, *Atmos. Environ., 17,* 111 (1983).

6. S. Jiang, W. De Jonghe, and F. Adams, *Anal. Chim. Acta, 136,* 186 (1982).

7. D. Tanzer and K. G. Heumann, *Anal. Chem., 63,* 1984 (1991).

8. N. Oyamada and M. Ishizaki, *Anal. Sci., 2,* 365 (1986).

9. T. Cooke and K. Bruland, *Environ. Sci. Technol., 21,* 1214 (1987).

10. L. Ebdon, S. J. Hill, G. E. Millward, and A. P. Walton, in *The Biological Alkylation of Heavy Elements* (P. J. Craig and F. Glockling, eds.), Special Publication No. 66, Royal Society of Chemistry, London, 1988, p. 258 ff.

11. G. A. Cutter and K. W. Bruland, *Limnol. Oceanogr., 29,* 1179 (1984).

12. D. Tanzer and K. G. Heumann, *Atmos. Environ., Part A, 24A,* 3099 (1990).

13. F. Challenger and H. E. North, *J. Chem. Soc.,* 68 (1934).

14. T. J. Ganje and E. I. Whitehead, *Proc. S.D. Acad. Sci., 37,* 81 (1958).

15. G. M. Abu-Erreish, E. I. Whitehead, and O. E. Olson, *Soil Sci.*, *106*, 415 (1968).

16. A. A. Hamdy and G. Gissel-Nielsen, *Z. Pflanzenernaehr. Bodenkd.*, *6*, 671 (1976).

17. V. V. Koval'skii, V. V. Ermakov, and S. V. Letunova, *Zhurnal Obshchei Biologii, 26,* 634 (1965).

18. L. Barkes and R. W. Fleming, *Bull. Environ. Contam. Toxicol.*, *12,* 308 (1974).

19. A. J. Francis, J. M. Duxbury, and M. Alexander, *Appl. Microbiol.*, *28,* 248 (1974).

20. J. W. Doran and M. Alexander, *Soil Sci. Soc. Am. J.*, *41,* 70 (1977).

21. O. E. Olson, E. E. Cary, and W. H. Allaway, *Agron. J.*, *68,* 839 (1976).

22. D. C. Reamer and W. H. Zoller, *Science, 208,* 500 (1980).

23. S. G. Jiang, H. Robberecht, and F. Adams, *Appl. Organomet. Chem.*, *3,* 99 (1989).

24. R. W. Fleming and M. Alexander, *Appl. Microbiol., 24,* 424 (1972).

25. H. G. Smith, *J. Gen. Microbiol., 21,* 61 (1959).

26. A. A. Razak, S. E. Ramadan, and K. El-Zawahry, *Biol. Trace Elem. Res., 25,* 187 (1990).

27. F. Challenger, C. Higginbottom, and L. Ellis, *J. Chem. Soc.*, 95 (1933).

28. F. Challenger and P. T. Charlton, *J. Chem. Soc.*, 424 (1947).

29. F. Challenger, D. B. Lisle, and P. B. Dransfield, *J. Chem. Soc.*, 1760 (1954).

30. D. P. Cox and M. Alexander, *Microb. Ecol., 1,* 136 (1974).

31. Y. K. Chau, P. T. S. Wong, B. A. Silverberg, P. L. Luxon, and G. A. Bengert, *Science, 192,* 1130 (1976).

32. J. W. Doran and M. Alexander, *Appl. Environ. Microbiol., 33,* 31 (1977).

33. R. Zieve and P. J. Peterson, *Sci. Tot. Environ., 19,* 277 (1981).

34. P. J. Peterson, personal communication.

35. R. Zieve, P. J. Ansell, T. W. K. Young, and P. J. Peterson, *Trans. Br. Mycol. Soc., 84,* 177 (1985).

36. V. A. Baltrusaitis and M. R. Speyer, Animal repellent and a method of protecting plants against animal browsing, Eur. Pat. Appl., EP 200,297 (1986).

37. U. Karlson and W. T. Frankenberger, Jr., *Soil Sci. Soc. Am. J.*, *53,* 749 (1989).

38. E. T. Thompson-Eagle, W. T. Frankenberger, and U. Karlson, *Appl. Environ. Microbiol.*, *55*, 1406 (1989).

39. T. G. Chasteen, G. M. Silver, J. W. Birks, and R. Fall, *Chromatographia*, *30*, 181 (1990).

40. G. Burton, T. Giddings, P. Debrine, and R. Fall, *Appl. Environ. Microbiol.*, *53*, 185 (1987).

41. A. A. Razak, S. E. Ramadan, and K. El-Zawahry, *Biol. Trace Elem. Res.*, *25*, 193 (1990).

42. P. M. Haygarth, K. C. Jones, and A. F. Harrison, *Sci. Total Environ.*, *103*, 89 (1991).

43. L. Wu and Z. Z. Huang, *Ecotoxicol. Environ. Saf.*, *22*, 267 (1991).

44. B. G. Lewis, C. M. Johnson, and C. C. Delwiche, *J. Agric. Food Chem.*, *14*, 638 (1966).

45. C. Evans, C. J. Asher, and C. M. Johnson, *Aust. J. Biol. Sci.*, *21*, 13 (1968).

46. B. Radziuk and J. Van Loon, *Sci. Tot. Environ.*, *6*, 251 (1976).

47. C. Asher, C. Evans, and C. M. Johnson, *Aust. J. Biol. Sci.*, *20*, 737 (1967).

48. R. Zieve and P. J. Peterson, *Sci. Tot. Environ.*, *32*, 197 (1984).

49. B. A. G. Lewis, in *Environmental Biogeochemistry*, *Vol. 1, Carbon, Nitrogen, Phosphorus, Sulfur and Selenium Cycles* (J. O. Nriagu, ed.), Ann Arbor Science, Ann Arbor, Michigan, 1976, p. 389 ff.

50. B. G. Lewis, C. M. Johnson, and T. C. Broyer, *Biochim. Biophys. Acta*, *237*, 603 (1971).

51. B. G. Lewis, C. M. Johnson, and T. C. Broyer, *Plant Soil*, *40*, 107 (1974).

52. V. V. Rendig, M. B. Jones, and R. G. Burau, Cycling of Selenium in a Soil-Plant System, 1986-87 Technical Progress Report, UC Salinity/Drainage Task Force, University of Calif., Davis, California, 1987.

53. J. Bender, J. P. Gould, Y. Vatcharapijarn, and G. Saha, *Water Air Soil Pollut.*, *59*, 359 (1991).

54. E. T. Thompson-Eagle and W. T. Frankenberger, Jr., *Water Res.*, *25*, 231 (1991).

55. E. T. Thompson-Eagle and W. T. Frankenberger, Jr., *J. Environ. Qual.*, *19*, 125 (1990).

56. E. T. Thompson-Eagle and W. T. Frankenberger, Jr., *Environ. Toxicol. Chem.*, *9*, 1453 (1990).

57. N. Oyamada, G. Takahashi, and M. Ishizaki, *Eisei Kagaku*, *37*, 83 (1991).

58. A. Foda, J. H. Vandermeullen, and J. J. Wrench, *Can. J. Fish Aquat. Sci.*, *40*, 215 (1983).

59. G. Gassmann, F. Schorn, D. G. Müller, and E. Fölster, Flüchtige Selenverbindungen aus marinen Algen, Biolog. Anstalt Helgoland, Jahresber. 1988, Hamburg, Germany, 1989, p. 83.

60. N. R. Bottino, C. H. Banks, K. J. Irgolic, P. Micks, A. E. Wheeler, and R. A. Zingaro, *Phytochemistry, 23,* 2445 (1984).

61. D. J. Velinsky and G. A. Cutter, *Geochim. Cosmochim. Acta, 55,* 179 (1991).

62. B. W. Mosher and R. A. Duce, *J. Geophys. Res., D: Atmos., 92,* 13289 (1987).

63. F. Hofmeister, *Arch. Exptl. Pathol. Pharmakol., 33,* 198 (1893/ 1894).

64. K. P. McConnell, *J. Biol. Chem., 145,* 55 (1942).

65. D. F. Petersen, H. L. Klug, and R. D. Harshfield, *Proc. S. Dakota Acad. Sci., 30,* 73 (1951).

66. H. E. Ganther, O. A. Levander, and C. A. Baumann, *J. Nutr., 88,* 55 (1966).

67. K. P. McConnell and D. M. Roth, *Proc. Soc. Exp. Biol. Med., 123,* 919 (1966).

68. O. E. Olson, B. M. Schulte, and E. I. Whitehead, *J. Agric. Food Chem., 11,* 531 (1963).

69. S. Jiang, H. Robberecht, and D. van den Berghe, *Experientia, 39,* 293 (1983).

70. S. Jiang, H. Robberecht, F. Adams, and D. van den Berghe, *Toxicol. Environ. Chem., 6,* 191 (1983).

71. H. E. Ganther, *Biochemistry, 5,* 1089 (1966).

72. H. E. Ganther and H. S. Hsieh, in *Trace Element Metabolism in Animals,* Vol. 2 (W. G. Hoekstra, ed.), Butterworths, London, 1974, p. 339 ff.

73. H. S. Hsieh and H. E. Ganther, *Biochim. Biophys. Acta, 497,* 205 (1977).

74. J. Yonemoto, M. Webb, and L. Magos, *Toxicol. Lett., 24,* 7-14 (1985).

75. J. Yonemoto, *Proc. ICMR Semin., 8,* 253 (1988).

76. D. H. Groth, in *Effects and Dose-Response Relationships of Toxic Metals* (G. F. Nordberg, ed.), Elsevier, Amsterdam, 1976, p. 121 ff.

77. J. Parizek, I. Benes, A. Babicky, and J. Benes, *Physiol. Bohemoslov, 18,* 105 (1969).

78. J. L. Byard, *Arch. Biochem. Biophys., 130,* 556 (1969).

79. R. J. Kraus, S. J. Foster, and H. E. Ganther, *Anal. Biochem., 147,* 432 (1985).

80. A. T. Nahapetian, V. R. Young, and M. Janghorbani, *Anal. Biochem.*, *140*, 56 (1984).

81. I. S. Palmer, R. P. Gunsalus, A. W. Halverson, and O. E. Olson, *Biochim. Biophys. Acta*, *208*, 260 (1970).

82. L. Barkes and R. W. Fleming, *Bull. Environ. Contam. Toxicol.*, *15*, 504 (1976).

83. J. M. Janda and R. W. Fleming, *J. Environ. Sci. Health, Part A*, *13*, 697 (1978).

84. F. Challenger, *Chem. Rev.*, *36*, 315 (1945).

85. J. W. Doran, in *Advances in Microbial Ecology*, Vol. 6 (G. F. Nordberg, ed.), Plenum Press, New York, 1982, p. 1 ff.

86. M. L. Bird and F. Challenger, *J. Chem. Soc.*, 574 (1942).

87. A. Drotar, L. R. Fall, E. A. Mishalanie, J. E. Travernier, and R. Fall, *Appl. Environ. Microbiol.*, *53*, 2111 (1987).

88. W. P. Ridley, L. J. Dizikes, and J. M. Wood, *Science, 197*, 329 (1977).

89. M. Berman, T. Chase, and R. Bartha, *Appl. Environ. Microbiol.*, *56*, 298 (1990).

90. B. C. McBride and R. S. Wolfe, *Biochemistry*, 4312 (1971).

91. Y. T. Fanchiang, W. P. Ridley, and J. M. Wood, in *Advances in Inorganic Biochemistry* (G. L. Eichhorn and L. G. Marzilli, eds.), Elsevier, New York, 1979, p. 147 ff.

92. H. S. Hsieh and H. E. Ganther, *J. Nutr.*, *106*, 1577 (1976).

93. J. L. Hoffman and K. P. McConnell, *Arch. Biochem. Biophys.*, *254*, 534 (1987).

94. S. J. Foster, R. J. Kraus, and H. E. Ganther, *Arch. Biochem. Biophys.*, *247*, 12 (1986).

95. X. Zhang, G. Yang, and L. Gu, *Yingyang Xuebao, 13*, 32 (1991).

96. S. J. Calderone, W. T. Frankenberger, Jr., D. R. Parker, and U. Karlson, *Soil Biol. Biochem.*, *22*, 615 (1990).

97. W. T. Frankenberger, Jr. and U. Karlson, *Soil Sci. Soc. Am. J.*, *53*, 1435 (1989).

98. U. Karlson and W. T. Frankenberger, Jr., unpublished.

99. W. T. Frankenberger, Jr. and U. Karlson, in *Dissipation of Soil Selenium by Microbial Volatilization at Kesterson Reservoir* (S. E. Hoffman, ed.), Final Report, U.S. Department of the Interior, Bureau of Reclamation, Sacramento, California, 1988, p. 1 ff.

100. L. H. Sørensen, *Soil Biol. Biochem.*, *6*, 287 (1974).

101. H. G. Schlegel, *Allgemeine Mikrobiologie*, Georg Thieme Verlag, Stuttgart, 1985, p. 176.

102. G. Gissel-Nielsen, *Agric. Food Chem., 19*, 1165 (1971).

103. P. H. Masscheleyn, R. D. Delaune, and W. H. Patrick, *J. Environ. Qual., 20*, 522 (1991).

104. P. H. Masscheleyn, R. D. Delaune, and W. H. Patrick, *J. Environ. Sci. Health, Part A: Environ. Sci. Eng., 26*, 555 (1991).

105. T. Ylaranta, *Ann Agric. Fenn., 21*, 103 (1982).

106. P. H. Masscheleyn, R. D. Delaune, and W. H. Patrick, Jr., *Environ. Sci. Technol., 24*, 91 (1990).

107. W. D. Grant and B. J. Tindall, in *Microbes in Extreme Environments* (R. A. Herbert and G. A. Codd, eds.), Special Publications of the Society for General Microbiology, Vol. 17, Academic Press, London, 1986, p. 25 ff.

108. R. H. Reed, in *Microbes in Extreme Environments* (R. A. Herbert and G. A. Codd, eds.), Special Publications of the Society for General Microbiology, Vol. 17, Academic Press, London, 1986, p. 55 ff.

109. U. Karlson and W. T. Frankenberger, Jr., *Soil Sci., 149*, 56 (1990).

110. U. Karlson and W. T. Frankenberger, Jr., *Sci. Tot. Environ., 92*, 41 (1990).

111. E. T. Thompson-Eagle and W. T. Frankenberger, Jr., unpublished.

112. W. T. Frankenberger, Jr., Microbial Transformations in Seleniferous Soils, 1985-86 Technical Progress Report, UC Salinity/Drainage Task Force, University of California, Davis, California, 1986.

113. Z. Wang, L. Zhao, L. Zhang, and J. Sun, *J. Environ. Sci., 3*, 113 (1991).

114. H. Yamada and T. Hattori, *Soil Sci. Plant Nutr., 35*, 553 (1989).

115. R. Fujii and S. J. Deverel, in *Selenium in Agriculture and the Environment* (L. W. Jacobs, ed.), SSSA Spec. Publ. Nr. 23, Soil Science Society of America Inc., Madison, Wisconsin, 1989, p. 195 ff.

116. O. Weres, A. Jaouni, and L. Tsao, *Appl. Geochem., 4*, 543 (1989).

117. R. Neal, G. Sposito, K. Holtzclaw, and S. Traina, *Soil Sci. Soc. Am. J., 51*, 1161 (1987).

118. H. R. Geering, E. E. Cary, L. H. P. Jones, and W. H. Allaway, *Soil Sci. Soc. Am. Proc., 32*, 35 (1968).

119   M. A. Elrashidi, D. C. Adriano, and W. L. Lindsay, in *Selenium in Agriculture and the Environment* (L. W. Jacobs, ed.), SSSA Spec. Publ. No. 23, Soil Science Society of America Inc., Madison, Wisconsin, 1989, p. 51 ff.

120. U. Karlson and W. T. Frankenberger, *Soil Sci. Soc. Am. J., 52,* 1640 (1988).

121. A. Sugimae, *Taiki Osen Gakkaishi, 22,* 278 (1987).

122. E. T. Thompson-Eagle and W. T. Frankenberger, Jr., *Adv. Soil Sci., 17,* 261 (1992).

123. S. V. Letunova, V. V. Koval'skii, and V. V. Ermakov, *Tr. Biogeokhim. Lab. Akad. Nauk SSSR, 12,* 238 (1968).

124. U. Karlson and K. N. Timmis, unpublished.

125. P. Doelman, in *Microbial Communities in Soil* (V. Jensen, A. Kjøller, and L. H. Sørensen, eds.), Elsevier, London, 1986, p. 369 ff.

126. W. T. Frankenberger, Jr., unpublished.

127. R. Zieve and P. J. Peterson, *Soil Biol. Biochem., 17,* 105 (1985).

128. R. S. Oremland and J. P. Zehr, *Appl. Environ. Microbiol., 52,* 1031 (1986).

129. R. Zieve and P. J. Peterson, *Trace Subst. Environ. Health, 18,* 262 (1984).

130. R. Zieve and P. J. Peterson, *Toxicol. Environ. Chem., 11,* 313 (1986).

131. R. Zieve and P. J. Peterson, *Planta, 160,* 180 (1984).

132. R. A. Duce, G. L. Hoffman, and W. H. Zoller, *Science, 187,* 59 (1975).

133. H. V. Weiss, M. Koide, and E. D. Goldberg, *Science, 172,* 261 (1971).

134. J. Låg and E. Steinnes, *Geoderma, 20,* 3 (1978).

135. C. McDonald and H. J. Duncan, *Atmos. Environ., 13,* 413 (1979).

136. J. Kubota, E. E. Cary, and G. Gissel-Nielsen, Selenium in Rainwater of the United States and Denmark, Trace Substances in Environmental Health, IX, Proc. Univ. Missouri Conf., 1975, pp. 123-130.

137. C. Boutron, M. Leclerc, and N. Risler, *Atmos. Environ., 18,* 1947 (1984).

138. R. Atkinson, S. M. Aschmann, D. Hasgawa, E. T. Thompson-Eagle, and W. T. Frankenberger, *Environ. Sci. Technol., 24,* 1326 (1990).

139. B. W. Mosher, and R. A. Duce, in *Occurrence and Distribution of Selenium* (M. Ihnat, ed.), CRC Press, Boca Raton, Florida, 1989, p. 295 ff.

140. H. F. Mayland, H. F. James, K. E. Panter, and J. L. Sondregger, in *Selenium in Agriculture and the Environment* (L. W. Jacobs, ed.), SSSA Spec. Publ. No. 23, Soil Science Society of America Inc., Madison, Wisconsin, 1989, p. 15 ff.

141. J. O. Nriagu, *Nature*, *338*, 47 (1989).

142. Z. Wang, L. Zhao, and A. Peng, *Huanjing Hauxue*, *8*, 7 (1989).

143. W. T. Frankenberger, Jr. and U. Karlson, Selenium detoxification, Patent No. 4,861,482, U.S. Patent Office (1989).

144. W. T. Frankenberger and U. Karlson, in *On-Site Bioreclamation* (R. E. Hinches and R. F. Olfenbuttel, eds.), Butterworth-Heinemann, Stoneham, Massachusetts, 1991, p. 239 ff.

145. K. P. McConnell and O. W. Portman, *Proc. Soc. Exptl. Biol. Med.*, *79*, 230 (1952).

146. I. Ostadalova and A. Babicky, *Arch. Toxicol.*, *45*, 207 (1980).

147. O. G. Raabe and M. A. Al-Bayati, in *Dissipation of Soil Selenium by Microbial Volatilization at Kesterson Reservoir* (S. E. Hoffman, ed.), Final Report, U.S. Department of the Interior, Bureau of Reclamation, Sacramento, California, 1988, pp. B3 ff.

148. B. D. Obermeyer, I. S. Palmer, O. E. Olson, and A. W. Halverson, *Toxicol. Appl. Pharmacol.*, *20*, 135 (1971).

149. K. W. Franke and A. L. Moxon, *J. Pharm. Exp. Therapeut.*, *58*, 454 (1936).

150. K. W. Franke and A. L. Moxon, *J. Pharm. Exp. Therapeut.*, *61*, 89 (1937).

151. M. I. Smith, E. F. Stohlman, and R. D. Lillie, *J. Pharm. Exp. Therapeut.*, *60*, 449 (1937).

152. H. L. Klug, D. F. Petersen, and A. L. Moxon, *Proc. S. Dakota Acad. Sci.*, *28*, 117 (1949).

153. L. Fishbein, in *Advances in Modern Toxicology, Vol. 2, Toxicology of Trace Elements* (R. A. Goyer and M. A. Mehlman, eds.), John Wiley and Sons, New York, 1977, p. 191 ff.

154. R. J. Kraus and H. E. Ganther, *Biol. Trace Elem. Res.*, *20*, 105 (1989).

155. J. Parizek, J. Kalouskova, V. Kuronova, J. Benes, and L. Pavlik, *Physiol. Bohemoslov*, *25*, 573 (1976).

156. C. Nebbia, J. Fink-Gremmels, and S. M. Gennaro, *Res. Commun. Chem. Pathol. Pharmacol.*, *67*, 117 (1990).

157. C. Nebbia, S. M. Gennaro, E. Zittlau, and J. Fink-Gremmels, *Res. Vet. Sci.*, *50*, 269 (1991).

158. K. Schwarz, L. A. Porter, and A. Fredga, *Bioinorg. Chem.*, *3*, 145 (1974).

159. S. Monti, Dialkyl diselenides having an antineoplastic activity, Eur. Pat. Appl., EP 331,917 (Cl. A61K31/095) (1989).

160. M. L. Bird and F. Challenger, *J. Chem. Soc.*, 163 (1939).

161. H. A. Schroeder, J. Buckman, and J. J. Balassa, *J. Chronic Dis., 20,* 147 (1967).

162. R. H. DeMeio, *J. Ind. Hyg. Toxicol., 28,* 229 (1946).

163. F. Challenger, *Adv. Enzymol., 12,* 429 (1951).

164. J. M. Wood, *Science, 183,* 1049 (1974).

165. A. Jernelöv and A. L. Martin, *Ann. Rev. Microbiol., 29,* 61 (1975).

166. I. V. Kozik, N. P. Novikova, L. A. Sedova, and E. N. Stepanova, *Gig. Tr. Prof. Zabol.,* 51 (1981).

167. C. A. Hendrick, W. P. Haskins, and A. K. Vidaver, *Appl. Environ. Microbiol., 48,* 56 (1984).

8

Making and Breaking the Co-Alkyl Bond in B$_{12}$ Derivatives

John M. Pratt

Department of Chemistry
University of Surrey
Guildford GU2 5XH, England

1. INTRODUCTION

1.1. Aims and Organization of the Review

The Co corrinoid complexes (derivatives of vitamin B_{12}) have been
developed by nature's R&D department as catalysts for unusual assign-
ments requiring the use of organometallic chemistry, namely, the
making and breaking of metal-alkyl bonds.   The parent B_{12} (the prefix
"vitamin" will be omitted throughout this chapter) or cyanocobalamin
is a diamagnetic six-coordinate Co(III) complex with the structure
shown in Figs. 1 and 2.   All the currently known reactions of B_{12}-
dependent enzymes (where "B_{12}" in this phrase is accepted as meaning
"Co corrinoid") involve the making and breaking of a Co-alkyl bond
(see Sec. 5).   The alkyl ligand is either the simple methyl or the
complex deoxyadenosyl ligand shown in Fig. 3.   These alkyl-Co corrinoids can formally be considered as complexes of the Co(III) ion
with a coordinated carbanion (R^-), although the electron distribution corresponds more closely to a covalent bond formed between the
radical R· and Co(II) (see Sec. 3.4).   The main forms present in
humans are the methyl, deoxyadenosyl, and aquo complexes, though the
last is probably reduced to Co(II) in vivo.   The cyanide complex is
mainly an artefact of the isolation procedure, though it can also be
formed in vivo from traces of cyanide derived from cigarette smoke or
cyanogenic foods such as cassava [1].   A more sinister aspect of B_{12}
chemistry is the reaction of the Hg(II) ion with CH_3-Co corrinoids in
anaerobic bacteria (and in vitro) to form the toxic CH_3Hg^+ ion, which
may, as around Minamata Bay in Japan, cause serious human poisoning
by entering the food chain; this has focused attention on the potential

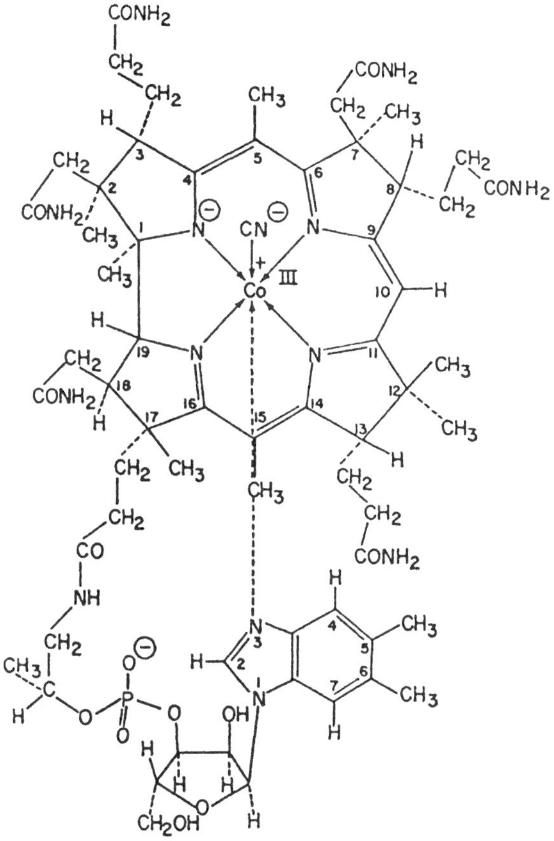

FIG. 1. Molecular structure of B$_{12}$. The positive charges of the Co(III) ion are balanced by the negative charges on the corrin ring, cyanide, and phosphate groups. [From Ref. 10 with permission from Academic Press Inc. (London) Ltd.]

role of B$_{12}$-dependent enzymes in the methylation of other elements with possible impact on the environment (see Chaps. 1 and 10 of this volume).

The presence of a simple Co-C bond to the ligand shown in Fig. 3 was established by X-ray analysis in 1961 [2]. The discovery of naturally occurring organometallic compounds which were stable (in the dark) toward both air and water upset the prevailing ideas about the general instability of the transition metal-to-carbon (alkyl) bond.

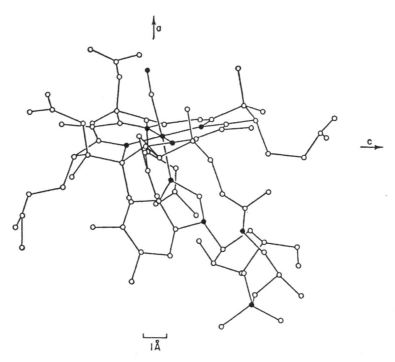

FIG. 2. Atomic positions in the molecule of B_{12} viewed parallel
with the crystallographic b axis in crystals of wet B_{12}. The solid
circles represent the Co, P, and some of the N atoms. (From Ref.
156 with permission.)

Coming at a time of rising interest in organometallic chemistry,
caused by the discovery of ferrocene exactly 10 years earlier (1951),
this triggered an explosive development of the organometallic and
coordination chemistry of the Co corrinoids and related "model" Co
complexes, which has continued unabated for 30 years. So much
information is now available on the making and breaking of Co-C
bonds in general that no review, even if restricted to B_{12}, can be
comprehensive. The Co corrinoids alone have been implicated in, or
convicted of, methyl transfer to 25 or more elements in vitro including many metals and metalloids (Cr, Ni, Pd, Pt, Cu, Au, Hg, Ga, In,
Tl, Ge, Sn, Pb, As, Sb, Bi, Se, Te, and I) as well as lighter elements
and Co in other complexes (see references in [3]). In addition, the

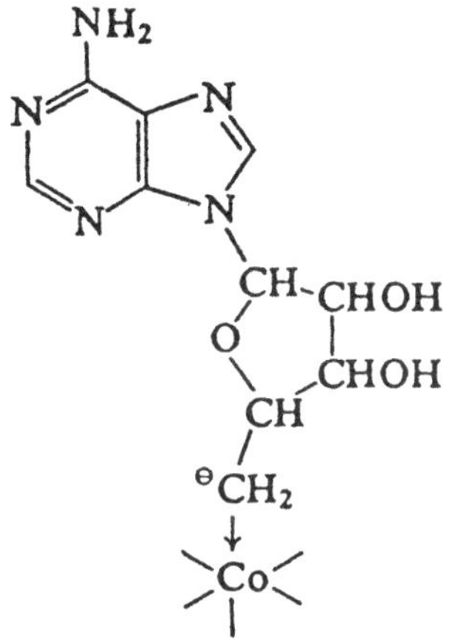

FIG. 3. The deoxyadenosyl (Ado) ligand present in the B$_{12}$ coenzymes (Ado corrinoids).

Co corrinoids are (or were until recently) the most complex known naturally occurring compounds which are not peptides or other poly-meric material. They display many features (diversity of nonbonded intramolecular interactions, conformational isomers, binding of ions and molecules, even the occurrence of extraordinary oscillating reactions) more usually associated with peptides, and B$_{12}$ has been called the "poor man's protein" [4,5]. Reactions involving the central Co ion cannot properly be understood without reference to the crowded and highly asymmetrical environment of mixed rigidity and flexibility and of mixed hydrophobic and hydrophilic character provided by the organic part of the corrinoid structure (and, ulti-mately, by the protein).

In comparison with other Co complexes, the corrinoids probably offer (1) a simpler basic set of reaction paths (see Sec. 4.8) and (2) a larger "database" of redox potentials and equilibrium and rate constants (see Sec. 3) as a framework for discussing possible reaction mechanisms, but (3) a wider range of variants on the basic themes due to the complex corrin structure which promotes the binding of metal ions and complexes and the formation of adducts and dimers (see Sec. 2)

and probably (4) more controversy over reaction mechanisms. There
is an urgent need for a critical review of in vitro Co-C bond formation and cleavage in the corrinoids. This chapter will focus specifically on the corrinoids, on Co-C bonds to alkyl ligands without
functional groups (from CH_3 to cyclohexyl, benzyl, etc.), and on
those examples which best serve to clarify the mechanistic pathways
and the factors involved; it aims to be critical, not comprehensive.
Short summaries of structural data, nonbonding interactions and ion
binding (Sec. 2), coordination chemistry (Sec. 3), and enzymatic
reactions (Sec. 5) are also included.

1.2. Main Sources of Information

A useful coverage of much of the B_{12} field was provided by the two-
volume multiauthor B_{12}, edited by Dolphin and published in 1982 [6];
this covered both the more chemical aspects (15 chapters in Vol. 1)
and the biochemical/medical aspects (14 chapters in Vol. 2). A book
entitled *Comprehensive* B_{12} by Schneider and Stroinsky, with greater
emphasis on the biological and medical aspects, appeared in 1987
[1]; see also the 1991 review by Ellenbogen and Cooper [7]. Two
aspects not covered in the two-volume B_{12} are the electrochemistry
(redox potentials) of Co corrinoids, which has subsequently been
reviewed [8,9], and their general coordination chemistry (ligand
substitution, redox reactions, etc.), which has not been treated
since the author's 1972 book [10]. Co-C bond formation and cleavage,
particularly in noncorrinoids, was thoroughly reviewed by Toscano
and Marzilli in 1984 [3]; the present review on B_{12} can be considered complementary. Rules for the nomenclature of the corrinoids
were approved by the IUPAC-IUB Commission on Biochemical Nomenclature in 1975 [11]; see also Sec. 2.

Although the Co corrinoids constitute the only major role for
Co in enzymes, it should be noted that Co porphyrins of the isobacteriochlorin series but with no known function have been isolated
from *Desulphovibrio* species [12-14] and a few enzymes of known

function (glucose isomerase [15] and nitrile hydratase [16]), which
require "inorganic" Co (i.e., not as a porphyrin or corrinoid complex) from other organisms.

2. STRUCTURES AND NONBONDING INTERACTIONS

B$_{12}$ is the parent of a large family of naturally occurring Co corrinoids. They all possess the same conjugated equatorial ring as in
B$_{12}$ itself, but may differ as to whether the side chains terminate
in carboxylic acid or amide groups, whether the side chain on C17
carries a nucleotide (phosphate + ribose + base), and, if so, what
is the nature of the base. Corrinoids which possess the same side
chains as B$_{12}$ itself, including the base *5,6-dimethylbenziminazole*
(here abbreviated to *Bzm*), and differ only in the nature of the axial
ligands are termed *cobalamins (Cbl)* while the *cobinamides (Cbi)*
differ in the absence of the whole nucleotide. The cobalamins are
termed "base-on" when Bzm is coordinated and "base-off" when the Bzm
is displaced (e.g., by protonation; see Sec. 3.1); B_{12a}, B_{12r}, and
B_{12s} are often used to designate the Co(III) aquo (or hydroxo),
Co(II) and Co(I) Cbl, respectively. Only the Cbl's are used (i.e.,
are actively absorbed from the gut) by mammals; hence their importance. Almost all work on the general coordination and organometallic chemistry has been carried out with the Cbl's (providing a
series where one axial ligand is Bzm) and the Cbi's (where the axial
ligand is H$_2$O/HO$^-$ or can be readily replaced by others such as CN$^-$).
Bacteria produce and use a bewildering range of corrinoids, differing
mainly in the nature of the nucleotide base. The forms most commonly
found in organisms possess methyl or *deoxyadenosyl (Ado)* as the axial
ligand. All naturally occurring corrinoids are very hydrophilic due
to the presence of amide and acid side chains, but can be made more
soluble in organic solvents by esterification of the side chains to
give the so-called *cobesters* (usually the methyl esters). The structures of over 30 corrinoids have now been determined, mostly six-
coordinate Co(III) corrinoids; for a summary up to 1982 see [17].

The Co corrinoids are far more crowded and distorted than the
Fe porphyrins. The corrin ring is nonplanar due to (1) the presence
of the C-C bond between the tetrahedral C1 and C19 and (2) the need
to minimize the repulsions between the substituents on the exocyclic
ring. Stereoisomers are possible when the two axial ligands are
different; ligands are designated α or β, respectively, when they are
below or above the corrin ring as presented in Fig. 2. Both axial
sites are surrounded by a hydrophobic cylinder composed of four
"sentinel" methyl or methylene groups, which may offer serious steric
hindrance to both the approach and eventual coordination of an incoming ligand; see the diagrams in Refs. 17 and 18. Methyl C20 causes
a distinct tilting away of both the base Bzm in the Cbl's and of CN^-
in cobyric acid; the coordinated Bzm is further distorted by contact
with the corrin ring between C5 and C6 [10]. The corrin ring can,
however, exhibit a variable degree of folding (2-22°) along the Co-C10
line [18].

Nonsteric factors are equally important. Except for side chain
g on C18 which appears to be held by interaction between its amide
(or ester) O and the C20 methyl [17,19], all the amide side chains
are very flexible [17]; that on C7 can form H bonds to β ligands [10,
20]. The very exposed nucleotide side chain (see Fig. 2) is, as
might be expected, involved in various nonbonded interactions. The
role of the nucleotide in molecular recognition by the Cbl-specific
protein *intrinsic factor* has recently been elucidated; it catalyzes
some conformation change of the protein [21]. The uv-vis and nuclear
magnetic resonance (NMR) spectra indicate that both the protonated
and unprotonated (but uncoordinated) forms of Bzm interact in some
way with the corrin structure, i.e., the nucleotide side chain does
not "dangle" free [22-24]. Tetracyanoethylene provides the first
clear-cut example of adduct formation, forming adducts with CH_3-Cbl
(K = 4 x 10^2 M^{-1}, charge transfer band at 420 nm), CH_3-Cbi, and
H_2O-Cbl [25].

The ability of the side chains to bind metal ions and even
whole complexes is well known [10]; for the coordination (and protonation) of Bzm see Sec. 3.1. Equilibrium constants of 3 x 10^3 and

and 4 x 10^3 M^{-1} have now been determined for the binding of $PtCl_4^{2-}$ and $Pt(CN)_4^{2-}$, respectively, by CH_3-Cbl, which are directly relevant to methyl transfer from Co to Pt [26] (see Sec. 4.7.2). The reduction of H_2O-Cbl by thiols is catalyzed by traces of firmly bound metal ions [27]. Many others in the B_{12} field (including ourselves) have experienced cases of unreproducible reactions, very likely due to trace metal catalysis, but have hesitated to publish the results; see p. 449 of Ref. 28 for an example involving Co-C fission in studies designed to model the B_{12}-dependent enzymatic isomerizations and Ref. 29 for an example in Co-C bond formation.

NMR studies have revealed that corrinoids can form dimers in solution. The dimerization constant of CH_3-Cbl in unbuffered solution with no added salts is 2-4 x 10^2 M^{-1}. Formation of the dimer is promoted by NaCl (K_D = 1-2 x 10^4 M^{-1} in 2.5 N NaCl) and by lowering the temperature, but not by $NaClO_4$, and is suppressed by adding CH_3OH or $(CH_3)_2SO$. Et-Cbl and protonated CH_3-Cbl show a lower tendency to dimerize, and Pr-, Ado-, CN-, and H_2O-Cbl none at all. All these Cbl's, whether forming dimers or not, showed NaCl-specific shifts of some of the resonances [30]. CH_3-Cbl also forms some adduct with diaquo-Cbi [31] but not with aquo-Cbl [30]. The NMR data suggest a β-face-to-β-face interaction with roughly parallel corrin rings [30], in agreement with the X-ray data for the iodide-bridged Co(II)-cobester dimer where the coplanar rings are about 5.8 Å apart [10,17]. The formation of these dimers opens up reaction paths (see Secs. 4.4 and 4.6.1) which are apparently not available to noncorrinoid complexes and emphasizes that the reactions of the Co cannot be understood in isolation from its organic matrix.

3. COORDINATION CHEMISTRY

3.1. General Comments

The Co corrinoids exhibit oxidation states from the stable Co(III) down to the unstable Co(I); in the presence of the strong donor CH_3 a very unstable Co(IV) is also attainable. A variety of chemical reagents may be used to reversibly reduce and oxidize the Co corrinoids; borohydride, for example, will rapidly reduce the red aquo-Cbi/Cbl to the yellow Co(II) and then far more slowly to the greenish Co(I). The distinctive changes in color and uv-vis spectra provide an easy way to monitor these reactions. In the presence of air, the Co(I) corrinoids are instantaneously oxidized to Co(II) and more slowly to Co(III), but under nitrogen will even liberate H_2 to reach Co(II) and undergo extraordinary oscillating reactions [32]. The kinetics of the reaction of Co(II) with O_2 [33] are also unusual (see Sec. 4.6.1); a Co(II)-O_2 adduct has been isolated and its structure determined [34]. The simplest way to prepare R-Co corrinoids is by reaction of Co(I) with RI or RBr (see Sec. 4.5). Where R is a primary alkyl ligand the R-Cbl's and R-Cbi's are very stable (in the dark) toward O_2, excess BH_4^-, acid and base, but very light-sensitive in solution and must be handled accordingly. Most work has been carried out in aqueous solution with the Cbl's and Cbi's and, in the case of R-Cbi's, almost exclusively with the upper (β) isomer (but see Sec. 4.5); some additional points have been established with cobesters in organic solvents.

The Cbl's possess the nucleotide side chain terminating in the base Bzm and a major feature of their solution chemistry is protonation and displacement of Bzm from coordination in acid solution. The observed pK is the result of competition between the two Lewis acids (Co^{3+} and H^+) for the Bzm and is very dependent on the nature of the trans ligand. Ignoring the complications (see Sec. 2) due to interaction of the noncoordinated Bzm (whether protonated or not) with the corrin structure, it is useful to adopt a "working" pK of 5.0 for noncoordinated Bzm [22,24,35]. The closer the observed pK for displacement and protonation of Bzm to 5.0 (H_2O-Cbl -2.1 to -2.4 [10,36], CN-Cbl +0.1 [10,24], CH_3-Cbl ∼2.7 [10,24], Ado-Cbl 3.3-3.5 [10], Co(II)-Cbl 2.9-3.2 [9,37]), the weaker the Co-N bond; cf. the trans effect (Sec. 3.4). The base-off Cbl's are very similar to the corresponding Cbi's in both physical properties (e.g., uv-vis spectra) and chemical properties (e.g., rates of Co-C bond cleavage, redox potential). The reversible conversion of base-on to base-off R-Cbl's (six-

and five-coordinate, respectively; see Secs. 3.2 and 3.4) with pKs
in the region 3-5 can both complicate mechanistic studies and provide
an easy way to probe the "trans effect" (actually the effect of a
change in coordination number) on the rate of reaction. Further complications may arise when the attacking reagent is a metal ion such
as Hg(II) which may itself coordinate the N atom of Bzm (see Sec.
4.7.1). Far greater acidity is required to protonate the phospho-
diester of the nucleotide (pK -0.1 in H$_2$O-Cbl [36]) and the corrin
ring (pK ~-1.5, probably on C10 [38]). The Co-CH$_3$ bond is stable even
in concentrated H$_2$SO$_4$ [39].

3.2. Valencies and Coordination Numbers

All the d^6 Co(III) corrinoids with typical ligands (H$_2$O, CN$^-$, Cl$^-$,
nitrogenous bases) are diamagnetic and six-coordinate with colors
ranging from orange to purple [10]. The Co(III) diaquo-Cbi, which
can be considered as the parent corrinoid, exhibits pK values of
5.9 and 10.3 linking the diaquo, aquohydroxo, and dihydroxo forms
[40], and aquo-Cbl a pK ~7.8 [8-10]. The monomeric d^7 Co(II) corrinoids are yellow, low-spin (i.e., one unpaired electron), and five-
coordinate, as established with two crystal structures [41,42]; a
green, iodide-bridged, diamagnetic, and dimeric Co(II) cobester is
also five-coordinate [10,17]. The extended X-ray absorption fine-
structure spectroscopy (EXAFS) spectrum of the very unstable Co(I)-
Cbl fairly definitely establishes a slightly distorted square planar
coordination [43], in agreement with the fact that gray-green Co(I)-
Cbi and Cbl have identical spectra even in neutral solution, i.e.,
Bzm is never coordinated. It is generally accepted that the square
planar d^8 Co(I) will, like the square planar d^8 Ni(CN)$_4^{2-}$ complex,
be diamagnetic.

All R-Cbl's and R-Cbi's studied are diamagnetic. X-ray
diffraction has established six coordination for the red base-on
forms of CH$_3$-, Ado-, and two other organo-Cbl's [2,20,44-46]. The
yellow R-Cbi's and base-off R-Cbl's are almost certainly five-

coordinate [47,48] (see also Sec. 3.4), but a structure determination is clearly desirable. The reduction of both H_2O-Cbl and -Cbi in glacial acetic acid gives very unstable yellow species with spectra similar to that of a base-off Et-corrinoid and probably the hydrido-Co(III) complex [49]; a pK of ~1 has been observed in aqueous solution by cyclic voltammetry and assigned to the equilibrium $Co(I) + H^+ \rightleftharpoons Co(III)-H^-$ [37]. Evidence for CH_3-Co(II) and CH_3-Co(IV) corrinoids is given below (Sec. 3.3).

3.3. Redox Potentials

The electrochemistry of the Co corrinoids has been well studied; values of ~-0.04 and -0.85 V vs. SCE have been reported for the two successive steps in the reduction of H_2O-Cbl to Co(I) in the pH-independent region 4.7-7.8, and +0.27 and -0.74 V vs. SCE for the analogous two steps in the reduction of diaquo-Cbi and the acidified "base-off" Cbl [8,9]. Values of -1.46 and -1.60 V vs. SCE have been determined for the reversible one-electron reduction of CH_3-Cbi and CH_3-Cbl, respectively, in 1:1 DMF:nPrOH to give very unstable products [50]; radiolysis at low temperature suggests an initial electron capture by the corrin ring, followed by the formation on annealing of a CH_3-Co(II)-Cbl with the extra electron in a $d_{x^2-y^2}$ orbital [51]. Values of +1.07 V vs. SCE have been determined for the reversible one-electron oxidation of CH_3- and Et-cobester in acetonitrile, while chemical oxidation with Ce(IV) at low temperature gives a product which retains the Co-CH_3 bond and shows an electron spin resonance (ESR) signal characteristic of other CH_3-Co(IV) complexes [52].

These potentials show that reduction from Co(III) to Co(II) is far easier, and from Co(II) to Co(I) slightly easier, with Cbi than Cbl; this reflects the increasing stability of the Co-N(Bzm) over the Co-OH_2 bond in the order Co(I) (neither coordinated) < Co(II) < Co(III). They also show that introduction of the far stronger donor CH_3^- further stabilizes Co(III) over Co(II) (CH_3-Cbi -1.46 V) and even stabilizes Co(IV) sufficiently to bring it within the detection range (+1.07 V). In the past it has been suggested that the Co(I) required for alkylation (in vivo or in vitro) could be formed by the disproportionation of Co(II)-Cbl, but comparison of the potentials shows that the disproportionation constant for 2 Co^{2+} $\rightleftharpoons$ Co^{3+} + Co^{1+} for the Cbl is so small (K$_d$ ~10^{-15} between pH 4.7 and 7.8) that this mechanism can probably be ignored [53,54].

3.4. Trans Effects

The ability to vary one axial ligand X in the Co(III) corrinoids systematically in an order of increasing donor power from H$_2$O through CN$^-$ to methyl and ethyl provides a unique opportunity to study the "trans effect," i.e., the effect of changing X on another "probe" ligand in the trans position (ground state effect) or on the equilibrium and rate constants for the substitution of one ligand by another (thermodynamic and kinetic trans effect, respectively). The data [48] reveal a remarkably simple pattern in which (1) the ligands X occur in the same trans effect order of increasing σ-donor power from H$_2$O to Et (viz. H$_2$O < Bzm < HO$^-$ < CN$^-$ < -C≡CH < -CH=CH$_2$ < -CH$_3$ < -CH$_2$CH$_3$) for all the parameters observed at all three levels (ground state, thermodynamic, and kinetic) and (2) as the donor power of the ligand increases and the positive charge on the Co decreases, so (a) the binding constant for CN$^-$ falls by ~10^{16} (and those for the less covalently bound imid and Bzm by 10^6-10^8), (b) the coordination even of H$_2$O eventually becomes so weak that the five-coordinate complex becomes the ground state, and (c) the rate constant for substitution of coordinated H$_2$O by CN$^-$ and other ligands rises by over 10^5 to approach the diffusion-controlled limit.

EXAFS spectra suggest a similar electron density on the Co in CH$_3$-Cbl, Ado-Cbl, and Co(II)-Cbl [43], in agreement with the similar pK values for protonation of Bzm (see Sec. 3.1). It appears that the Co(III) + CH$_3^-$ unit is comparable to Co(II) as far as both the Co and the rest of the complex are concerned and that the Co-CH$_3$

bond can be considered basically as a nonpolar (i.e., covalent) bond, but also (see next section) a very polarizable one, which can be made and broken by reactions involving transition states which may correspond to $(Co^{1+} + CH_3^+)$, $(Co^{2+} + CH_3^{\bullet})$, or $(Co^{3+} + CH_3^-)$.

4. ORGANOMETALLIC CHEMISTRY: MAKING AND BREAKING THE Co-C BOND

4.1. General Comments

The oxidation states from Co(I) to Co(III) are all readily accessible in simple corrinoids, while the electron density on the Co in a Co-alkyl corrinoid approximates to Co(II) (see Sec. 3.4); the result is a fairly covalent but very polarizable Co-C bond which can be both made and broken by transfer reactions in which the transition state corresponds to any of the three oxidation states as shown in (1). The only unassisted path (i.e., not involving the transfer of an alkyl group to or from another atom) is the reversible reaction involving Co(II) and free radicals. Additional paths can be opened up by an initial electron transfer to a strong oxidant [to give a paramagnetic R-Co(IV) ion] or from a strong reductant (to reduce the corrin ring and/or the Co ion) before the actual transfer step; no example of Co-C bond formation via such paths has been reported and the thermodynamics of such steps may, in fact, always be unfavorable.

$$Co^{1+} \cdots R^+ \rightleftharpoons Co\text{-}R \rightleftharpoons Co^{3+} \cdots R^-$$
$$\Updownarrow$$
$$Co^{2+} \cdots R^{\bullet} \tag{1}$$

We consider in turn Co-C bond fission via an initial oxidation or reduction (Sec. 4.2), unassisted reactions involving Co(II) and free radicals (Sec. 4.3), where Co-C formation and fission can be studied in the absence of complications due to donor or acceptor, and then the recently discovered family of methyl transfers between Co corrinoids (Sec. 4.4), followed by other alkyl transfers (Secs. 4.5-4.7). Wherever possible we examine the effects of varying the nature of the group R being transferred, the atom or group acting as

donor or acceptor and the trans ligand, also the possible involvement of any adducts or additional redox centers. The last is of
particular interest in the light of ESR studies of intermediates of
the B$_{12}$-dependent isomerase reactions which show overlap between the
atomic orbitals of Co and a C-centered radical at ≥ 10 Å (see references and discussion in Ref. 47) and the "linear electric field
effect" which can distinguish a radical-metal ion "charge transfer
center" (as in cytochrome c peroxidase compound I) from discrete
radicals and/or metal ions by the polarizability of its electron
density under the influence of the electric field [55]. This points
to the possibility of electronic interaction between species normally
treated as discrete entities, leading to an enhanced polarizability
of the system through "mixing in" additional states [4,47].

4.2. Co-C Bond Fission via Initial Electron Transfer

The stable R-Co(III) corrinoids can be reversibly oxidized to unstable
R-Co(IV) complexes and reversibly reduced to give unstable R-Co(II)
complexes (see Sec. 3.3). Vol'pin and coworkers reported that the
CH$_3$- and Et-Co(IV) cobesters formed by reversible oxidation in non-
aqueous solvents at low temperature (see Sec. 3.3) were not decomposed by pyridine (i.e., not susceptible to attack by nucleophiles)
and produced some methane and ethane (indicating formation of CH$_3$
radicals), while the electrochemical results suggested subsequent
oxidation and modification of the corrin ring with retention of the
Co-C bond [52]. The irreversible oxidation of CH$_3$-Cbl by IrCl$_6^{2-}$ in
aqueous solution gives H$_2$O-Cbl (in either the presence or absence of
O$_2$) and methane/ethane or CH$_3$Cl (with a deficiency or excess of oxidant, respectively), again indicating decomposition of the presumed
CH$_3$-Co(IV) intermediate to give Co(III) and methyl radicals which can
dimerize or abstract Cl from a second IrCl$_6^{2-}$; competitive inhibition
by IrCl$_6^{3-}$ provided evidence for the preequilibrium binding of IrCl$_6^{2-}$
and the observed (i.e., composite) rate constants were higher for the
base-on than the protonated base-off form of CH$_3$-Cbl (83 and 1.1
M^{-1} sec^{-1}, respectively) [56]. CH$_3$OH and H$_2$O-Cbl were, however,

identified as the products (involving nucleophilic attack by the
solvent) from the irreversible two-electron electrochemical oxidation of CH_3-Cbl in aqueous solution [57].

Lexa and Savéant [50] obtained rate constants, calculated
from the sweep rate required to obtain reversibility, for the
decomposition of CH_3-Cbi and CH_3-Cbl reduced at the cathode in
dimethylformamide (DMF)/PrOH at low temperature (see also Sec. 3.3):
reduced CH_3-Cbi from 14 sec^{-1} at -20°C to 2500 sec^{-1} at +19° and
reduced CH_3-Cbl 1200 sec^{-1} even at -30°C, indicating a trans-
labilizing effect of ~300 by Bzm. The products were not analyzed
and the reaction was assumed to involve decomposition to give
Co(I) + methyl radicals; Co(I) and ethane (from the dimerization
of methyl radicals) were, however, identified as the products of
irreversible two-electron reduction of CH_3-Cbl in aqueous solution
[57]. Lexa and Savéant also noted that even reduced CH_3-Cbi was
far more unstable than noncorrinoid analogs and ascribed this to the
cis electronic effect of the equatorial ligands; their data suggest
that the corrin (namely, reduced CH_3-Cbi) is ~10^5 more labile than
the tetraphenylporphyrin analog [50]. The same one-electron reduction of R-Cbl's at the electrode to give Co(I) and the radical R was
proposed in 1971 as the key step in the Cbl-catalyzed reduction of
alkyl halides to alkanes and alkylmercury compounds (from further
reactions of the radicals with the mercury electrode, etc.) [58].
The reaction of CH_3- and Pr-Cbl with zinc amalgam (the authors
merely mentioned "Hg^{2+} in the presence of zinc dust", but this would
have reacted immediately to produce zinc amalgam) to give alkylmercury derivatives, which was reported in 1969 [59], is probably
analogous.

4.3. Reactions Involving Co(II) and Free Radicals

4.3.1. Reaction Pathways

Sufficient experimental information is now available to form a
relatively simple and self-consistent picture of the potentially
reversible reactions (2) which are involved in the unassisted decompo-

sition of the Co-C(alkyl) bond. The same pattern of reactions is
observed for both five- and six-coordinate alkyl corrinoids and
whether initiated by heat or light (photolysis). The alkyl groups
R can be placed in an order of increasing steric bulk and increasing
labilization of the Co-C bond (Sec. 4.3.2) such that the Co-CH$_3$ and
Co-Et bonds break slowly even at 100° but may be readily dissociated
by photolysis, while Cbl's with bulky ligands such as iPr, cyclohexyl, and neopentyl [-CH$_2$C(CH$_3$)$_3$], together with benzyl (-CH$_2$C$_6$H$_5$)
where the radical is more stable, decompose even at room temperature;
the six-coordinate base-on Cbl's are more labile than the five-
coordinate Cbi's or analogous base-off Cbl's. Most work has been
done with Cbl's possessing the ligands methyl, benzyl, and neopentyl
which exemplify overall homolytic fission (upper path) and Et, iPr,
and cHx which exemplify β elimination (lower path). Homolytic
fission is here abbreviated to h.f.:

$$\text{Co-R} \rightleftharpoons (\text{Co}^{2+} \cdots R^{\bullet}) \rightleftharpoons \text{Co}^{2+} + R^{\bullet}$$

$$\Updownarrow$$

$$\text{Co-H} \rightleftharpoons \text{Co}^{1+} + H^{+} \tag{2}$$

$$+ \text{ olefin}$$

The potentially reversible steps of scheme (2) comprise h.f.
of the Co-R bond to give a caged (Co^{2+} + radical) pair which can
then (1) diffuse away to form the separated Co^{2+} and radical, (2)
isomerize, or (3) lose H from the β position to give an olefin (hence
the term "β elimination") and the hydride Co-H which, in aqueous solution, will lose H^{+} to give the Co(I) complex (pK ~1; see Sec. 3.2).
Subsequent reactions are essentially irreversible: the radical R$^{\bullet}$
may dimerize or disproportionate, abstract a H atom (e.g., from
thiols) to give the alkane, or be oxidized (e.g., by O$_2$); Co^{1+} and
Co-H are instantaneously oxidized by O$_2$ but will also decompose even
under N$_2$ (Sec. 3.1); Co^{2+} is rapidly oxidized by products such as RO$_2^{\bullet}$
from the reactions of R$^{\bullet}$ with O$_2$. Evidence for the various steps in
the scheme can be built up as follows.

A study of CH$_3$-Cbl and Ado-Cbl by picosecond flash photolysis
has shown that the immediate product is a caged (Co^{2+} + R$^{\bullet}$) pair

with kinetic properties distinct from those of the separated Co^{2+}
and free radical, and that the free methyl radical and Co^{2+} recom-
bine with a second-order rate constant of ~2 x 10^9 M^{-1} sec^{-1}, i.e.,
close to the rate of dimerization of methyl radicals (~10^{10} M^{-1} sec^{-1})
[60]. It seems reasonable to assume that the immediate product of
homolytic fission by thermal activation is also the caged pair
(structure discussed in Ref. 61). The isomerization of cyclopropyl-
methyl- to but-3-enyl-Cbl, which occurs on heating to 60°C, provides
an example of rearrangement; because the rate is unaffected by the
presence of O_2, it probably proceeds within the caged pair [62]. It
can be shown that the lack of any significant effect of O_2 on reactions of the caged pair does not necessitate any special mechanism
to protect the radical against attack by O_2 but follows from the low
solubility of O_2 in water (~10^{-3} M) compared to the very high effective local concentrations of the partners in the caged pair [61].

    Good examples of overall h.f. without complications due to β
elimination are provided by the photolysis of CH_3-Cbl and the thermal
decomposition of benzyl- and neopentyl-Cbl at ambient temperature.
All these reactions show the same marked acceleration of rate in the
presence of radical scavengers such as O_2, thiols, and iPrOH but
little reaction in their absence, which provides indirect evidence
for the reversible formation of Co^{2+} and a free radical [10,63,64].
Direct evidence for radical formation has been obtained by radical
trapping from both the photolysis and thermolysis of CH_3-Cbl and
Ado-Cbl [65-67]. The presence of O_2 has a far smaller effect on
the rate of decomposition where β elimination predominates (see,
for example, Ref. 64); cHx-Cbi is decomposed by O_2 in the presence
of either vanadyl ions or cobalt boride as catalyst, but the products
and mechanism have not been established [61]. Evidence for a common
first step for both β elimination and overall h.f. was obtained by
showing that varying R and changing from R-Cbl to R-Cbi had similar
effects on the rates of both overall reactions [61]. The β elimination is pH-independent [68]; there is evidence that Co-H and Co(I)
are intermediates [61,68]. Unstable Co-H complexes can be prepared

by reduction in glacial acetic acid [49] and will react with olefins
to give alkyl corrinoids [49,69].

In the above examples of Co-C bond formation the radicals have
been prepared by either (1) thermal decomposition or photolysis of
a preexisting Co-C bond or (2) abstraction of the H atom from a Co-H
complex by an olefin. Other known examples include (3) an ingenious
use of mixed "oxidizing-reducing" conditions provided by V^{3+} ions
and O_2 to generate radicals in the presence of Co(II) from the appropriate alkane, alkene, ether, alcohol, aldehyde, or carboxylic acid
(see Ref. 3 for references) and (4) the reactions of Co(II)-Cbl (B$_{12r}$)
with various alkyl halides RX in aqueous or methanolic solution
(*except* for RI in water) which show the same kinetics (first-order
dependence on both [Co^{2+}] and [RX]) as previously found for other
low-spin Co^{2+} complexes and have been interpreted in terms of the
same mechanism, i.e., rate-determining abstraction of X by the Co^{2+}
to give Co^{3+}X$^-$ + R$^\bullet$, followed by the rapid reaction of R$^\bullet$ with Co^{2+}
to give R-Co and the substitution of X$^-$ by H$_2$O to give H$_2$O-Co^{3+} [70].
For the very different kinetics and mechanism observed for Co(II)
and RI, see Sec. 4.6.1.

4.3.2. *Steric Effects of the Alkyl Ligands*

There has been great interest in exploring the steric effects of simple
alkyl ligands on the stability of the Co-C bond in both corrinoid and
noncorrinoid complexes, driven primarily by interest in trying to understand the mechanism of Co-C bond labilization in the Ado corrinoids in
the B$_{12}$-dependent isomerases (see Sec. 5.1). The result is an unusually
detailed and self-consistent picture of steric effects which, as in the
case of electronic trans effects (see Sec. 3.4), can be traced back from
kinetic through thermodynamic to ground state effects.

The effects of increasing steric distortion around the coordinated
carbon atom (C$_\alpha$) have been systematically studied by varying the
alkyl ligand R in the three series (1) from -CH$_3$ to -CH(CH$_3$)$_2$ [the
Co-C(CH$_3$)$_3$ complex is too unstable to study], (2) from -CH$_2$CH$_3$ to
-CH$_2$C(CH$_3$)$_3$ and (3) from cyclopropyl to cyclohexyl; these serve to

TABLE 1

Steric Effects in Alkyl Corrinoids

		R =
Ground State Effects (Structures and Physical Properties)		
1. R-Cbl	Co-C bond length (Å)	
2. R-Cbl	Co-$\hat{C}$-C	
3. R-Co(DH)$_2$py	Co-C bond length (Å)	
4. R-Co(DH)$_2$py	Co-$\hat{C}$-C	
5. R-Cbi	Ratio A_{460} : A_{440} in visible spectra	
Thermodynamic Effects		
6. R-Cbl	pK (Bzm + H$^+$)	
7. R-Cbl	Co-C BDE (kcal/mol)	
8. R-Cbi	Co-C BDE (kcal/mol)	
Kinetic Effects (Rates of Co-C Bond Fission)		
9. R-Cbl	k (sec^{-1}) at 25°	
10. R-Cbl	$t_{\frac{1}{2}}$ at ca 25°	
11. R-Cbl	$t_{\frac{1}{2}}$ at 95°	
12. R-Cbi	k (sec^{-1}) at 25°	
13. R-Cbi	$t_{\frac{1}{2}}$ at ca 25°	
14. R-Cbi	$t_{\frac{1}{2}}$ at 95°	

Note: Except where otherwise indicated, data are taken from references in right-hand column (some of which are summaries of data).
[a]Ref. 153; [b]Ref. 154; [c]Ref. 67; [d]Ref. 155.

distinguish the effects of substitution on C_α and C_β and the effects of increasing and decreasing the Co-C-C bond angle. Bulkier ligands studied include cyclooctyl, -CHEt$_2$, and -CHCH$_3$·C(CH$_3$)$_3$ [68]. As will be seen below, all three series of alkyl ligands, together with Ado, can be placed in virtually a single order of increasing steric compres-

Ado	CH_3	cPr	Et	iPr	np	cHx	Ref.
{ 2.00 { 2.04	1.99						151
{ 124° { 121°	—						151
	1.998		2.065	2.085	2.060		152
	—		122°	114°	130°		152
0.89	0.90	0.97	1.03	1.23	1.22	1.27	61,74
3.3-3.5	2.5-2.7	2.8	3.9	⇌ 4.5	4.7	4.7	74
{ 30±2[a] { 26±2[b]	37±3[c]			19	21		64 154
34.5±1[d]				26	30		64
$10^{-9±1}$	$10^{-12±3}$						66,67
	v.long	6 mo	6 mo	2.8 min	75 min	44 min	64,68
180 min	3½ days		9 min				4
$10^{-11±1}$							66
				130 hr	73 days	67 days	64,68
	~28 days		380 min				4

sion, i.e., $CH_3 \backsim$ cPr $\backsim$ Ado < Et $\backsim$ Pr < iBu $\backsim$ cBu < iPr, np, cPe, cHx.
Only the end members of the three series and Ado have been included
in Table 1. Most work has been carried out on R-Cbi's and R-Cbl's
without other axial ligands. The R-Cbi's and protonated base-off
R-Cbl's can be treated as entirely five-coordinate (see Sec. 3.2) and,

with few exceptions, the two series appear to show identical uv-vis spectra and rates of Co-C bond fission; one exception is that the rate of Co-C(cHx) fission is sensitive to the added acid in the protonated base-off Cbl (~10 times faster in HCl than in H_3PO_4) but not in the Cbi [68], presumably due to some interaction with the nucleotide side chain.

Steric effects on the following experimental observables have been studied and the results summarized in Table 1:

1. Structural properties such as Co-C bond lengths and Co-C-C bond angles. The few relevant data available for the corrinoids (rows 1-2) can be supplemented by a much larger body of data from the cobaloximes R-Co(DH)$_2$py. The observation of marked ground state steric effects with the planar cobaloximes (rows 3-4) allows one to focus attention on the immediate coordination sphere of the Co and to ignore, at least initially, any additional direct steric interaction between the coordinated alkyl ligand and upward-projecting substituents on the corrin ring.

2. Other physical properties such as the uv-vis spectra of the R-Cbi's and base-off R-Cbl's, where the relative intensities of the two most intense vibrational bands (~440 and 460 nm) of the first electronic transition show a systematic variation and act as a molecular strain gauge (see row 5).

3. Equilibrium constants for coordination by the five-coordinate R-Cbi's of, e.g., imidazole or CN$^-$, which are usually too small to be detected (but see Ref. 61 for a summary of the few binding constants known for imidazole), and of Bzm in the R-Cbl's, where a more extended series can be observed (row 6); the closer the pK (for protonation of Bzm and displacement from coordination) to 5.0, the weaker the Co-N bond (see Sec. 3.1).

4. Equilibrium constants for the reversible dissociation of the Co-C bond to give Co(II) + free radical, here given

(rows 7-8) as the bond dissociation energy or enthalpy
(BDE) determined by kinetic methods (for details, see Ref.
71). BDEs for Co-iPr and Co-np have here been derived
from the published activation enthalpies [64] by subtract-
ing 2 kcal/mol to allow for the activation enthalpy of
viscous flow in water [71]. Similar trends are observed
in the noncorrinoid field; cf. Co-C BDEs of 34.6 ± 1.4
and 21.3 ± 2.0 kcal/mol for R = CH_3 and iPr in R-Co(DH)$_2$py
[72] and 27.1, 21.8, and 20.3 kcal/mol for R = Pr, iPr,
and np in R-Co(saloph)py [71].

5. Rate constants and/or half-times for Co-C fission at 25°C
(or 95°C) in the R-Cbl's (rows 9-11) and R-Cbi's (12-14);
because they involve a common first step, β elimination
and overall h.f. can be treated on a common basis (see
Sec. 4.3.1). Since the rate constant depends on the
entropy as well as enthalpy of activation, it may not
show a simple relationship to the BDE; there does, however, appear to be some relationship between the entropy
and enthalpy of activation in Co-C bond fission [71].

The data of Table 1 show that:

1. Significant steric effects can be caused by substitution
on C_β (but not on C_γ [68]) as well as C_α and there are
unexpectedly large differences even between methyl and
ethyl. Similar steric effects are seen in the equilibrium constants for the coordination of substituted amines
by H_2O-Cbl [73], i.e., ligands with a tetrahedral C atom
and with a tetrahedral N atom show parallel effects.

2. The area around C_α is very easily distorted by increasing
the Co-C bond length and/or increasing the Co-C-C bond
angle (to values even greater than the trigonal C-C-C
bond angle of 120°C). This is not, however, accompanied
by any significant deviation in the Co-C bond from 90 ± 3°
to the plane of the corrin ring and this has been ascribed
to strong repulsion from the cylinder of high electron

density around the Co-C axis provided by the lone pairs on
the four equatorial N atoms [74].

3. All the simple alkyl ligands can be placed in virtually a
single order for steric effects at all three (ground state,
thermodynamic, and kinetic) levels and in both five- and
six-coordinate complexes, though the correlation between
the kinetic and thermodynamic effects may be upset where
the stability of the radical is significantly affected by
steric (cf. cPe and cHx) or electronic (e.g., $PhCH_2$)
effects; such additional effects will not be discussed here.

The observed pattern of equilibrium and rate constants can be
explained [47,74] by assuming that steric compression can be increased
through either (1) increasing the effective steric bulk of the alkyl
ligand R or (2) converting a five-coordinate corrinoid (with Co probably displaced out of the plane toward the single axial ligand) into
a six-coordinate complex (with Co more in the plane), such that increasing the steric bulk of R in the six-coordinate R-Cbl's will
increasingly destabilize the six- relative to the five-coordinate R-Co
and Co(II) forms (decrease in BDE and log K for coordination of imidazole [61], rise in pK for protonation of Bzm, increase in rate of
Co-C bond fission) and, conversely, converting from a six- to a five-
coordinate R-corrinoid will relieve some of the steric compression
(rise in BDE, fall in rate of Co-C fission).

Comparisons can also be made of Bzm with imidazole and CN^- as
the trans ligand. First, it has been suggested that labilization of
the Co-C bond in the R-Cbl's is due (partly or even mainly) to steric
compression of the Bzm against the C5-C6 region of the corrin ring;
since both np- and cHx-Cbi are also labilized by imidazole, though
the binding constants are too low to be detected [63,68], the role of
compression of Bzm against the corrin ring can be ignored as a significant factor in labilizing the Co-C bond. Second, a kinetic study of
Co-C(benzyl) fission in the Cbi, Cbl, and cyanide complex showed that
CN^- as the trans ligand causes a striking decrease in the BDE (~27,
20, and 9 kcal/mol in the three complexes named), but a smaller

increase in the rate constant at 25° (9.3 x 10^{-6}, 2.4 x 10^{-3}, and
1.2 x 10^{-2} sec^{-1}, respectively) due to a much larger negative value
of $\Delta S^{\neq}$ (~14, 2, and -29 e.u., respectively) [75]. This appears to
be the only direct study so far reported on the trans effect on
homolytic Co-C bond fission, which is an area obviously needing
attention.

4.4. Methyl Transfer Between Co Corrinoids

Reactions involving the direct transfer of methyl from CH$_3$-Cbl to
the Co(III) diaquo-Cbi reported by Fanchiang, Bratt, and Hogenkamp
in 1984 [31] and to Co(II)-cobester and Co(II)- and Co(I)-Cbi by
Kräutler shortly afterward [76,77] are reversible and provide a
unique opportunity for establishing the thermodynamic trans effect
(i.e., effect of the trans ligand on the equilibrium constants) in
methyl transfer. The reactions between CH$_3$-Cbl and the three Cbi's
can be written schematically as in (3)-(5), where N represents the
nucleotide base Bzm and only the coordinated axial ligands are given:

$$N\text{-Co-CH}_3 + Co^{1+} \rightleftharpoons Co^{1+} + CH_3\text{-Co} \tag{3}$$

$$N\text{-Co-CH}_3 + Co^{2+}\text{-OH}_2 \rightleftharpoons N\text{-Co}^{2+} + CH_3\text{-Co} \tag{4}$$

$$N\text{-Co-CH}_3 + H_2O\text{-Co}^{3+}\text{-OH}_2 \rightleftharpoons N\text{-Co}^{3+}\text{-OH}_2 + CH_3\text{-Co} \tag{5}$$

Equilibrium constants of 4 x 10^{-3} and 0.6 were calculated for
reactions (3) and (4), respectively, from the composition of the
equilibrated mixtures [76,77] and a value of ≥ 20 can be calculated
for (5) from the data in [31], i.e., log K = -2.4, -0.2, and ≥ 1.3
for (3)-(5). As Kräutler points out [77,78], the positions of these
equilibria are determined primarily by the fact that the Co-N bond
strength increases relative to that of Co-OH$_2$ (as shown by the decrease
in pK for protonation and displacement of Bzm) in the order Co^{1+} (Bzm
not coordinated) < Co^{2+}-N ~ N-Co-CH$_3$ < N-Co^{3+}-OH$_2$. As between Co(III)
corrinoids, the very strong donor ligand CH$_3^-$ is destabilized by increasing the donor power of the trans ligand from H$_2$O to Bzm. Extend-
ing this trend to other ligands in the known trans effect order (i.e.,

H_2O < Bzm $\leq$ HO^- < CN^-; see Sec. 3.4) will explain the facts that converting diaquo- to aquohydroxo-Cbi by raising the pH reduces the rate and/or extent of transfer from CH_3-Cbl and that CN-Cbi cannot be methylated [31].

A half-time of ~9 min was reported for reaction (4) in either direction, but replacing Co(II)-Cbi by Co(II)-cobester increased the half-time to ~3 days (i.e., very approximately by a factor of 500) while leaving the equilibrium constant unchanged (0.6); the rate of (3) was merely noted as higher than that of (4) [77]. For reaction (5) under comparable conditions one can deduce a half-time of 30-60 min from figure 1 of Ref. 31, i.e., only slightly greater than for (4), but replacing CH_3-Cbl by the epimer in which side chain e on C13 (see Fig. 1) points up reduces the rate by almost an order of magnitude; the NMR spectrum also shows the formation of some complex between CH_3-Cbl and diaquo-Cbi before the transfer step [31]. It is clear that the rate-determining step is the formation of some productive dimeric complex and that either its rate of formation and/or its geometry around the active site (the Co-Co axis) are very sensitive to changes in the nature and conformation of the side chains. One cannot therefore say anything about the rates of the different reactions (3)-(5), except that they are all facile. The whole process is very reminiscent of an enzymatic reaction—formation of a nonbonded adduct (the dimer), followed by enclosure of the reactive center in a hydrophobic environment before the reaction proper; another example of B_{12} behaving as the "poor man's protein."

Fanchiang et al. [31] showed that the rate of reaction (5) was unaffected by the presence of O_2, H_2O_2, or N_2O which would trap Co(II) and Co(I) formed as reaction intermediates or traces of adventitious Co(II) acting as catalyst. They pointed out that the direct methylation of the Co(III) diaquo-Cbi required the prior formation of the five-coordinate intermediate through loss of a coordinated H_2O and that, because of the known trans effect (see Sec. 3.4), this was more likely to be generated from the aquohydroxo complex. Since they found that the rate of transfer decreased with pH as the diaquo was converted into the aquohydroxo form, they proposed a mechanism

involving an initial electron transfer to give CH$_3$-Co(IV)-Cbl and
Co(II)-Cbi in a discrete step before methyl transfer [31]. However,
the decrease in rate as diaquo- is converted to aquohydroxo-Cbi would
be expected on thermodynamic grounds alone (see above) and the rate
of loss of coordinated H$_2$O from diaquo-Cbi, as reflected in the rate
constants of 0.2-2×10^3 M^{-1} sec^{-1} for ligand substitution reactions
[40], is more than adequate to cope with the relatively slow rate of
9.6×10^{-2} M^{-1} sec^{-1} which they reported for reaction (3).

Computer modeling indicates that steric hindrance between the
side chains in the dimer will occur when the Co-Co distance falls
below 5.8 Å [77], in good agreement with the Co-Co distance of 5.65 Å
actually observed in the iodide-bridged Co(II)-cobester dimer, where
the Co is five-coordinate and displaced by 0.13 Å out of the plane
toward the I$^-$ [10,17]. The reaction therefore involves stretching
the Co-C bond from 2.0 Å, as observed in CH$_3$-Cbl [44], to a transition state in which the C atom of the CH$_3$ unit is weakly bonded to
two Co atoms at ∼2.8 Å. The structure of the dimer also reveals
that one of the side chains swings round to place its ester methyl
group in the otherwise empty sixth coordination site at a distance
of 4.1 Å from the Co; as Glusker points out, "it may indicate an
affinity of five-coordinate Co(II) for methyl groups" [17]. Any such
long-range interaction would obviously serve to reduce the energy of
the transition state and facilitate methyl transfer, while the hydrophobic environment would minimize any reorganization energy in the
surrounding medium.

4.5. Alkyl Transfer Involving Co(I)

The alkylation of Co(I)-Cbl by reagents such as CH$_3$I according to
(6) was discovered and developed in 1962-1963 by the groups of Smith,
Johnson, and Bernhauer, by Müller and Müller, and by Zagalak (see
[10]). There has been considerable interest in methyl transfer both
to and from Co(I) because of its possible relevance to the enzymatic
reactions (see Sec. 5.2). We now know a fair amount about alkyl

(especially CH_3) transfer to Co(I), but have no confirmed examples of transfer to another nucleophile to leave Co(I).

$$Co(I) + CH_3I \longrightarrow Co-CH_3^+ + I^- \tag{6}$$

$$Co(I) + RX \longrightarrow Co\cdots R\cdots X \longrightarrow Co-R + X^- \tag{7}$$

$$Co(I) + RX \longrightarrow Co(II) + R\cdot + X^- \longrightarrow Co-R + X^- \tag{8}$$

The Co(I) ion in reaction (6) behaves like a typical organic nucleophile in a S_{N2} substitution reaction, but Schrauzer and co-workers showed in 1968 that Co(I)-Cbl and the Co(I) cobaloximes were more reactive than any previously known nucleophile; cf. the second-order rate constants (in M^{-1} sec^{-1}) for the reaction of CH_3I in CH_3OH at 25°C with CH_3OH (i.e., solvolysis, 1.3×10^{-10}), NH_3 (4×10^{-5}), I^- (i.e., exchange with labeled I^-, 3.4×10^{-3}), (Bu_3P)-Co(I)-cobaloxime (2.5×10^3), and Co(I)-Cbl (3.4×10^4) [79,80]. Co(I)-Cbl and Cbi both react with PrCl at the same rate, in agreement with their four-coordinate structure, i.e., there is no trans effect. In studies with a wider range of alkyl halides RX Schrauzer and Deutsch always observed second-order kinetics and, from the variation of rate with R and X, concluded that the mechanism involved the one-step S_{N2} substitution (7) rather than the two-step electron transfer reaction (8). They also noted that increasing steric strain around the trans-ferred C atom had a similar effect on the rates of both Co(I)-Cbl and Co(I)-cobaloximes but an apparently far greater destabilizing effect on the Co-C bond of the product in the case of the Cbl's [80].

At the same time Friedrich and coworkers (see Ref. 10) were able to prepare and identify the two separate α/β (lower/upper) isomers of CH_3-Co corrinoids and to interconvert them by heating or photolysis in solution in the absence of O_2. They observed that, in the case of cobyric acid (with no Bzm to coordinate to the lower site), the same proportion (8%) of the lower isomer was obtained both as the immediate product of methylation and after equilibration of the separate isomers, i.e., under both kinetic and thermodynamic control. Kräutler subse-quently observed that the α/β ratio at equilibrium changed from 1:12 in 20% aqueous CH_3OH to 8.5:1 in toluene (i.e., the equilibrium constant

changes by 10^2), that similar ratios were obtained from methylation
by CH_3I in the two solvents (i.e., the same ratios are obtained
under conditions of kinetic and thermodynamic control), but that
methylation by the tosylate in toluene gave the equilibrium ratio
observed in aqueous CH_3OH [81]. The solvent dependence of the rate
constants has not been determined. It has also been found that Co(I)-
Cbl in 50% aqueous CH_3OH will react with the cyclododecyl iodide but
not with the tosylate (the data suggest a difference in rate of $\geq$ 200)
[82]. The similar solvent-dependence observed for the α/β ratio under
both thermodynamic and kinetic control suggests similarities in the
interaction between the corrin structure and both the incoming CH_3I
and the coordinated CH_3 (perhaps also the solvent and counterion);
cf. the weak $Co\cdots CH_3OCOR$ interaction observed at 4.1 Å in the structure of the Co(II)-cobester dimer [17], already discussed in Sec. 4.4.
Both examples of the differing behavior of iodide and tosylate as
leaving group were explained by assuming that the tosylate reacted
via the usual S_{N2} mechanism (7) and the iodide via the electron
transfer and radical path (8) [81,82] but, in view of the possibility
of "charge transfer centers" interacting over distances up to at
least 10 Å (see Sec. 4.1), it may be more realistic to consider transition states with partial charge transfer (the amount depending on R,
the leaving group, and the solvent) than to discuss mechanisms solely
in terms of discrete, noninteracting species such as those in Eq. (8).

Other classes of alkylating agents which have been used [most
frequently as the CH_3 derivative with Co(I)-Cbl in aqueous solution]
are given in the following list (from table 13.3 of Ref. 10 except
where another reference is given) and arranged according to the nature
of the donor atom: RCl, RBr, RI; Et_3O^+ [83], R_2SO_4, R_3PO_4, R-tosylate,
$CH_3OOC\cdot COOR$ (but not ROH, ROR); $R_2SCH_3^+$ (but not $RSCH_3$); CH_2N_2 (but not
$N(CH_3)_4^+$ or $pyCH_3^+$ [4]). Other reagents which will alkylate the Co
include ethylene oxide, ethyleneimine, and propylenesulfide (but not
cyclopropane) and tetrahydrofuran; they can be considered as analogs
of the otherwise inert ethers ROR, amines RNH_2, and thioethers RSR,
where ring opening provides a thermodynamic driving force and demon-

strate that there is nothing intrinsically wrong with simple N, O, or S as the leaving atom. Although there have been several reports of methyl transfer from CH_3-Cbl to, say, CN^- or thiols, none of these reactions have been confirmed under well-defined conditions where one could exclude the faster reactions observed in the presence of some oxidizing agent (e.g. O_2) and which probably involve Co(II) (see Sec. 4.6.5). No reactions were observed in either the forward direction between Co(I)-Cbl and the $N(CH_3)_4^+$, N-CH_3-pyridinium, and protonated N-CH_3-imidazole ions (or several other N-CH_3 compounds) or in the reverse direction between CH_3-Cbl and $N(CH_3)_3$, pyridine and imidazole [4]; this excludes the possibility that failure to observe a reaction is due to adverse thermodynamics. There appears to be no simple and facile path for methyl transfer between Co and N.

These results highlight both the remarkable reactivity of the Co(I) ion as a nucleophile and our ignorance of the main factors which govern the rate of such group transfers, especially to and from N.

4.6. Alkyl Transfer Involving Co(II)

Reactions which formally involve Co(II) are of interest because they include several examples of mechanisms not observed for noncorrinoids. Excluding transfer between CH_3-Co and Co(II) corrinoids (Sec. 4.4), there appears to be no example of simple transfer to Co(II) and only one other example of simple transfer from CH_3-Co to leave Co(II). We consider in turn the two more complex examples of transfer to Co(II) and then three examples of increasing complexity of transfer from Co(II).

4.6.1. *Co(II)-Assisted Alkylation of Co(II)*

The kinetics of the slow reaction of Co(II)-Cbl with various RX in aqueous solution and in CH_3OH have been studied by Blaser and Halpern [70]. They found that several RI (but no RBr or RCl; see Sec. 4.3.1) in aqueous solution (but not in CH_3OH) exhibited a rate law not observed

with any noncorrinoid Co(II) complexes, namely, first-order in [RI]
but second-order in [Co^{2+}]. The rate was pH-independent from 4 to 7,
but showed an inflection with a pK ~3 corresponding to the pK for
protonation and displacement of Bzm in Co(II), with the protonated
base-off form having a reactivity ≤1% of that of the base-on form.
A previously assumed mechanism involving disproportionation of 2
Co^{2+} ⇌ Co^{3+} + Co^{1+} followed by alkylation of the Co(I) could be
eliminated because of the known, very low value of the disproportionation constant (10^{-14}-10^{-15} M^{-1}) (see Sec. 3.3). They therefore
considered two mechanisms involving an initial rapid and reversible
formation of either (1) a Co(II) dimer or (2) a Co(II)·RI adduct,
followed by the irreversible reaction with (1) RI or (2) the second
Co(II) to give R-Co + H$_2$O-Co(III) + X$^-$. They decided against (1),
commenting that "while such a mechanism cannot be discounted, the
absence of independent evidence in other contexts for formation of
a Co(II) dimer is disturbing."

There is now sound evidence for the role of dimers (irrespective
of the oxidation state of the Co) in the Co-to-Co methyl transfers
(see Sec. 4.4) and anomalous kinetics, which may indicate a common
denominator, had already been reported for the reaction of Co(II)
with O$_2$ [33]. The rate of oxidation of Co(II)-Cbl by O$_2$ in aqueous
solution at 0°C showed a second-order dependence on [Co^{2+}] but above
a partial pressure of 0.4 atm. O$_2$ was independent of [O$_2$]; such a
rate law is unknown for the oxidation of any other metal complexes
by O$_2$. An initial disproportionation can also be excluded here and
the rate law reinterpreted in terms of an initial rate-determining
formation of a Co(II) dimer, followed by rapid reaction with O$_2$.
Postulating the same Co(II) dimer as an intermediate in the reaction
with RI would require that its rate of reaction with RI be sufficiently slow for the initial dimer formation to be treated as a
rapidly reversible preequilibrium; use of the listed [70] value of
k$_3$ = 1 x 10^4 M^{-2} sec^{-1} for the most reactive RI (namely, ICH$_2$COOCH$_3$)
and ~5 x 10^{-4} M RI (deduced from the text) gives a pseudo-second-
order rate constant of k$_2$ ~5 M^{-1} sec^{-1} at 25°C, which would not
conflict with the higher substrate-independent rate of proposed

dimer formation (k_2 = 70 M^{-1} sec^{-1} at 0°C) achieved with O_2. Blaser
and Halpern [70] found that the values of k_3 varied by almost 10^2
from ICH_2COOCH_3 down to CH_3I and $ICH_2CO_2^-$ but offered no comment and
did not discuss the geometry or movement of the nuclei and electrons
in their proposed mechanism. The higher rate observed with the bulky
ICH_2COOCH_3 probably precludes reaction in the constrained space
between the two Co atoms in the dimer (which might be accessible to
CH_3I) and requires reaction at the side (whether α or β) which is
most exposed to the solvent. The reaction could then be written as
in (9), where the dimers are enclosed in square brackets, with the
two Co(II) ions (separated by, say 5-6 Å; see Sec. 4.4) forming a
"charge transfer center" (see Sec. 4.1) polarized to act like
Co(III) + Co(I) under the influence of the attacking reagent.

$$[Co(II)\cdot\cdot Co(II)] + RI \longrightarrow [Co(III)\cdot\cdot Co-R] + I^-$$

$$\longrightarrow Co(III) + Co-R + I^- \qquad (9)$$

4.6.2. Thiol-Assisted Alkylation of Co(II)

Two groups reported in 1963-1964 that CH_3-Cbl can be formed from the
reaction of CH_3I with H_2O-Cbl in the presence of thiols or sulfide
(see Ref. 10). Seven different thiols and a variety of alkylating
agents (CH_3/Et/PrI, $ClCH_2COOH$, CH_3-tosylate) have been used but few
mechanistic details given. Thiols will readily reduce H_2O-Cbl to
the Co(II) but not the Co(I) level. The highest rate was apparently
observed with Co(II) + H_2S, while no significant reaction was observed
with Co(II) prepared in the absence of thiols [84]. Since alkylation
was observed at pH 2.3 where the reducing power of thiols is repressed
($2RSH = RSSR + 2e^- + 2H^+$), it seems unlikely that the reaction involves
the intermediate formation of Co(I). There is, however, ESR and uv-vis
evidence for the formation of Co(II)-thiol or thiolate complexes [85]
and H_2S appears to form some product with Co(II) at room temperature
[84]. It is possible that, just as the combination of Co(III) + CH_3^-
has some characteristics of Co(II) (see Sec. 3.4), so the combination
of Co(II) + RS^-/HS^- might show characteristics of Co(I). A possible

mechanism is therefore reaction (10) in which the Co(II) + RS⁻ could be polarized to act like Co(I) + radical under the influence of the attacking reagent. Further work is obviously needed.

$$[RS^--Co(II)] + CH_3I \longrightarrow RS^\cdot + Co-CH_3 + I^- \text{ (or } RS^- + I^\cdot)$$

$$\text{followed by } 2\ RS^\cdot \longrightarrow RSSR \tag{10}$$

4.6.3. Alkylation of Cr(II) by R-Co

Espenson and Sellers [86] studied the reaction of Cr(II) with CH_3- and Et-Cbl in acid (pH 0-2.3), where the R-Cbl's exist in the protonated base-off form. The reaction proceeds according to (11). The rate was independent of pH, and showed a first-order dependence on the concentration of each reagent with $k_2 = 3.6 \times 10^2$ and $4.4\ M^{-1}\ sec^{-1}$ for CH_3- and Et-Cbl, respectively (i.e., ratio of 82) at 25°C. They suggested the rather uncommon process of S_{H2} displacement at the saturated C, i.e., involving the transfer of a radical from Co(II) to Cr(II), and commented that the alternative mechanism of electron transfer to generate Cr(III), followed by transfer of a carbanion, was "only subtly different." The trans effect has not been studied.

$$Co^{III}R^- + Cr(II) \longrightarrow Co(II) + [(H_2O)_5Cr^{III}R^-]^{2+} \tag{11}$$

4.6.4. Alkylation of Sn(II) by R-Co

Fanchiang and Wood [87] studied the reaction of $SnCl_2$ with CH_3- and Et- (also $ClCH_2$-) Cbl in aqueous HCl pH 0-1 where the R-Cbl's are present in their protonated base-off form. The rate shows a first-order dependence on the concentrations of both $SnCl_2$ and R-Co and proceeds only in the presence of O_2 or H_2O-Cbl as oxidant but (above a minimum concentration not determined) is zero-order in each; Fe^{3+} can also act as the oxidant [88]. The observed rate constant is the sum of a pH-independent k_1 and a proton-dependent k_2, i.e., $k_{obs} = k_1 + k_2[H^+]$, where $k_1 = 0.32\ M^{-1}\ sec^{-1}$ and $k_2 = 0.85\ M^{-2}\ sec^{-1}$. In 1 N HCl at 23°C k_{obs} was $1.04\ M^{-1}\ sec^{-1}$ (same with O_2 or H_2O-Cbl) for CH_3-Cbl and $1.66 \times 10^{-2}\ M^{-1}\ sec^{-1}$ for Et-Cbl (ratio CH_3:Et = 65). They proposed a mechanism involving the initial reversible transfer

of a radical according to (12), followed by oxidation of the organotin complex by O_2 or H_2O-Cbl such that the first step becomes rate
limiting on increasing the concentration of the oxidant. Reaction
(12) is obviously analogous to reaction (11) with Cr(II), except that
it is reversible and the absolute rate constants are lower (by ~400).
The ratio of rates of the CH_3- and Et-Cbls are similar; cf. 65 for
Sn(II) and 82 for Cr(II). The authors were unable to pinpoint the
origin of the proton dependence of k_2, and the trans effect has not
been studied. Wood subsequently suggested that oxidation of Sn(II)
to Sn(III) precedes transfer [89], which would still involve transfer
of the CH_3 radical to leave Co(II); it is, however, difficult to see
how the rate-determining step could then become independent of the
nature and concentration of the oxidant.

$$Co^{III}R^- + Sn^{II} \longrightarrow o^{II} + Co^{II} + R^-Sn^{III} \tag{12}$$

4.6.5. Alkylation of Thiols by R-Co

Methyl transfer to thiols is of particular interest because of its
relevance to the enzyme-catalyzed transfer to thiols such as homocysteine and coenzyme M (see Sec. 5.2). Woods and coworkers reported
in 1962 [90] that protein-free (as well as enzyme-bound) CH_3-Cbl (but
not Et-Cbl) reacted slowly with homocysteine (but not cysteine) to
give methionine at 37°C and pH 7.8 under H_2 in the presence of mer-
captoethanol (ME); the possible role of ME as CH_3 acceptor or source
of RSSR (see below) was not investigated. Very slow reactions of
CH_3-Cbl with other thiols have also been found at pH 14 [91,92]. In
1974 Agnes et al. [93] reported a much more rapid O_2-dependent methyl
transfer from CH_3-Cbl to ME at 25°C and pH 7.8 to give the methyl
thioether (identified by GLC); in the absence of O_2 the reaction proceeded for a short while before it stopped, but in O_2 the rate of
reaction built up to a steady-state after an induction period. A
similar O_2-dependent reaction showing an induction period has also
been found for the reaction between CH_3-Cbl and coenzyme M and shown
to be independent of pH from 7 to 14 [94]; the product was later

identified as the methyl thioether by its ability to act as substrate for the enzymatic formation of methane [95].

Co corrinoids are excellent catalysts for the autoxidation of thiols to disulfides [10]. The induction period could therefore represent the buildup of simple Co(II) and Co(III) Cbl's from the demethylation of CH$_3$-Cbl and the attainment of a steady state then explained by the increasing rate of removal of RS$^{\cdot}$ radicals through dimerization. Reactive species such as RS$^{\cdot}$, the anion radicals RSSR^{-}, and their conjugate acids RSSRH are now well established as intermediates in the reactions of thiols but, unfortunately, little is known about the pK values and other equilibrium constants (e.g., RS$^{\cdot}$ + RS^{-} $\rightleftharpoons$ RSSR^{-}), rate constants, and redox potentials of relevant thiols. The simplest and most likely mechanism for methyl transfer to thiols is therefore probably reaction (13) [94,96], though more complex schemes are conceivable [4,93], and it has been suggested that other reagents such as disulfides could act as the oxidant required to generate RS$^{\cdot}$ in the absence of O$_2$ [94], perhaps through the simple equilibrium (14).

$$RS^{\cdot} + CH_3\text{-}Co \longrightarrow RSCH_3 + Co(II) \tag{13}$$

$$RS^{-} + RSSR \rightleftharpoons RS^{\cdot} + RSSR^{-} \tag{14}$$

A more detailed study has been made of the slow methyl transfer from CH$_3$-Cbl to ME and dithiothreitol (DTT) at 43°C in the absence of O$_2$ but in the presence of apparently significant concentrations of the corresponding disulfides [97]. The rate with ME increased from pH 9.0 to 10.2 but the expected leveling off required to establish a correlation with the pK (9.7) of ME was not reached. The reactions are very slow even with 10^{-2} M CH$_3$-Cbl and 10^{-1} M DTT; $t_{1/2}$ ~1.5 hr at pH 9.7 and 43°C and over 5 days required at room temperature. The products were identified as RSCH$_3$ (NMR) and Co(II) (uv-vis). Similar pseudo-first-order rate constants of 0.54, 0.24, and 0.21 x 10^{-4} sec^{-1} were found for the reactions of CH$_3$-, Et-, and Pr-Cbl, respectively, at pH 9.7 and 43°C [98], i.e., there is no significant steric effect. They proposed [97] a mechanism involving initial attack by RS^{-}

according to (15) to leave Co(I), which was then oxidized by RSSR
to Co(II) according to (16).

$$RS^- + CH_3-Co \longrightarrow RSCH_3 + Co(I) \tag{15}$$

$$1/2 \text{ RSSR} + Co(I) \longrightarrow RS^- + Co(II) \tag{16}$$

$$1/2 \text{ RSSR} + CH_3-Co \longrightarrow RSCH_3 + Co(II) \tag{17}$$

Several comments can be made. First, the simple addition of reactions (15) and (16) gives (17), which represents the net overall
reaction. This strongly supports the earlier suggestion [94] that
disulfides can provide the oxidant needed to drive the reaction,
but leaves unanswered whether the mechanism involves direct attack
by RS^- according to (15) or reaction according to (14) to generate
a low concentration of the reactive $RS^{\cdot}$, which then attacks the
CH_3-Co according to (13). Second, since CH_3-Cbl is now known to
dimerize with $K_D = 10^2$-10^4 M^{-1} (see Sec. 2), it may be dangerous to
try to correlate reactions of 10^{-2} M CH_3-Co in the presence of RSSR
with reactions of 10^{-4} M CH_3-Co in the presence of O_2. The lack of
any steric effect recalls the reaction of RI with the dimeric Co(II)-
Cbl (see Sec. 4.6.1). One can conclude (i) that the only facile path
available for methyl transfer from a CH_3-corrinoid to a thiol so far
discovered involves some reactive intermediate which can provisionally
be considered to be the thiyl radical $RS^{\cdot}$ reacting according to (13),
and (ii) that no example of transfer to the anion RS^- according to
(15) has yet been conclusively established.

The dealkylation of CH_3- and Et-corrinoids by CN^- shows parallels
(need for O_2, induction period) but the reaction has not been studied
in detail [10].

4.7. Alkyl Transfer Involving Co(III)

Only two examples of the transfer of a carbanion (e.g., CH_3^-) *to* a
Co(III) corrinoid, in addition to that from CH_3-Cbl to H_2O-Cbi (see
Sec. 4.4), have been reported, i.e., from $LiCH_3$ and CH_3MgI to dicyano-

cobester in Et_2O-THF by Wagner and Bernhauer in 1964 [99] and from a dimethyl-Co(III) complex to H_2O-Cbl in CH_3OH or H_2O-THF by Costa et al. in 1971 [100]; these reactions have not been further studied or exploited. Although the slow reaction of I_2 and ICl with CH_3/Et/ Ado-Cbl to give RI, reported by Bernhauer and Irion in 1964 [101], may involve the transfer of a carbanion (but see Ref. 3), the first clear-cut case of the transfer of a carbanion *from* Co(III) to an electrophile was the reaction with Hg(II) reported by Hill et al. in 1970 [102]. Reactions with many metal ions and complexes have now been studied and we focus here on the two most thoroughly studied, namely, (1) with Hg(II) as the type reaction and (2) with Pt(II) + Pt(IV) complexes combined, as a more complex example where the mechanism is still controversial.

4.7.1. *Alkylation of Hg(II)* [102-111]

CH_3-Cbl reacts with many Hg(II) complexes to form H_2O-Cbl and CH_3Hg^+ complexes; the uv-vis spectra give good isosbestic points and the rate shows a first-order dependence on the concentrations of both CH_3-Cbl and Hg(II) [102,106-110], unaffected by the presence or absence of O_2 [105]. The reaction can therefore be written as in (18), involving an S_{E2} substitution or carbanion transfer. No significant demethylation occurs with the Hg_2^{2+} ion [106] and no reaction has been reported with metallic Hg. Confusion has been caused by the report in 1968 that CH_3- and Pr-Cbl can be dealkylated by Hg^{2+} in the presence of zinc dust, i.e., with Zn amalgam [59]; this reaction does not exhibit the large steric effect ($CH_3 >>$ Et,Pr, see below) found with Hg(II) and probably involves initial reduction by the Zn (see Sec. 4.2).

$$Co-CH_3 + Hg(II) \longrightarrow Co(III)-OH_2 + CH_3Hg^+ \qquad (18)$$

The observed pattern of equilibria and reactions may be complex because the Hg(II) can (1) remove the CH_3^- (rates given here as k_2/M^{-1} sec^{-1} at or near 25°C), (2) coordinate the N atom of Bzm and convert the base-on to the more inert base-off form, and (3) coordinate added

ligands to form a mixture of complexes (e.g., from $HgCl^+$ to $HgCl_4^{2-}$)
though the data apparently refer to the HgX_4^{2-} complex at least where
$X = Cl^-$, Br^- [111], and CH_3COO^- [109]. As the binding constants
toward the added ligand fall (in the order $CN^- > SCN^- > Br^- > Cl^- >$
CH_3COO^- [107,111]), so the reactivity toward base-on CH_3-Cbl rises,
as shown by the following values of k_2 (calculated from their table 1
in the case of Ref. 107) in the presence of: CN^- ($<10^{-3}$ [107]) < SCN^-
(0.54 [107]) ~ Br^- (0.44 [111], 0.35 [107]) < Cl^- (4.0 [110], 3.6
[111], 2.8 [107]) < CH_3COO^- (350-380 [102,109,111], 300 [104], $\geq 10^3$
[107]) but H_2O (~10 [109]). No coordination of Bzm was observed
[107,110,111] except with the CH_3COO^- complex (K_N = 70-72 [106,110])
and the aquo complex (log K_N = 4.8 [109]). The value of k_2 for the
simple aquo complex appears anomalous, but the reason is not known
(? different charge and/or stereochemistry, coordination to amide
side chains). $(CH_3)_2Hg$ may also be produced [103,108] from the much
slower reaction of CH_3Hg^+ with CH_3-Cbl: cf. k_2 ~3.7 x 10^2 and 6.9 x
10^{-2} for Hg(II) and CH_3Hg^+, respectively, with acetate [111] and ~3.6
and 0.25, respectively, as the mixed aquo and hydroxo complexes at
pH 3 [108]. Demethylation also occurs with other organomercury ions
such as $PhHg^+$ [105]. Values of k_2 = 0.12 [102] and 0.06 [104]
reported for CH_3-Cbi and 0.11 [110] for base-off CH_3-Cbl at pH 1
indicate a trans labilizing effect of Bzm (and six coordination) of
~3000. Transalkylation by Hg(II) also shows a significant steric
effect ($CH_3 \gg$ Et, Pr, iPr); cf. $k_2 < 10^{-5}$ [102] and 6 x 10^{-5} [104]
for Et-Cbi, indicating CH_3/Et $\geq 10^3$ in the Cbi's. The differences
in k_2 reported by the two groups are, however, greater than might
be expected and may reflect some involvement of trace metals in the
slower reactions.

4.7.2. Alkylation of Pt [26,111-119]

The reaction of CH_3-Cbl with a mixture of $PtCl_4^{2-}$ and $PtCl_6^{2-}$ to give
CH_3Cl via some Pt-CH_3 intermediate was first reported by Williams
and coworkers in 1971 [112]. The main contributions have been made
by Taylor and Wood and their coworkers. This very interesting trans-

alkylation, unique to the corrinoids, was reviewed by Fanchiang in 1985 [119] and deserves to be analyzed again in detail.

CH_3-Cbl is rapidly demethylated by a mixture of $PtCl_4^{2-}$ and $PtCl_6^{2-}$; the immediate, but not fully characterized, product is probably $CH_3PtCl_5^{2-}$ [114,115,117], which then gives a better characterized $CH_3PtCl_3^{2-}$ [114]. A far slower reaction with Pt(IV) alone gives CH_3Cl without any detectable Pt-CH_3 intermediate (i.e., may involve a different path), but no reaction could be observed with Pt(II) alone [113,116,118]. $PtCl_4^{2-}$ can also be consumed stoichiometrically (product not identified) if free radicals are liberated from CH_3-Cbl by photolysis [113]. The thermal reaction is stoichiometric in added Pt(IV) and the Pt(II) appears to act as a catalyst [113], but the Pt in the Pt-CH_3 product was shown (NMR evidence using ^{13}C and ^{195}Pt) to originate from the added Pt(II) [117]. The uv-vis spectrum indicates conversion of CH_3-Cbl to H_2O-Cbl with good isosbestic points [113] and no effect of O_2 on rate or products [116]. This particular overall reaction is therefore described by (19), but $PtCl_6^{2-}$ can be replaced by many other six-coordinate Pt(IV) complexes and $PtCl_4^{2-}$ by several other square planar Pt(II) complexes [111,113,116-118]. Equation (19) serves to highlight the movement of the Pt atoms; deleting $PtCl_4^{2-}$ from both sides of the equation demonstrates that the driving force of the reaction is a ligand substitution reaction of the Pt(IV) complex.

$$Co-CH_3 + {}^*PtCl_4^{2-} + PtCl_6^{2-} \longrightarrow Co-OH_2 + CH_3^*PtCl_5^{2-} + PtCl_4^{2-} + Cl^-$$

(19)

$$Co-CH_3 + Pt^{2+} + Pt^{4+} \rightleftharpoons [Co-CH_3 \cdot Pt^{2+}] + Pt^{4+} \rightleftharpoons$$

$$[Co-CH_3 \cdot Pt^{2+} \cdot Pt^{4+}] \longrightarrow products$$

(20)

The rate shows a first-order dependence on the concentrations of all three reagents at low concentrations; as concentrations are increased, the rate becomes zero order in $[Pt^{2+}]$ for all Pt(IV) complexes tested but zero order in $[Pt^{4+}]$ for only some Pt(II) complexes. Wood and coworkers [118] therefore proposed the sequence shown in (20), where CH_3-Cbl reacts reversibly first with Pt(II) to form the binuclear adduct A [with kinetically determined binding constants K_A = 2-6 x 10^3 M^{-1} for both $PtCl_4^{2-}$ and $Pt(CN)_4^{2-}$] and then with Pt(IV) to form the trinuclear adduct B (with kinetically determined K_B = 1 x 10^3 M^{-1} for $Pt(CN)_5Cl^{2-}$ and 8 x 10^3 M^{-1} for $Pt(CN)_4Cl_2^{2-}$ [116,118]) before the rate-determining step. The slight changes in the uv-vis spectra allowed the direct determination of K_A = 3 x 10^3 and 4 x 10^3 M^{-1} for $PtCl_4^-$ and $Pt(CN)_4^{2-}$, respectively, and K_B = 1 x 10^3 M^{-1} for $Pt(CN)_5Cl^{2-}$ in the presence of $Pt(CN)_4^{2-}$ [26,118], in good agreement with the kinetically determined values. The scheme of Eq. (20) is therefore self-consistent, but the structures of A and B have not been established.

Wood and coworkers first [116] suggested that the Pt(II) complex in A was held by coordination of the amide side chains but later [118] proposed a structure in which the Pt(II) becomes five-coordinate by forming a π complex with some part of the corrin ring "similar to the well-known synergic donor-acceptor bond between transition metal complexes, including Pt(II), and olefins and poly-enes"; and for B they proposed that the incoming Pt(IV) uses one ligand to form a bridge to the Pt(II), loses a second ligand to the solvent, and undergoes a two-electron exchange with the Pt(II) to give a structure (see scheme 1 of Ref. 118, axial ligands only given here) which can be written as [corrin·Pt^{4+}·X·Pt^{2+}]. Taylor and co-workers had earlier proposed a bridged [Pt^{2+}·X·Pt^{4+}·X] complex, held near the Co-CH$_3$ by some corrin N···Cl interaction [115]. Several comments can be made.

Consideration of the likely steric interactions with sub-stituents on the corrin ring (see figures in Ref. 17 and comments in Sec. 2) indicates that the Pt ion could not approach close enough to coordinate to any part of the conjugated ring or to allow a coor-dinated Cl to interact with a corrin N; it also suggests that the disposition of the valleys between the upward-projecting "sentinel" groups could provide a "receptor" site for a planar four-lobed unit such as $Pt(CN)_4^{2-}$ with the Pt situated above the Co-CH$_3$ group. It is therefore interesting to compare the coordination of a ligand such as [Co(CN)$_6$]$^{3-}$ since the Co-C-N-Co and Co-CH$_3$··Pt distances would be comparable and place the four lateral cyanides at similar heights above the corrin ring. The equilibrium constants for the coordination of [Co(CN)$_6$]$^{3-}$ by H$_2$O-Cbi (log K$_1$ > 6.3, log K$_2$ = 5.2 [40]) are unexpectedly high and could possibly reflect some (? entropically) favorable interaction between the corrin structure and a ligand of this stereochemistry. There is, in fact, no need to postulate any π bonding in A and B and alternative modes of interaction can be suggested. Furthermore, the liberation of X$^-$ required in the reversible formation of B was not confirmed by equilibrium studies and the required suppression of the rate by added X$^-$ conflicts with the finding that "reaction rates (with PtCl$_4^{2-}$ + PtCl$_6^{2-}$) in NaClO$_4$-HClO$_4$ solution were much slower than those with chloride" [116]. One cannot, however, exclude the possibility that the final rate-determining step is not associated with electron transfer or ligand substitution but with some conformation change of the corrin structure (e.g., to bring the Pt(IV) complex into the position required for the next step).

A comparison of the rates of dealkylation by PtCl$_4^{2-}$ + PtCl$_6^{2-}$ at pH 2 gave the following relative rates: CH$_3$-Cbl 1.0, Et-Cbl 0.012, Pr-Cbl 0.006, CH$_3$-Cbi 0.02, Pr-Cbi 0.001 [113]. A more detailed comparison in the light of Eq. (20) showed that K$_A$ and the composite kK$_B$ were 15 and 2 times greater, respectively, for CH$_3$-Cbl than for CH$_3$-Cbi with one pair of Pt complexes [118] and 10 and 120 times greater for the base-on than the base-off forms of CH$_3$-Cbl with another pair [119]. These reactions therefore show steric and trans effects of the same direction (CH$_3$ > Et and Cbl > Cbi) as with Hg(II) but of smaller magnitude.

Possible mechanisms can be discussed in the light of the above. In his 1985 review [119] Fanchiang pointed out that earlier arguments against the simple transfer of a carbanion [as with Hg(II)], which were based on the relatively small trans effect observed for Bzm [118], were invalidated by subsequent findings (see above). He stated that "there is no concrete evidence to rule out or support any of the 3 mechanisms," i.e., (1) transfer of CH_3^- to Pt(IV) to leave Co(III), (2) interaction of Pt(II) + Pt(IV) to give Pt(III) + Pt(III), followed by transfer of a methyl radical to one Pt(III) and oxidation of the Co(II) by the second Pt(III), and (3) an initial oxidation of CH_3-Cbl by Pt(IV) to give CH_3-Co(IV) + Pt(III), followed by transfer of a CH_3 radical to the Pt(III) to leave Co(III). Two comments should be made. First, as already pointed out [115,118], Pt(II) complexes are known to catalyze ligand substitution reactions of the very inert Pt(IV) complexes [120,121] via a bridged inter-mediate as shown in (21), where only the axial ligands are given. This provides presumptive evidence for mechanism (1), i.e., the observed complexities of the demethylation of CH_3-Cbl reflect the mechanism required to "activate" the Pt(IV) complex for ligand sub-stitution involving the transfer of a carbanion ligand, as proposed by Taylor and coworkers [115]. Second, one would expect the rates of any reaction involving the redox mechanisms (2) or (3) to show some correlation with the relevant $Pt^{2+}/^{4+}$ redox couples but these are, unfortunately, not known. It is however, striking that all the complexes PtX_4 where X may be Cl^-, CN^-, SCN^-, or even NH_3 (in the presence of $PtCl_6^{2-}$ and CH_3-Cbl) show "comparable reaction rates" [116]. It is difficult to see how such a range of ligands could fail to cause a sufficient change in potential and hence in rate *if* mechanism (2) or (3) was operating.

$$X-Pt^{4+}-Y + Pt^{2+} + Z \rightleftharpoons X-Pt-Y-Pt-Z \longrightarrow Pt^{2+} + Y-Pt^{4+}-Z + X$$

$$(21)$$

Until there is evidence to the contrary, the reactions of alkyl corrinoids with Pt(II) and Pt(IV) should be considered an unusual and interesting variant of the carbanion transfer reaction, caused by the kinetic inertness of Pt complexes. It appears that the initially formed CH_3-Pt(IV) complexes may be converted into CH_3-Pt(II) complexes, presumably by reactions analogous to (21). $PtCl_4^{2-}$ will react, however, directly with the $(CH_3)_3Pb^+$ ion to give $CH_3PtCl_3^{2-}$ [122] and CH_3-Cbl will react with the more labile $PdCl_4^{2-}$ to give CH_3Cl and palladium metal via a presumed $CH_3PdCl_3^{2-}$ [110,111,123].

4.8. Summary

The above survey illustrates, without pretending to be comprehensive, how much variety (and controversy!) can be found even within the limited field of corrinoids with simple alkyl ligands. Comparison with the noncorrinoids [3] emphasizes the following points:

1. There are six *basic pathways* common to both corrinoids and noncorrinoids. These involve: initial one-electron oxidation or reduction of the R$^-$Co(III) unit; unassisted homolytic fission of the Co-C bond; and the transfer of formal R$^-$, R$^·$ and R$^+$ between another donor/acceptor and Co(III), Co(II), and Co(I) respectively. Examples of reactions in either direction are known for all but the first two. The corrin ring is less susceptible to one-electron oxidation/ reduction or to irreversible covalent modification than many other conjugated equatorial ligands; this simplifies the pattern observed for corrinoids. A major gap in our knowledge for both corrinoids and noncorrinoids is whether or not there exists any path of the type RX + Co-H $\longrightarrow$ X$^-$ + R-Co + H$^+$.

2. Of the *ancillary pathways* observed for noncorrinoids some are shared by corrinoids (e.g., further reaction of the caged Co(II)$\cdot\cdot$R$^·$ to give Co-H + olefin), others are not (e.g., insertion of O$_2$ into Co-R to give Co-O-O-R [3], reversible transfer of R from the Co to the equatorial N atom in the porphyrins on oxidation [124]), probably mainly for steric reasons (see Sec. 2).

3. The corrinoids show a large number of *variants* on these basic pathways, connected with their ability to form dimers (see Secs. 4.4 and 4.6.1) and nonbonded "adducts" (i.e., not involving coordination to Bzm) with, e.g., IrCl$_6^{2-}$ and IrCl$_6^{3-}$ (Sec. 4.2), Pt(II) and Pt(IV) complexes (Sec. 4.7.2), also Fe(III) [125], AuCl$_4^-$ and AuBr$_4^-$ [126]. The reactions of CH$_3$-Cbl with Pt(II) and Pt(IV) and the unusual kinetics

of Co(II)-Cbl with RI and O_2 (Sec. 4.6.1) are unique to the
corrinoids. As shown in the case of Pt, it is often difficult to establish the mechanism of transalkylation reactions,
especially where some adduct is formed in a preequilibrium.
For other less thoroughly studied transalkylations see [3,
111,122].

4. One fascinating aspect to emerge from work on the corrinoids
    is the apparently *long range of interaction*; cf. the distance
    of 4.1 Å actually observed between the Co ion and the methyl
    group (of an ester side chain) located in the vacant coordi-
    nation site and 5.6 Å as the probable overall Co··CH_3··Co dis-
    tance in the transition state for CH_3 transfer between two
    corrinoids (see Sec. 4.4). The fact that electronic inter-
    action can occur over greater distances (see ESR evidence in
    Sec. 4.1) means that it may be more realistic to discuss mech-
    anisms involving partial charge transfer in the transition
    state (e.g., to explain differences between RI and R-tosylate,
    see Sec. 4.5, or the unusual kinetics observed for Co(II) +
    RI or O_2, see Sec. 4.6.1) and not just complete electron
    transfer and discrete noninteracting species such as CH_3^--
    Co(IV) complexes, and even to abandon the distinction between
    transfer of CH_3^-, $CH_3^·$, or CH_3^+ (except for the purposes of
    classification) in favor of viewing the reaction as compris-
    ing the common transfer of a fairly neutral CH_3 group *and* a
    varying degree and direction of electron transfer directly
    between donor and acceptor (depending on the relative redox
    potentials).

5. The corrinoids offer a good "database" of structures, redox
    potentials, equilibrium and rate constants as a framework
    against which to assess the properties and reactions of the
    alkyl corrinoids; one serious gap is the lack of X-ray struc-
    tural data on a yellow five-coordinate alkyl corrinoid.
    There is also now sufficient evidence on steric effects (cf.
    CH_3/Et) and trans effects (cf. Cbi's/Cbl's, though this may

reflect the changes in coordination number rather than in
the trans ligand) and changes in the donor/acceptor to
indicate the emerging patterns, though further studies
would be welcome on, for example, the trans effect (without
changing the coordination number), the effect of solvent
(nucleophilic substitution reactions are generally very
solvent-dependent [127]), and particularly the cis effect
(cf. the effect of C10 substitution on sulfite transfer
from the Co(III)-SO$_3^{2-}$ complex to β-naphthylamine to give
the sulfamic acid [10,128]).

5. B$_{12}$-DEPENDENT ENZYMATIC REACTIONS

The currently known B$_{12}$-dependent reactions all fall into two main
groups: methyl transfer between two substrates, involving the alter-
nate formation of Co(I) and CH$_3$-Co intermediates, and unusual isomerizations of a single substrate (coupled with their further reduction in the case of the ribonucleotides), involving reversible homolytic fission of the Co-C (Ado) bond in the coenzyme to give Co(II)
and the Ado radical, which then attacks the substrate. We briefly
summarize those aspects directly related to making and breaking the
Co-C bond. Various proteins are also needed to absorb and transport
the Co corrinoids [1,7,129-132] and to reduce them and form the Ado-Co
link [132-134]. Et-corrinoids are apparently also found in nature but
little is known about them [135].

5.1. Isomerization and Ribonucleotide Reduction

The isomerases or mutases [1,6] catalyze the isomerization of various
substrates (10 currently known), exemplified by the reversible C-
skeleton rearrangement shown in Fig. 4a (a type of reaction previously
unknown in organic chemistry) and the irreversible rearrangement of
diols to aldehydes in Fig. 4b; they all involve the interchange of H

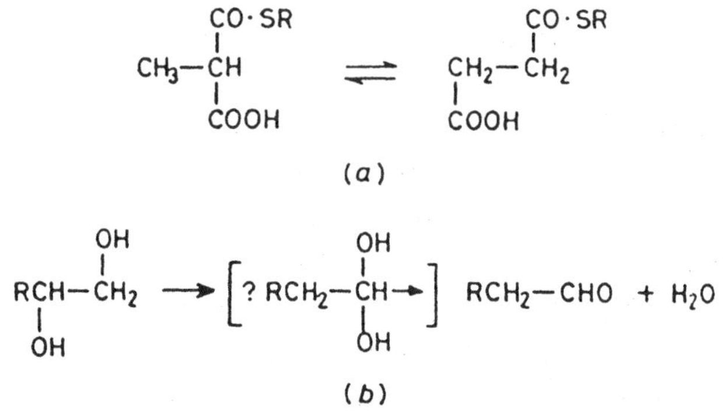

$$
\begin{array}{ccc}
\begin{array}{c} CO \cdot SR \\ | \\ CH_3-CH \\ | \\ COOH \end{array}
& \rightleftharpoons &
\begin{array}{c} CO \cdot SR \\ | \\ CH_2-CH_2 \\ | \\ COOH \end{array}
\end{array}
$$

$$(a)$$

$$
\begin{array}{c} OH \\ | \\ RCH-CH_2 \\ | \\ OH \end{array}
\longrightarrow
\left[? \begin{array}{c} OH \\ | \\ RCH_2-CH \rightarrow \\ | \\ OH \end{array} \right]
\quad RCH_2-CHO \ + \ H_2O
$$

$$(b)$$

FIG. 4. Examples of B_{12}-dependent enzymatic rearrangements
(a) between methylmalonyl and succinyl derivatives (RSH =
coenzyme A) and (b) of diols to aldehydes (R = H, CH_3, CH_2OH).

with C, N, or O between neighboring C atoms. Two such enzymes
occur in humans: methylmalonyl-CoA mutase and leucine-2,3-amino-
mutase. The ribonucleotide reductases [1,136] catalyze reduction
of the -CHOH- of ribose to the -CH_2- of 2-deoxyribose but, by analogy
with the diol/aldehyde rearrangement, this probably involves the
initial rearrangement of a diol to a carbonyl followed by reduction
of the carbonyl [47]. All these enzymes require an Ado-Co corrinoid
and it is generally agreed that the first step involves reversible
homolytic fission of the Co-C bond to form Co(II) and the Ado radical
which then abstracts an H atom from the substrate; the substrate
radical undergoes rearrangement, followed by return of the H atom
and reformation of the Co-C bond. One can therefore focus attention
on the simple equilibrium (22) and its manipulation by the protein.

$$
\text{Ado-Co} \rightleftharpoons \text{Ado}^{\cdot} + \text{Co(II)} \tag{22}
$$

It was suggested in 1975 that the protein uses steric distortion of
the Co coordination sphere (most probably of the Co-C-C bond angle),
caused by a substrate-induced conformation change, to displace equilibrium (22) to the right [137]. We now know that the rate of Co-C
bond fission observed in protein-free Ado-Cbl must be increased by a
massive factor of $\geq 10^{11}$ [138,139] to reach that observed in the

enzymatic reaction, that this is achieved in two stages of ~10^6 each
when the Ado-Cbl is bound by the apoenzyme and when the substrate is
bound, and that this represents a displacement of the equilibrium
constant for reaction (22) [138]. The general concept of steric
labilization of the Co-C bond is fully supported by the data in
Sec. 4.3, but details of the magnitude and direction of distortion
must obviously await crystallographic data on B_{12}-dependent isomerases.

5.2. Methyl Transfer

The methyltransferases involve transfer of CH_3 groups from a N or O
atom to Co(I) to form Co-CH_3 for onward transmission to a S atom.
The source of the CH_3 is usually folic acid, which catalyzes the
de novo synthesis of CH_3 groups by reduction from the HCOOH level
down to N^5-CH_3-tetrahydrofolic acid which can be considered as an
amine R_2NCH_3; CH_3 transfer between N and Co is reversible [140].
Some bacteria can also use CH_3OH as the methyl donor [141,142],
though whether directly or via some intermediate is not known. The
CH_3 groups can then be directed along three paths as follows: (1)
Methylation of the thiol homocysteine ($RSCH_2CH_2CHNH_2COOH$, where R = H)
to give the essential amino acid methionine (R = CH_3); this enzyme
occurs in most or all animals (including humans), many bacteria, but
no plants [140,143]. The enzyme catalyzing the much slower direct
transfer from tetrahydrofolate to homocysteine was isolated in 1970
[144] but no further results reported since then. (2) Methylation
of the thiol coenzyme M ($HSCH_2CH_2SO_3^-$), followed by reduction of the
CH_3-SR group to methane by a Ni-containing enzyme (see Chap. 9).
This enzyme has been found only in the methanogenic Archaebacteria.
(3) Transfer of the CH_3 group to the Ni/Fe-containing enzyme called
carbon monoxide dehydrogenase (CODH) which, in the presence of CO and
the thiol coenzyme A (here written RSH), catalyzes the formation of
CH_3COSR and some CH_3COOH, i.e., inserts CO between CH_3^+ transferred
from the Co and the nucleophiles RS^- and HO^-. There is more than
one methyl receptor on CODH which may include the thiol cysteine

and/or a metal ion [145]. CODH has been found only in the primitive acetogenic and sulfate-reducing Eubacteria and the methanogenic Archaebacteria.

6. SUMMARY: B_{12} IN HISTORICAL PERSPECTIVE

Much of the driving force for development of the organometallic chemistry of B_{12} has been our response to the shocks given to our preconceived ideas (a transition metal-alkyl bond stable to air and water!) and to insistant questions as to how (e.g., how does the protein labilize the Co-C bond) and why (e.g., why the rare metal Co?). The unexpected discovery of a thermally (but not photochemically) stable Co-C bond in Ado-Cbl in 1961 [2] triggered the development of B_{12} coordination and organometallic chemistry across a broad front in the 1960s (see Ref. 10), but further jolts around 1970 channeled much of this activity in two directions. The detection of organic radicals as intermediates in the isomerase reaction (suggesting homolytic rather than the previously assumed heterolytic fission of the Co-C bond), coupled with discovery of the far greater lability of Co-C bonds to secondary alkyl ligands, led to the idea of steric labilization of the Ado-Co bond by the protein and to experimental studies of such steric effects (see Sec. 4.3 and Refs. 137 and 138), while discovery of the methylation of mercury by CH_3-Cbl in vivo and in vitro prompted wider studies on methyl transfer reactions in general (see Secs. 4.2, 4.5-4.7, and Refs. 89 and 122).

The first strand has provided a comprehensive picture of unassisted Co-C fission and the associated steric factors (Sec. 4.3) and the data on protein-free alkyl corrinoids against which to compare the protein-bound Ado-Cbl and hence identify the role of the protein. The B_{12}-dependent isomerases are the first family of redox-active metalloenzymes where we can claim to understand what the protein does and, at least in outline, how it does it [138]; they provide an example where the protein serves to "activate" the coenzyme (not the substrate) and to change an equilibrium (not a rate) constant.

These steric effects, like the electronic trans effects (see Sec. 3.4), provide a very satisfying picture linking together effects at all three levels (kinetic; thermodynamic; structural and ground state). The second strand has given us a broader picture comprising both the basic pathways shared with noncorrinoids and the additional complexities expected from B_{12} as the "poor man's protein." It exposes our ignorance: the protein can apparently use simple donors such as CH_3OH to methylate Co(I); we cannot. Does the protein "activate" the coenzyme and/or the substrate? One can expect an increasing shift of interest from the isomerases and steric factors (now a "mature" subject) toward methyl transfer and other (e.g., electronic, solvent) factors.

$$M-R + H_2O \rightleftharpoons M-OH + RH \quad (R = H \text{ or alkyl}) \tag{23}$$

The fundamental question remains the nature of the Co-C bond and "why Co?". Until recently it was assumed that σM-H/C bonds (from transition metals to H or alkyl ligands) were thermodynamically unstable toward reaction (23) and that any observed stability toward water was due merely to kinetic inertness. In 1982 [47] it was suggested that Co-H and some Co-C bonds are thermodynamically stable toward hydrolysis; this was demonstrated experimentally (with actual or minimum equilibrium constants determined) first for the Co-H bond in $[Co(CN)_5H]^{3-}$ [146] and then for corrinoids with H [49] and the substituted alkyls $-CH(CN)_2$ and $-CH_2NO_2$ [147]. Treating the covalent M-C bond as formed from Co(II) and a C-centered radical, it was also suggested [47] that nature had selected Co(II) because of the combination of relatively low 3d-4s/4p promotion energy (required for significant overlap with the C orbital) and accessibility of both the radical Co(II) and nucleophilic Co(I) states (required for enzymatic activity) with biologically available ligands (contrast Cr). The importance of the 4s and 4p orbitals has been amply confirmed by recent experimental and theoretical work on the electronic structures and bond dissociation energies (BDEs) of the binary M-H molecules and $M-H^+$ ions in the gas phase (see summary in Ref. 148). This demonstrates the very high s character in such M-H bonds (up to ~75%) and the remarkably good inverse correlation between BDEs and the d-s promotion energy; one can assume similar electronic effects for M-C bonds.

In developing the Co corrinoids nature has therefore exploited the remarkable combination of what is probably the most stable M-C bond available in a biological system (Co-C), one of the most active nucleophiles known (Co(I)), and a highly reactive free (though protein-bound) radical. There is, however, nothing unique about B_{12}. We now have one example each of an isomerase and an anaerobic ribonucleotide reductase [149], where the Ado-corrinoids can be replaced as cofactor by S-adenosylmethionine (SAM) together with a transition metal; the common denominator is presumably reversible formation of the Ado radical according to (24), where SAM is written as $AdoSR_2^+$. Two other classes of ribonucleotide reductase are also known where the cofactors are Fe (as in humans) or Mn together with a tyrosine radical, but these are confined to aerobic organisms [149].

$$AdoSR_2^+ + M^{n+} \rightleftharpoons Ado^\bullet + R_2S + M^{(n+1)+} \tag{24}$$

The replacement of RNA by DNA was probably the key step at the base of the whole evolutionary tree as we now know it (except for some viruses). Since SAM and B_{12} both appear to have evolved before DNA [150] and since the only currently known enzymes for the reduction of ribonucleotides to form the building blocks for DNA under anaerobic conditions use SAM or B_{12} as cofactors, it would appear that the key chemical innovation required to consolidate the DNA revolution was the controlled application of free radical chemistry (in ribonucleotide reduction) and it is quite conceivable that metal-alkyl derivatives of B_{12} played an active role in one of the most momentous events in evolution around 4×10^9 years ago.

ABBREVIATIONS AND SYMBOLS

Ado   deoxyadenosyl radical or ligand (see Fig. 3)
BDE   bond dissociation energy or enthalpy
Bu   butyl

Bzm	5,6-dimethylbenziminazole (present in the nucleotide side chain of B$_{12}$; see Fig. 1)
cBu	cyclobutyl
cHx	cyclohexyl
cPe	cyclopentyl
cPr	cyclopropyl
Cbi	cobinamides (see Sec. 2)
Cbl	cobalamins (see Sec. 2)
CODH	carbon monoxide dehydrogenase
DH	dimethylglyoximate anion
DMF	dimethylformamide
DTT	dithiothreitol
ESR	electron spin resonance
Et	ethyl
EXAFS	extended X-ray absorption fine-structure spectroscopy
GLC	gas-liquid chromatography
h.f.	homolytic fission
iBu	isobutyl
iPr	isopropyl
k	rate constant
K	equilibrium constant
ME	mercaptoethanol
NMR	nuclear magnetic resonance spectroscopy
np	neopentyl
nPr	*n*-propyl
Ph	phenyl
Pr	propyl
Py	pyridine
R	alkyl radical, hence R-Cbl, etc., with R as the ligand on the Co
saloph	*N,N'*-disalicylidene-*o*-phenylenediamine dianion (ligand)
SAM	*S*-adenosylmethionine
SCE	saturated calomel electrode
THF	tetrahydrofuran
uv	ultraviolet

vis   visible

X   Cl, Br, and I, hence RX, X⁻, etc.

X, Y, Z   used to denote ligands in general in equilibria, etc.

REFERENCES

1. Z. Schneider and A. Stroinski, *Comprehensive B_{12}*, Walter de Gruyter, Berlin, 1987.

2. P. G. Lenhert and D. C. Hodgkin, *Nature, 192,* 937 (1961); P. G. Lenhert, *Proc. Soc., A303,* 45 (1968).

3. P. J. Toscano and L. G. Marzilli, *Progr. Inorg. Chem., 31,* 105 (1984).

4. J. M. Pratt, J. Aron, S. M. Chemaly, and H. M. Marques, in Ref. 5, p. 46.

5. P. J. Craig and F. G. Glockling (eds.), *The Biological Alkylation of Heavy Elements,* Royal Society of Chemistry, London, Special Publication No. 66, 1988.

6. D. Dolphin (ed.), *B_{12},* 2 Vols., John Wiley and Sons, New York, 1982.

7. L. Ellenbogen and B. A. Cooper, *Food Sci. Technol., 40,* 491 (1991).

8. D. Lexa and J. Savéant, *Acc. Chem. Res., 16,* 235 (1983).

9. K. A. Rubinson, H. V. Parekh, E. Itabashi, and H. B. Mark, *Inorg. Chem., 22,* 458 (1983).

10. J. M. Pratt, *Inorganic Chemistry of Vitamin B_{12}*, Academic Press, London, 1972.

11. Nomenclature of Corrinoids, *Pure Appl. Chem., 48,* 497 (1976).

12. J. J. G. Moura, I. Moura, M. Bruschi, J. Le Gall, and A. V. Xavier, *Biochem. Biophys. Res. Commun., 92,* 962 (1980).

13. E. C. Hatchikian, *Biochem. Biophys. Res. Commun., 103,* 521 (1981).

14. A. R. Battersby and Z. Sheng, *Chem. Commun., 1982,* 1393.

15. J. Hemker, L. Kleinschmidt, and H. Witzel, *Rec. Trav. Chim., 106,* 350 (1987).

16. T. Nagasawa, K. Takeuchi, and H. Yamada, *Biochem. Biophys. Res. Commun., 155,* 1008 (1988) and *Eur. J. Biochem., 196,* 581 (1991).

17. J. P. Glusker, in Ref. 6, p. 23.

18. V. B. Pett, M. N. Liebman, P. Murray-Rust, K. Prasad, and J. P. Glusker, *J. Am. Chem. Soc., 109,* 3207 (1987).

19. A. J. Markwell, J. M. Pratt, M. S. Shaikjee, and J. G. Toerien, *J. Chem. Soc., Dalton Trans., 1987,* 1349.

20. N. W. Alcock, R. M. Dixon, and B. T. Golding, *Chem. Commun., 1985,* 603.

21. E. R. Andrews, J. M. Pratt, and K. L. Brown, *FEBS Lett., 281,* 90 (1991).

22. S. M. Chemaly and J. M. Pratt, *J. Chem. Soc. Dalton Trans., 1980,* 2267.

23. P. K. Mishra, R. K. Gupta, P. C. Goswami, P. N. Venkatasubramanian, and A. Nath, *Biochim. Biophys. Acta, 668,* 406 (1981).

24. K. L. Brown and S. Peck-Siler, *Inorg. Chem., 27,* 3548 (1988).

25. Y. Fanchiang, *Chem. Commun., 1982,* 1369.

26. Y. Fanchiang and J. J. Pignatello, *Inorg. Chim. Acta, 91,* 147 (1984).

27. D. Cavallini, R. Scandurra, E. Barboni, and M. Marcucci, Atti Accad. Naz. Lincei, *Rend. Sci. fis. mat. e nat., 45,* 390 (1968); idem, *FEBS Lett., 1,* 272 (1968).

28. J. Rétey, in *Vitamin B$_{12}$,* Proc. 3rd Eur. Symp. on Vitamin B$_{12}$ and Intrinsic Factor (B. Zagalak and W. Friedrich, eds.), de Gruyter, Berlin, 1979, p. 439.

29. D. A. Baldwin, E. A. Betterton, and J. M. Pratt, *J. Chem. Soc. Dalton Trans., 1983,* 225.

30. J. J. Pignatello and Y. Fanchiang, *J. Chem. Soc. Dalton Trans., 1985,* 1381.

31. Y. Fanchiang, G. T. Bratt, and H. P. C. Hogenkamp, *Proc. Natl. Acad. Sci., USA, 81,* 2698 (1984).

32. S. M. Chemaly, R. A. Hasty, and J. M. Pratt, *J. Chem. Soc. Dalton Trans., 1983,* 2223.

33. E. W. Abel, J. M. Pratt, R. Whelan, and P. J. Wilkinson, *S. Afr. J. Chem., 30,* 1 (1977).

34. E. Hohenester, C. Kratky, and B. Kräutler, *J. Am. Chem. Soc., 113,* 4523 (1991).

35. W. W. Reenstra and W. P. Jencks, *J. Am. Chem. Soc., 101,* 5780 (1979).

36. K. L. Brown, J. M. Hakimi, and D. W. Jacobsen, *J. Am. Chem. Soc., 106,* 794 (1984).

37. D. Lexa and J. Savéant, *Chem. Commun., 1975,* 872.

38. K. L. Brown and S. Peck, *Inorg. Chem., 26,* 4145 (1987).

39. J. M. Pratt, *J. Chem. Soc., 1964,* 5154.

40. D. A. Baldwin, E. A. Betterton, and J. M. Pratt, *J. Chem. Soc., Dalton Trans., 1983,* 217.

41. B. Kräutler, W. Keller, M. Hughes, C. Caderas, and C. Kratky, *Chem. Commun.*, *1987*, 1678.

42. B. Kräutler, W. Keller, and C. Kratky, *J. Am. Chem. Soc.*, *111*, 8936 (1989).

43. M. D. Wirt, I. Sagi, E. Chen, S. M. Frisbie, R. Lee, and M. R. Chance, *J. Am. Chem. Soc.*, *113*, 5299 (1991).

44. M. Rossi, J. P. Glusker, L. Randaccio, F. M. Summers, P. J. Toscano, and L. G. Marzilli, *J. Am. Chem. Soc.*, *107*, 1729 (1985).

45. H. J. Savage, P. F. Lindley, J. L. Finney, and P. A. Timmins, *Acta Crystalogr.*, *B43*, 280 (1987).

46. T. G. Pagano, L. G. Marzilli, M. M. Flocco, C. Tsai, H. L. Carrell, and J. P. Glusker, *J. Am. Chem. Soc.*, *113*, 531 (1991).

47. J. M. Pratt, in Ref. 6, p. 325.

48. D. A. Baldwin, E. A. Betterton, and J. M. Pratt, *S. Afr. J. Chem.*, *35*, 173 (1982).

49. S. M. Chemaly and J. M. Pratt, *J. Chem. Soc. Dalton Trans.*, *1984*, 595.

50. D. Lexa and J. Savéant, *J. Am. Chem. Soc.*, *100*, 3220 (1978).

51. D. N. Ramakrishna Rao and M. C. R. Symons, *J. Chem. Soc. Far. Trans. I*, *79*, 269 (1983).

52. I. Ya. Levitin, A. L. Sigan, A. I. Prokof'ev, and M. Ye. Vol'pin, *Doklady, Akad. Nauk SSSR*, *311*, 370 (1990).

53. R. L. Birke, G. A. Brydon, and M. F. Boyle, *J. Electroanal. Chem.*, *52*, 237 (1974).

54. D. Lexa, J. Savéant, and J. Zickler, *J. Am. Chem. Soc.*, *99*, 2786 (1977).

55. K. Lerch, W. B. Mims, and J. Peisach, *J. Biol. Chem.*, *256*, 10088 (1981).

56. Y. Fanchiang, *Organometallics*, *2*, 121 (1983).

57. K. A. Rubinson, E. Itabashi, and H. B. Mark, *Inorg. Chem.*, *21*, 3571 (1982).

58. H. A. O. Hill, J. M. Pratt, M. P. O'Riordan, F. R. Williams, and R. J. P. Williams, *J. Chem. Soc. (A)*, *1971*, 1959.

59. J. M. Wood, F. S. Kennedy, and C. G. Rosen, *Nature*, *220*, 173 (1968).

60. J. F. Endicott and G. J. Ferraudi, *J. Am. Chem. Soc.*, *99*, 243 (1977).

61. D. A. Baldwin, E. A. Betterton, S. M. Chemaly, and J. M. Pratt, *J. Chem. Soc. Dalton Trans.*, *1985*, 1613.

62. S. M. Chemaly and J. M. Pratt, *Chem. Commun.*, *1976*, 988.

63. S. M. Chemaly and J. M. Pratt, *J. Chem. Soc. Dalton Trans.*, *1980*, 2274.

64. G. N. Schrauzer and J. H. Grate, *J. Am. Chem. Soc.*, *103*, 541 (1981).

65. K. N. Joblin, A. W. Johnson, M. F. Lappert, and B. K. Nicholson, *JCS Chem. Commun.*, *1975*, 441.

66. B. P. Hay and R. G. Finke, *J. Am. Chem. Soc.*, *109*, 8012 (1987).

67. B. D. Martin and R. G. Finke, *J. Am. Chem. Soc.*, *112*, 2419 (1990).

68. J. H. Grate and G. N. Schrauzer, *J. Am. Chem. Soc.*, *101*, 4601 (1979).

69. G. N. Schrauzer and R. J. Holland, *J. Am. Chem. Soc.*, *93*, 4060 (1971).

70. H. Blaser and J. Halpern, *J. Am. Chem. Soc.*, *102*, 1684 (1980).

71. J. Halpern, *Polyhedron*, *7*, 1483 (1988).

72. P. J. Toscano, A. L. Seligson, M. T. Curran, A. T. Skrobult, and D. C. Sonnenberger, *Inorg. Chem.*, *28*, 166 (1989).

73. D. A. Baldwin, E. A. Betterton, and J. M. Pratt, *J. Chem. Soc. Dalton Trans.*, *1983*, 2217.

74. S. M. Chemaly and J. M. Pratt, *J. Chem. Soc. Dalton Trans.*, *1980*, 2259.

75. F. Nome, M. C. Rezende, C. M. Saboia, and A. C. da Silva, *Can. J. Chem.*, *65*, 2095 (1987).

76. B. Kräutler, M. Hughes, and C. Caderas, *Helv. Chim. Acta*, *69*, 1571 (1986).

77. B. Kräutler, *Helv. Chim. Acta*, *70*, 1268 (1987).

78. B. Kräutler, in Ref. 5, p. 31.

79. G. N. Schrauzer, E. Deutsch, and R. J. Windgassen, *J. Am. Chem. Soc.*, *90*, 2441 (1968).

80. G. N. Schrauzer and E. Deutsch, *J. Am. Chem. Soc.*, *91*, 3341 (1969).

81. B. Kräutler and C. Caderas, *Helv. Chim. Acta*, *67*, 1891 (1984).

82. R. Breslow and P. L. Khanna, *J. Am. Chem. Soc.*, *98*, 1297 (1976).

83. G. J. King, C. Gazzola, R. L. Blakeley, and B. Zerner, *Inorg. Chem.*, *25*, 1078 (1986).

84. D. Dolphin and A. W. Johnson, *J. Chem. Soc.*, *1965*, 2174.

85. S. Cockle, H. A. O. Hill, S. Ridsdale, and R. J. P. Williams, *J. Chem. Soc. Dalton Trans.*, *1972*, 297.

86. J. H. Espenson and T. D. Sellers, *J. Am. Chem. Soc.*, *96*, 94 (1974).

87. Y. Fanchiang and J. M. Wood, *J. Am. Chem. Soc., 103,* 5100 (1981).

88. L. J. Dizikes, W. P. Ridley, and J. M. Wood, *J. Am. Chem. Soc., 100,* 1010 (1978).

89. J. M. Wood, in Ref. 5, p. 62.

90. J. R. Guest, S. Friedman, D. D. Woods, and E. L. Smith, *Nature, 195,* 340 (1962).

91. G. N. Schrauzer and R. J. Windgassen, *J. Am. Chem. Soc., 89,* 3607 (1967).

92. G. N. Schrauzer and E. A. Stadlbauer, *Bioinorg. Chem., 3,* 353 (1974).

93. G. Agnes, H. A. O. Hill, J. M. Pratt, S. C. Ridsdale, F. S. Kennedy, and R. J. P. Williams, *Biochim. Biophys. Acta, 252,* 207 (1971).

94. T. Frick, M. D. Francia, and J. M. Wood, *Biochim. Biophys. Acta, 428,* 808 (1976).

95. J. M. Wood, I. Moura, J. J. G. Moura, M. H. Santos, A. V. Xavier, J. Le Gall, and M. Scandellari, *Science, 216,* 303 (1982).

96. J. M. Pratt and P. J. Craig, *Adv. Organomet. Chem., 11,* 331 (1973).

97. H. P. C. Hogenkamp, G. T. Bratt, and S. Sun, *Biochemistry, 24,* 6428 (1985).

98. H. P. C. Hogenkamp, G. T. Bratt, and A. T. Kotchevar, *Biochemistry, 26,* 4723 (1987).

99. F. Wagner and K. Bernhauer, *Ann. New York Acad. Sci., 112,* 580 (1964).

100. G. Costa, G. Mestroni, and C. Cocevar, *Chem. Commun., 1971,* 706.

101. K. Bernhauer and E. Irion, *Biochem. Zeit., 339,* 521 (1964).

102. H. A. O. Hill, J. M. Pratt, S. Ridsdale, F. R. Williams, and R. J. P. Williams, *Chem. Commun., 1970,* 341.

103. N. Imura, E. Sukegawa, S. Pan, K. Nagao, J. Kim, T. Kwan, and T. Ukita, *Science, 172,* 1248 (1971).

104. G. N. Schrauzer, J. H. Weber, T. M. Beckham, and R. K. Y. Ho, *Tetrahedron Lett., 1971,* 275.

105. L. Bertilsson and H. Y. Neujahr, *Biochemistry, 10,* 2805 (1971).

106. R. E. De Simone, M. W. Penley, L. Charbonneau, S. G. Smith, J. M. Wood, H. A. O. Hill, J. M. Pratt, S. Ridsdale, and R. J. P. Williams, *Biochim. Biophys. Acta, 304,* 851 (1973).

107. H. Yamamoto, T. Yokoyama, J. Chen, and T. Kwan, *Bull. Chem. Soc. Jpn., 48,* 844 (1975).

108. V. C. W. Chu and D. W. Gruenwedel, *Z. Naturforschung, C31,*
     753 (1976).

109. V. C. W. Chu and D. W. Gruenwedel, *Bioinorg. Chem., 7,* 169
     (1977).

110. G. C. Robinson, F. Nome, and J. H. Fendler, *J. Am. Chem. Soc.,*
     99, 4969 (1977).

111. J. S. Thayer, *Inorg. Chem., 20,* 3573 (1981).

112. G. Agnes, S. Bendle, H. A. O. Hill, F. R. Williams, and R. J. P.
     Williams, *Chem. Commun., 1971,* 850.

113. R. T. Taylor and M. L. Hanna, *Bioinorg. Chem., 6,* 281 (1976).

114. R. T. Taylor, J. A. Happe, and R. Wu, *J. Environ. Sci. Health,*
     A13, 707 (1978).

115. R. T. Taylor, J. A. Happe, M. L. Hanna, and R. Wu, *J. Environ.*
     Sci. Health, A14, 87 (1979).

116. Y. Fanchiang, W. P. Ridley, and J. M. Wood, *J. Am. Chem. Soc.,*
     101, 1442 (1979).

117. Y. Fanchiang, J. J. Pignatello, and J. M. Wood, *Organometallics,*
     2, 1748 (1983).

118. Y. Fanchiang, J. J. Pignatello, and J. M. Wood, *Organometallics,*
     2, 1752 (1983).

119. Y. Fanchiang, *Coord. Chem. Rev., 68,* 131 (1985).

120. W. R. Mason, *Coord. Chem. Rev., 7,* 241 (1972).

121. L. I. Elding and L. Gustafson, *Inorg. Chim. Acta, 24,* 239 (1977).

122. J. S. Thayer and F. E. Brinckman, *Adv. Organomet. Chem., 20,* 313
     (1982).

123. W. M. Scovell, *J. Am. Chem. Soc., 96,* 3451 (1974).

124. J. Setsune and D. Dolphin, *Can. J. Chem., 65,* 459 (1987).

125. Y. Fanchiang, *Inorg. Chem., 23,* 3983 (1984).

126. Y. Fanchiang, *Inorg. Chem., 21,* 2344 (1982).

127. M. J. Kamlet, J. L. M. Abboud, and R. W. Taft, *Progr. Phys.*
     Org. Chem., 13, 485 (1981).

128. F. Wagner, *Proc. Roy. Soc., A288,* 344 (1965).

129. C. Bradbeer, in Ref. 6, Vol. 2, p. 31.

130. E. Nexø and H. Olesen, in Ref. 6, Vol. 2, p. 57.

131. H. Schjønsby, *Gut, 30,* 1686 (1989).

132. G. Neale, *Gut, 31,* 59 (1990).

133. F. M. Huennekens, K. S. Vitols, K. Fujii, and D. W. Jacobsen,
     in Ref. 6, Vol. 1, p. 145.

134. F. Watanabe and Y. Nakano, *Int. J. Biochem.*, *23*, 1353 (1991).

135. J. M. Wood, *Naturwissenschaften*, *62*, 357 (1975).

136. R. L. Blakeley, in Ref. 6, Vol. 2, p. 381.

137. J. M. Pratt, in *Techniques and Topics in Bioinorganic Chemistry* (C. A. McAuliffe, ed.), Macmillan, London, 1975, p. 109.

138. J. M. Pratt, *Chem. Soc. Rev.*, *14*, 161 (1985).

139. B. P. Hay and R. G. Finke, *Polyhedron*, *7*, 1469 (1988).

140. R. T. Taylor, in Ref. 6, Vol. 2, p. 307.

141. B. A. Blaylock and T. C. Stadtman, *Biochem. Biophys. Res. Commun.*, *17*, 475 (1964).

142. P. van der Meijden, C. van der Lest, C. van der Drift, and G. D. Vogels, *Biochem. Biophys. Res. Commun.*, *118*, 760 (1984).

143. R. V. Banerjee and R. G. Matthews, *FASEB J.*, *4*, 1450 (1990).

144. C. D. Whitfield, E. J. Steers, and H. Weissbach, *J. Biol. Chem.*, *245*, 390 (1970).

145. W. Lu, S. R. Harder, and S. W. Ragsdale, *J. Biol. Chem.*, *265*, 3124 (1990).

146. M. B. Mooiman and J. M. Pratt, *S. Afr. J. Chem.*, *35*, 171 (1982).

147. E. A. Betterton, S. M. Chemaly, and J. M. Pratt, *J. Chem. Soc. Dalton Trans.*, *1985*, 1619.

148. C. A. Tsipis, *Coord. Chem. Rev.*, *108*, 163 (1991).

149. R. Eliasson, M. Fontecave, H. Jörnvall, M. Krook, E. Pontis, and P. Reichard, *Proc. Natl. Acad. Sci. USA*, *87*, 3314 (1990).

150. S. A. Benner, A. D. Ellington, and A. Tauer, *Proc. Natl. Acad. Sci. USA*, *86*, 7054 (1989).

151. L. Randaccio, N. Bresciani Pahor, E. Zangrando, and L. G. Marzilli, *Chem. Soc. Rev.*, *18*, 225 (1989).

152. J. M. Pratt, *J. Mol. Cat.*, *23*, 187 (1984).

153. B. P. Hay and R. G. Finke, *J. Am. Chem. Soc.*, *108*, 4820 (1986).

154. S. H. Kim, H. L. Chen, N. Feilchenfeld, and J. Halpern, *J. Am. Chem. Soc.*, *110*, 3120 (1988).

155. B. P. Hay and R. G. Finke, *J. Am. Chem. Soc.*, *109*, 8012 (1987).

156. C. Brink-Shoemaker, D. W. J. Cruikshank, D. C. Hodgkin, M. J. Kamper, and D. Pilling, *Proc. Roy. Soc.*, *A278*, 1 (1964).

9

Methane Formation by Methanogenic Bacteria: Redox Chemistry of Coenzyme F430

Bernhard Jaun

Organic Chemistry Laboratory
ETH-Zürich, Universtätstrasse 16
CH-8029 Zürich, Switzerland

1. BIOCHEMISTRY OF METHANOGENESIS

1.1. Methanogenic Bacteria

The methanogens are a unique group of strictly anaerobic micro-organisms which produce methane as the final product of their catabolic activity. They are found in such diverse biotopes as the sediments of lakes and oceans, the rumen of cows, anaerobic digestors of wastewater treatment plants, and hot vents on the ocean floor. All known species of methanogenic bacteria are able to use hydrogen and carbon dioxide as their only sources of reducing equivalents and C_1 units, respectively (chemolithoautotrophic growth). While species from several families are able to use carbon monoxide or formate, the ability to grow on acetate, methanol, and methylamine is specific for the *Methanosarcinaceae*. In anaerobically fermenting biomass, the methanogenic bacteria live in tight symbiosis with the acidogenic bacteria which produce hydrogen and acetate, the substrates for the methanogens. In a typical anaerobic digester, ~70% of the methane formed originates from acetate, the rest from CO_2 and other C_1 units. For the cleavage of acetate, no external reducing equivalents are needed; the methyl group is reduced to methane while the carboxylate group is oxidized to CO_2 [1,2].

Our knowledge about the biochemistry of methanogenesis is
fairly recent. Because of their extreme sensitivity to oxygen, and
because they form a tight symbiotic system with the acidogens, methanogenic bacteria have been difficult to obtain in pure culture.
Although Schnellen isolated pure strains of both *Methanosarcina
barkeri* and *Methanobacterium formicium* in 1947 [3], the actual break-
through was achieved in 1958, when Hungate and coworkers developed
special techniques for anaerobic handling of cultures [4]. Since
then, about 20 different species of methanogens have been described.
The recent discovery of a new, extremely thermophilic methanogen by
Stetter et al. [5] indicates, however, that this list is still far
from being complete.

When it became possible to culture methanogenic bacteria in
large quantities in the laboratory, biochemical analysis revealed
that they differed from the classical bacteria regarding the composition of their cell walls [6], the lipids of their membranes [7],
and, in particular, the presence of several coenzymes not found in
other organisms. Sequence characterization of 16S ribosomal RNA of
a broad spectrum of microorganisms by Woese's group [8] revealed a
deep phylogenetic split between the methanogens and the classical
bacteria and led to the proposal of a new taxonomy of life in which
the methanogens (together with the thermoacidophiles and the extreme
halophiles) form a separate urkingdom, the Archaebacteria [9].

1.2. Biochemical Pathways of Methanogenesis

Our current knowledge of the pathways, enzymes, and coenzymes
involved in the conversion of CO_2 and H_2 to methane is largely due
to the pioneering work of the groups of Wolfe [10] and Thauer [11].
Scheme 1 summarizes the two major pathways starting from CO_2 and
acetate, respectively. The structural formulas of the participating
coenzymes are collected in Scheme 2.

Reduction of CO_2 to the level of formate leads to the intermediate formylmethanofuran, which has been isolated [12]. The enzyme

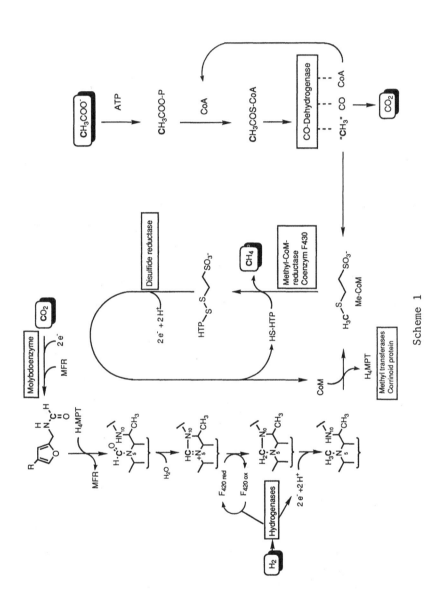

Scheme 1

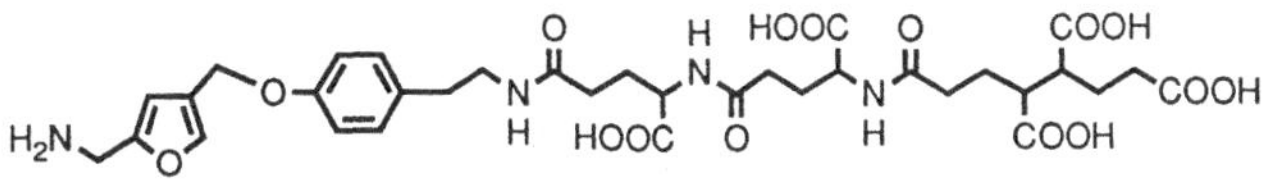

Methanofuran (MFR)

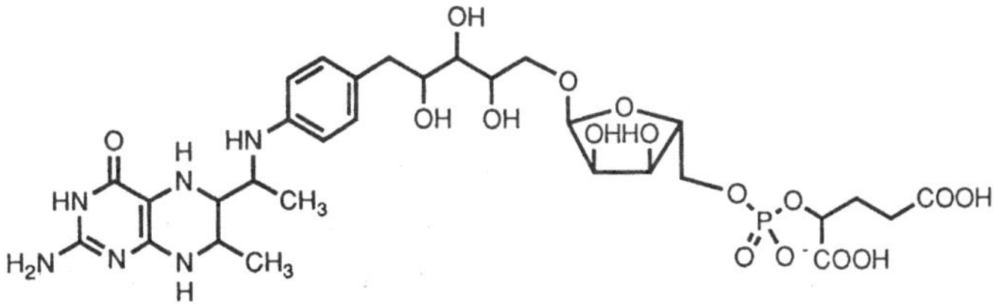

Tetrahydromethanopterin (H$_4$MPT)

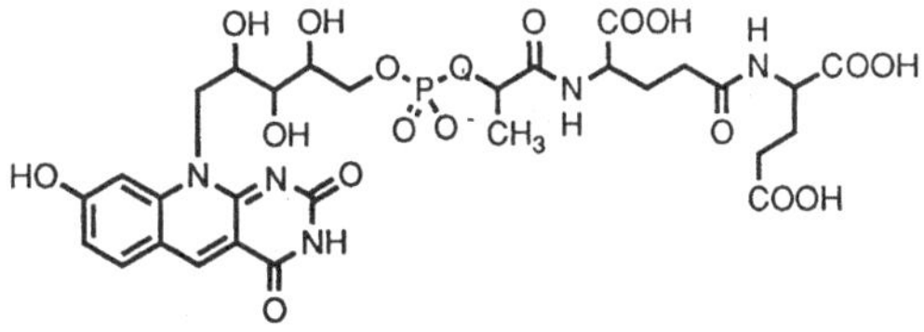

Factor F$_{420}$

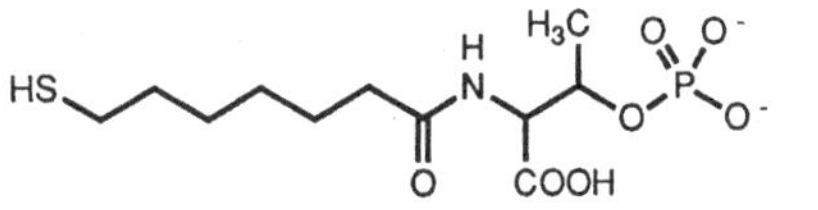

N-7-Mercaptoheptanoylthreonine-phosphate (HS-HTP)    Coenzyme M

Scheme 2

catalyzing this step contains molybdenum and a pterin cofactor [13]. From formylmethanofuran the formyl group is transferred to tetrahydromethanopterin (H$_4$MPT) [14,15], which serves as the carrier of the C$_1$ unit during its reduction to methylene-H$_4$MPT and further to methyl-H_4MPT. The source of reducing equivalents for the reductions is hydrogen, which is oxidized at two distinct hydrogenases, one of them using the deazaflavin F420 [16] as its primary electron acceptor. The currently available evidence suggests that methyl transfer from N^5-methyl-H_4MPT to 2-mercaptoethanesulfonate(coenzyme M) occurs in two steps, with a corrinoid-containing enzyme serving as the intermediate methyl carrier [17].

In *Methanosarcina*, acetate is first activated by phosphorylation and subsequent formation of acetyl-CoA [18]. Then it is cleaved into two one-carbon units by the enzyme complex carbon monoxide dehydrogenase which contains nickel, iron, and cobalt [19]. The nascent carbon monoxide unit is oxidized to carbon dioxide at the same enzyme complex, whereas the methyl group is transferred to CoM.

Methyl-CoM [20] is the first intermediate common to all pathways irrespective of the original substrate. It is converted to methane and the mixed disulfide of CoM and the thiol 7-mercaptoheptanoylthreonine phosphate (HS-HTP) [21] by the enzyme methyl-CoM reductase. Subsequently, the heterodisulfide is reduced back to the two thiols by a disulfide reductase [22]. The necessary electrons are again furnished by a hydrogenase and transferred to the disulfide reductase via an electron carrier, possibly a polyferredoxin [23]. It was discovered early by Gunsalus and Wolfe [24] that the first step of the cycle, reductive fixation of CO_2 to give formylmethanofuran, is coupled to the last step, the cleavage of methyl-CoM. More recent studies indicate that this "RPG" effect is linked to the concentration of the heterodisulfide CoM-SS-HTP [25].

1.3. Bioenergetics of Methane Formation

Compared with other catabolic pathways like photosynthesis or respiration, methanogenesis is an inefficient process. The fact that up to 15% w/w of the total cell protein consists of methyl-CoM reductase, one of the key enzymes of the process, shows that methanogens depend on high rates of methane formation. Estimated thermodynamic data for

TABLE 1

Estimated Free Energies for the Reactions of Methanogens

Reaction	$\Delta G'^{\circ a}$ (kJ/mol)	ΔG^{b} (kJ/mol)
$CH_3COO^-(aq) + H^+(aq) = CH_4(g) + CO_2(g)$	-36	-23
$CO_2(g) + 4H_2(g) = CH_4(g) + 2H_2O(l)$	-131	-17
$CO_2(g) + 3H_2(g) = CH_3OH(aq) + H_2O(l)$	-18	
$CH_3OH(aq) + H_2(g) = CH_4(g) + H_2O(l)$	-113	
$CH_3OH(aq) + HS\text{-}CH_2CH_2SO_3^- = CH_3SCH_2CH_2SO_3^- + H_2O(l)$	-49	
$CH_3SCH_2CH_2SO_3^-(aq) + H_2(g) = HS\text{-}CH_2CH_2SO_3^-(aq) + CH_4(g)$	-64[c]	
$CH_3\text{-}S\text{-}CH_2CH_2SO_3^- + HS\text{-}HTP = CH_4 + {}^-O_3SCH_2CH_2\text{-}S\text{-}S\text{-}HTP$	-43[c]	
${}^-O_3SCH_2CH_2\text{-}S\text{-}S\text{-}HTP + H_2 = HS\text{-}CH_2CH_2SO_3^- + HS\text{-}HTP$	-21[c]	

[a]At pH = 7.
[b]Estimated for typical concentrations in the biotope [26].
[c]Estimated from the experimental data for ethanethiol and methylethyl sulfide and diethyl disulfide (all in the liquid state).
Source: Compiled from references given in Ref. 26.

the major overall reactions are listed in Table 1 [26]. They show
that the two final steps, the generation of methane from methyl-CoM
and the reduction of the heterodisulfide, furnish the major part of
the free energy liberated in the process, whereas the early steps are
only slightly exergonic. Working with *Methanosarcina* and *M. thermoautotrophicum*, Gottschalk and coworkers showed that methanogenesis is
coupled to the synthesis of ATP via a chemiosmotic mechanism at several stages [27].

2. METHYL-COENZYME M REDUCTASE

2.1. Structural Properties of the Enzyme

Methyl-CoM reductase from different species has been purified to
homogeneity in the past years [28]. The protein part has a molecular
weight of ~300,000 and consists of two identical parts, each containing three different polypeptides ($\alpha_2\beta_2\gamma_2$). The operon coding for
these three peptide chains has been located and sequenced for several
species [29]. In the case of *Methanococcus voltae*, calculation of the
molecular weights from the genetically derived amino acid sequence
gives α = 60,443, β = 47,186, and γ = 28,726, leading to a value of
272,710 for the whole apoprotein.

   Firmly but not covalently bound to the protein are two molecules of coenzyme M, a not precisely known amount of the thiol HS-HTP
[21] and two molecules of coenzyme F430, the hydroporphinoid-nickel
complex which is the main subject of this chapter. Coenzyme F430 can
be dissociated from the protein under mild conditions. Wolfe and
coworkers observed partial recovery of methanogenic activity after
adding coenzyme F430, CoM, and Me-CoM to a mixture of the three
proteins α, β, and γ [30].

2.2. In Vitro Assays, Substrate Analogs and Inhibitors

Wolfe and coworkers demonstrated methane formation from methyl-CoM
and hydrogen in vitro, using an assay reconstituted from several
protein fractions (components A1, A2, A3a, A3b, and C), the low
molecular weight component B (later identified as HS-HTP), and
catalytic amounts of ATP, Mg^{2+}, NADPH, and coenzyme F_{420} [31].
The fact that the complex protein mixture A could be replaced by
an NADPH-dependent oxidoreductase revealed that component C is the
actual methyl-CoM reductase [32]. Ankel-Fuchs and Thauer used an
even simpler assay, consisting of component C, HS-HTP, catalytic
amounts of Cob(III)alamin and either dithiothreitol, $SnCl_2$, or
Ti(III)citrate as electron donors [33]. Finally, Thauer and coworkers
demonstrated that purified methyl-CoM reductase from *M. thermoauto-
trophicum* (strain Marburg) is able to catalyze the stoichiometric
conversion of HS-HTP and Me-CoM to CoM-SS-HTP and methane in the
absence of any external reducing agent, thus establishing that the
heterodisulfide is the product of the reaction catalyzed by methyl-
CoM reductase [34].

The highest activity observed in cell-free assays reached only
~5% of the activity for whole cells (as estimated based on the known
content of methyl-CoM reductase). The question of whether these low
in vitro activities are due to irreversible disactivation of the
enzyme during purification or to the destruction of the cell micro-
structure was resolved only recently by Thauer and coworkers. They
found that methyl-CoM reductase can be isolated in a much more active
form, showing ~50% of the in vivo activities, if the cells are pre-
conditioned reductively under H_2 and anaerobic purification techniques are used [35]. This clearly indicates that irreversible disactivation, presumably by reaction with oxygen, must be the major
cause for the low activity observed with the aerobically purified
enzyme (MCR_{ox}). Unfortunately, this recent result means that the

TABLE 2

Substrate Analogs and Inhibitors for Methyl-CoM Reductase (MCR_{ox})

| Substrate[a] | Activity as an analog | | Inhibitor |
	K_M (mM)	V_{max}[b]	I_{50} (mM)[c]
$CH_3SCH_2CH_2SO_3^-$ (native)	0.1	11	
$CH_3CH_2SCH_2CH_2SO_3^-$	1.3	7.4	
$CH_3SeCH_2CH_2SO_3^-$	0.3	35	
$CF_2HSCH_2CH_2SO_3^-$	2.5	20	
$CH_3SCH_2CH_2CO_2^-$	1.3	1.3	
$CH_2=CHCH_2SCH_2CH_2SO_3^-$			0.023[d]
$N\equiv C-SCH_2CH_2SO_3^-$			0.032[d]
$CF_3SCH_2CH_2SO_3^-$			7.0[d]
$CH_3OCH_2CH_2SO_3^-$			>20
$Br-CH_2CH_2SO_3^-$			0.004[d]
$Br-CH_2CH_2CH_2SO_3^-$			5×10^{-5}[d]
$N_3-CH_2CH_2SO_3^-$			0.001[d]

[a]Cyclopropyl-, vinyl-, and *n*-propyl-CoM are neither inhibitors nor analogs.
[b]$nmol \cdot hr^{-1} \cdot (mg\ protein)^{-1}$.
[c]Concentration of inhibitor giving 50% of the methane formation rate observed in the absence of the inhibitor with 0.1 M Me-CoM.
[d]Competitive inhibition.
Source: Compiled from Refs. 36-38.

biophysical and spectroscopic data reported so far, which were all obtained with MCR_{ox} (see Sec. 4), might not be relevant to the catalytically active state of the enzyme. At present, no method is available to reductively reactivate the enzyme after oxygen contact.

Even with the partly disactivated enzyme, valuable kinetic data on the selectivity of methyl-CoM reductase have been obtained by the groups of Wolfe [36], Walsh [37], and Thauer [38]. The known methyl-CoM analogs and inhibitors are listed in Table 2. Toward the thiol HS-HTP, the enzyme is highly selective: the analogs with aliphatic chains containing six or eight carbon atoms [34] and the nonnative D enantiomer [39] of HS-HTP are completely inactive.

2.3. Location of the Enzyme in the Cell

Initially, the question whether the methyl-CoM reductase system is a cytoplasmatic or a membrane-bound enzyme complex was controversial. In the isolation process, MCR is always found in the "soluble" cell fraction, not in the membranes. On the other hand, the finding of Gottschalk and coworkers [27] that the final steps of methanogenesis are coupled to a proton motive force did suggest a membrane-associated process. Immunoelectron microscopy using gold-labeled antibodies to MCR showed the label distributed over the entire cell interior in *M. thermoautotrophicum*, whereas in *Methanococcus voltae* it was located near the membranes [40]. Working with the methanogenic bacterium strain Göl, from which protoplasts can be obtained more easily than from other species, Mayer and coworkers [41] were able to generate inside-out membrane vesicles which had hollow spheric particles attached to the outer surface (originally the inner side) of the membrane via a short, rodlike structure. Gold labeling using antibodies to MCR revealed that several copies of the enzyme are located in the wall of the spheric part of these particles. The authors, who proposed the name *methanoreductosome* for these particles, interpreted their observations geometrically as shown in Fig. 1.

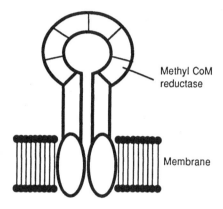

FIG. 1. Methanoreductosome. Geometric interpretation of electron microscopic results according to Ref. 41.

3.   STRUCTURE OF COENZYME F430

3.1.   Constitution and Configuration of Coenzyme F430

Factor 430 was first observed by LeGall [42] as a yellow band in chromatograms of cell extract and later described as a component of methyl-CoM reductase by Ellefson et al. [43]. Following the discovery that growth of methanogens critically depends on the presence of nickel in the culture medium [44], Thauer and coworkers found that a large part of the nickel present in the cells is incorporated into factor 430 [45]. Labeling experiments with [^{14}C]-δ-aminolevulinic acid confirmed the idea that factor 430 might be a nickel tetrapyrrole whereas incorporation of two [^{14}C]-methyl groups from methionine indicated a relationship to sirohydrochlorin [46]. At this stage, a collaboration between the laboratories in Marburg and at ETH in Zürich was initiated, which led to the determination of the constitution of coenzyme F430 and to the assignment of configuration for all but three of the stereogenic centers [47]. In fact, the structure of the hydroporphinoid ligand was determined for F430 M (formula 2), the derivative of the coenzyme obtained after acid-catalyzed methanolysis of crude isolates of factor 430. In contrast to F430 (formula 1) itself,

1 Coenzyme F430 (R = H)

2 F430M (R = CH$_3$, X = ClO$_4$)

the pentamethyl ester *2* is soluble in noncoordinating organic solvents. This property proved to be crucial for high-resolution nuclear magnetic resonance (NMR) studies because in the presence of donors the central nickel ion in both F430M and coenzyme F430 has a pronounced tendency to form five- and six-coordinate, paramagnetic complexes.

The series of biosynthetic incorporation experiments with selectively labeled [^{13}C]-δ-aminolevulinic acid and the spectroscopic studies that established the constitution of F430M as shown in formula *2* have been described in detail [47]. The remaining question was whether in the native coenzyme additional residues that might have been removed during methanolysis were covalently linked to the side chains of the hydroporphinoid macrocycle [48]. It was resolved by Livingston et al. [49] who showed that the native coenzyme, as isolated from the enzyme by very mild methods, was identical to the pentaacid *1* obtained from F430M by hydrolysis.

The relative configuration of the centers within rings A, B, and C as well as the correlation between rings A and B were determined by means of nuclear Overhauser difference spectroscopy (NOE) [47]. However, the configurational relationship across the ring junctions C-B and C-D could not be deduced from NOE's because the

sp^2-*meso* carbon atoms on both sides of ring C lie in the plane of
the macrocycle. The missing correlation followed from the absolute
configuration of the A-B part [50] and of ring C [47] which were
independently determined through chemical transformations and comparison of chiroptical properties with those of known compounds.
For technical reasons, mainly severe overlap in the ^{1}H-NMR spectrum
and strong coupling effects, the configuration of the three stereogenic centers in ring D could not be determined unambiguously by
one-dimensional NMR methods. Therefore, in Ref. 47, the configuration at centers C19 and C18 was only tentatively assigned from model
considerations, and it had to remain unassigned for C17.

Despite considerable efforts, no crystals of coenzyme F430 (*1*)
suitable for X-ray structure analysis have been obtained until now.
However, the bromide salt of 12,13-diepi-F430M (*3*, X = Br, see below)
did give crystals which, although somewhat disordered, were success-
fully analyzed with X-ray diffraction by Kratky and coworkers. The
resulting structure (see Fig. 8) showed without doubt that, in the
diepimer *3*, the configuration at C19 and C18 was opposite to our
earlier tentative assignment for F430M. Eventually, further studies
using 2D-NOE spectroscopy and deuterium incorporation experiments

3 12,13-diepi-F430M (R=CH$_3$)

demonstrated that 12,13-diepi-F430M and F430M do have the same configuration in ring D, thus completing the assignment of configuration
for all stereogenic centers in coenzyme F430 as shown in formula 1
[51].

        Summers and coworkers carried out a complete two-dimensional
NMR study of the native coenzyme dissolved in (D_3)-trifluoroethanol,
a polar but nonnucleophilic solvent in which F430 is diamagnetic [52].
Only minor chemical shift differences were observed relative to the
pentamethyl ester. The correlations obtained through homonuclear and
heteronuclear 2D methods confirmed the constitution and relative configuration reported for F430M and, in addition, allowed the assignment
of several carbon signals that had only been assigned as groups before.

        Besides being the first nickel-containing tetrapyrrole with a
known biological function, coenzyme F430 (1) exhibits several structural elements that deserve further comment. The π chromophore extends
only over three of the four nitrogens and makes F430 the most extensively reduced tetrapyrrole found in nature. Coenzyme F430 is a tetrahydrocorphin, *corphin* being the name proposed by Eschenmoser [53] for
this class of tetrapyrroles in which the carbon framework of a porphin
is combined with the linear chromophore typical for corrins. Two
additional rings are found in F430, the lactam ring fused to ring B
and the six-membered carbocycle formed through intramolecular acyla-
tion of meso position C15 by the propionic acid side chain at ring D.
Formation of this ring is the last step in the biosynthesis of coenzyme
F430, the direct precursor being $15,17^3$-seco-F430M [54]. As discussed
in Sec. 5 below, the fact that this annelation brings the carbonyl
function into conjugation with the π chromophore has a profound influence on the redox properties of coenzyme F430.

3.2. Partial Synthetic Derivatives and Stereoisomers
        of Coenzyme F430

It has been observed very early that after heat extraction from cells,
coenzyme F430 is usually found together with two other compounds which

have very similar absorption spectra but are chromatographically
distinct from the native coenzyme [55]. Furthermore, if neutral
or slightly basic aqueous solutions of the coenzyme were left
standing in contact with air, increasing amounts of a purple product,
called F560 according to its absorption maximum, were formed. All
four compounds gave pentamethyl esters if heated in acidic methanol.
Fast atom bombardment (FAB) mass spectra, NMR spectroscopy, and circular dichroism (CD) spectra of succinimides derived from ring C by
ozonolysis showed that the three compounds with absorption maxima
near 430 nm are stereoisomers differing in the configurations at C12
and C13 in ring C [56]. Regardless of the starting composition, heat
treatment of these isomers leads to an equilibrium mixture containing
4% of native F430 (1), 8% of 13-epi-F430, and 88% of 12,13-diepi-F430
(3). The fact that the diepimer is thermodynamically more stable
than the native stereoisomer has been interpreted as a consequence of
a nonideal puckering conformation enforced on ring C by steric interaction between the side chain at C13 and the carbonyl group of the
isocyclic ring.

The purple derivative (F560) formed from F430 upon contact
with oxygen is 12,13-didehydro-F430 (4), with an additional double
bond in ring C [56]. It can be reduced back to the tetrahydrocorphin

4 12,13-didehydro-F430 (R=H),F560

level with zinc/acetic acid, giving predominantly the native stereoisomer F430. Compared with other hydroporphyrins, coenzyme F430 is
in fact surprisingly stable toward dehydrogenation in ring C, which
is not blocked to oxidation by a pair of geminal substituents. This
(kinetic) stability may be attributed to the conformation of ring C,
in which, according to the NMR data, the two side chains point in the
quasi-axial directions whereas the two vicinal protons both occupy
quasi-equatorial positions. In order to acquire the antiperiplanar
conformation required for dehydrogenation, ring C has to flip into
a conformation where the side chain at C13 is quasi-equatorial and
may interact unfavorably with the carbonyl group of the isocyclic
ring.

The hydrocorphinoid macrocycle of coenzyme F430 can only be
oxidized to the level of a cyclic conjugated isobacteriochlorin if
the lactam ring is opened first. This transformation has been
carried out experimentally by Fässler [57] leading to the derivative
5 (for formulas of 5 and 6, see Scheme 5 below in Sec. 5.3). Isobacteriochlorin, 6, in which both the lactam ring and the carbocycle
have been opened, was synthesized from 12,13-didehydro-F430 by Kobelt
[58].

Recently, Hamilton et al. [59] reported the preparation of
pentaalkylamide derivatives of F430. Similar to the pentaesters,
these compounds are soluble in unpolar solvents in which they are
diamagnetic.

3.3. Coordination Chemistry of the Ni(II) Form

Compared to the exclusively four-coordinate Ni-corrinates, and to
the Ni-porphinates which bind axial ligands only weakly [60], coenzyme
F430 and F430M have a pronounced tendency to bind additional ligands
in the axial direction. Upon addition of the first axial ligand, the
electronic configuration changes from low-spin to high-spin. The
formation of five- and six-coordinate complexes of F430M with imidazole in methylene chloride has been studied by Fässler by means of

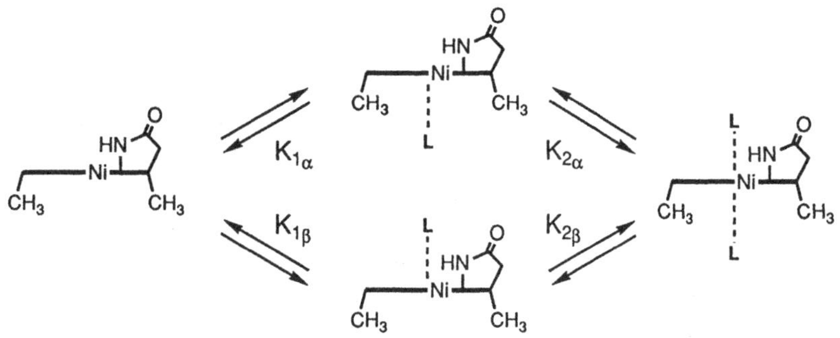

Scheme 3

UV-VIS spectroscopy and magnetic moment determination [57]. Upon
formation of five-coordinate complexes, the absorption maximum shifts
to shorter wavelengths and the band gains intensity. Transition from
five- to six-coordinate forms leads to a shift to longer wavelengths
and a further increase in extinction. Depending on the nature of the
axial ligands, the absorption maximum of the fully six-coordinate
form may be either slightly hypsochromic or bathochromic relative
to the four-coordinate complex. Since the α and the β sides of the
macrocycle are diastereotopic, two five-coordinate complexes have to
be considered for each monodentate ligand according to Scheme 3.

Therefore, the absorption spectrum of "the five-coordinate
form" is actually a weighted average for the two possible complexes,
and the constant K_1 as obtained by nonlinear computer fitting is the
sum of the two individual constants $K_{1\alpha} + K_{1\beta}$.

Axial complexation of F430M by several ligands other than
imidazole has been studied in our laboratory. Estimated values for
the dissociation constants obtained through computer fitting of
UV-VIS data are collected in Table 3.

The coordination chemistry of native F430 and of 12,13-diepi-
F430 has been investigated by several groups using UV-VIS, extended
X-ray absorption fine-structure spectroscopy (EXAFS), X-ray absorp-
tion spectroscopy (XAS), resonance Raman, and MCD techniques [61].
Unfortunately, the pentaacid is very sparingly soluble in solvents

TABLE 3

Estimated Dissociation Constants for Axial Ligand Binding
to F430M (Formula 2) in CH_2Cl_2

Exogenous ligand[a]	pK_1	pK_2
Cl^-	5.4	3.7
CN^-	6.2	3.2
Pyridine	2.2	0.8
Imidazole [57]	2.7	2.2
1,2-Dimethylimidazole	1.8	b

[a]Charged ligands were introduced as tetraethylammonium
salts.
[b]1,2-Dimethylimidazole was found to form only five-
coordinate complexes with F430M.
Source: Unpublished work by J. Lisowski, ETH-Zürich.

other than water, which competes with exogenous ligands for the axial
coordination sites. In pure water, at ambient temperature, the magnetic moment of 2.0 μ_B indicates that only part of the coenzyme is
in a high-spin form. Shiemke et al. [62] measured UV-VIS, EXAFS,
XAS, and Raman resonance spectra of native F430 in water and glycerol/
water at ambient and low temperatures which show that the fraction of
high-spin forms increases at lower temperatures. The same techniques
indicate that in 5 M aqueous cyanide, neat pyridine, or neat 1-methyl-
imidazole, the coenzyme is six-coordinate [63]. Interestingly, the
authors interpret their data as being consistent with an equilibrium
between the four-coordinate and six-coordinate forms, without inter-
mittent formation of five-coordinate complexes ($K_2 > K_1$). It remains
to be shown whether this apparent difference between the coordination
chemistry of the pentaester and the pentaacid is really due to the
different structures or rather to the different media—unpolar or
polar—used for the individual studies.

4. BIOPHYSICAL STUDIES

4.1. Axial Coordination of Additional Ligands in the Enzyme

Several research groups have investigated aerobically purified methyl-CoM reductase (MCR_{ox}), in which the Ni(II) valence state of coenzyme F430 predominates, by spectroscopic methods. In view of recent results [35], indicating that the resting state of the active enzyme might contain coenzyme F430 in the Ni(I) form and that the aerobically prepared enzyme (MCR_{ox}) is irreversibly disactivated, it is difficult to judge how relevant the information obtained on the coordination environment of F430 in MCR_{ox} may be with regard to the active enzyme. Therefore, these results will be discussed only briefly here.

Orme-Johnson [64] and Scott [65] and their coworkers carried out independent EXAFS studies of MCR_{ox}. Both research groups concluded that the nickel of coenzyme F430 in the enzyme is five- or six-coordinate. The data are consistent with either oxygen or nitrogen donor atoms as exogenous ligands and allow to exclude the presence of sulfur in the first coordination shell. Resonance Raman studies of hydrocorphinoid model complexes indicate that the separation of two high-frequency resonances at ~1540 and ~1630 cm^{-1} is a useful marker for the coordination number of the nickel ion [66]. In four-coordinate low-spin forms, this difference is ~95 cm^{-1}; in the high-spin complexes it decreases to 70-80 cm^{-1}. In the enzyme (MCR_{ox}), both marker lines were found at unusually high frequencies with a separation of 77 cm^{-1}, indicating five- or six-coordinate nickel in the enzyme [67]. X-ray absorption spectra of square planar Ni complexes show a typical pre-edge maximum which is absent in fully six-coordinate forms. Five-coordinate species show a less well-resolved shoulder [68]. Therefore, it is often difficult to decide whether a spectrum corresponds to a mixture of four- and six-coordinate or to a single five-coordinate species. The XAS spectrum of MCR_{ox} shows a straight edge without shoulder, as typically observed

with six-coordinate nickel having N_4O_2 or N_5O coordination spheres
[69]. Magnetic circular dichroism (MCD) spectroscopy of MCR_{ox} confirmed that the nickel of coenzyme F430 is in the $S = 1$ high-spin
state [61]. The value of the axial zero field splitting constant
deduced from the field and temperature dependence of the MCD spectra
is $D = +10$ cm^{-1}; near to the value of $D = +8$ cm^{-1} found for high-spin
F430 in water/glycerol 1:1.

4.2. EPR Detection of $S = 1/2$ Forms of Coenzyme F430 in Whole Cells and in the Enzyme

The aerobically purified enzyme MCR_{ox} exhibits an electron paramagnetic resonance (EPR) signal (now called MCR_{ox1}) of varying intensity
which is lost when F430 is dissociated from the apoenzyme [70].
Similar to the spectrum of Ni(I)F430M (see below), this in vivo signal
is of the axial type and shows partly resolved hyperfine coupling to
four nitrogen ligand atoms. However, the axial anisotropy of MCR_{ox1}
is much smaller than for Ni(I)F430M. With enzyme from bacteria grown
in the presence of ^{61}Ni, the lines of MCR_{ox1} were broadened due to
hyperfine coupling with the Ni nucleus, confirming the assignment of
MCR_{ox1} to the nickel ion of F430. The intensity of signal MCR_{ox1} did
vary considerably from one preparation to the other and no correlation
with the (low) activity of the enzyme could be established. In view
of the more recent results, MCR_{ox1} presumably corresponds to a form
of the coenzyme in the disactivated state. Its structure and valence
state, which must be either Ni(I) or Ni(III), remain to be determined.

More recently, Albracht et al. [71] reported that, in addition
to MCR_{ox1}, the EPR spectrum of whole cells of *Methanobacterium thermoautotrophicum* shows two new EPR signals after conditioning the cells
under H_2 at 60°C (Fig. 2). One of these signals, called MCR_{red1}, is
of the axial type with very pronounced anisotropy and is practically
identical to the EPR spectrum of Ni(I)F430M in tetrahydrofuran (THF).
Therefore, it was assigned to the square planar Ni(I) form of coenzyme
F430 by Albracht et al. The second signal, MCR_{red2}, is distinctly

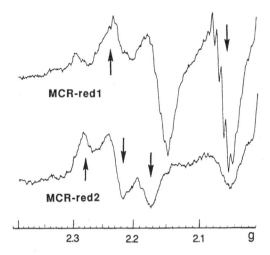

FIG. 2. EPR signals observed in whole cells of *M. thermoauto-trophicum* after treatment with H_2. (Reproduced with permission from Ref. 10).

rhombic with all three principal g values well above 2.1. If cells showing both EPR signals were conditioned at 60°C under 20% CO_2/80% N_2 and then frozen, the intensity of signal MCR_{red1} was reduced and MCR_{red2} disappeared completely. Since CO_2 is the natural substrate of the cells, this observation suggests that the species exhibiting signals MCR_{red1} and MCR_{red2} are both involved as intermediates in the enzymic reaction cycle. This hypothesis was corroborated by Krzycki and Prince [72], who observed the same signals in *Methanosarcina barkeri* and were able to show that their intensity correlates with methane formation from acetate.

In contrast to the aerobically prepared enzyme (MCR_{ox}), which showed only MCR_{ox1}, both signals MCR_{red1} and MCR_{red2} were detected in the highly active enzyme preparations described by Rospert et al. [35]. Because the nickel of coenzyme F430 is the only metal atom found in MCR, this observation proves that both signals must be due to coenzyme F430. In addition to the absorption of the Ni(II) form at 420 nm, the UV-VIS spectrum of the highly active enzyme shows bands at 386 nm and at ~720 nm that closely resemble the absorption spectrum of Ni(I)F430M (Fig. 3).

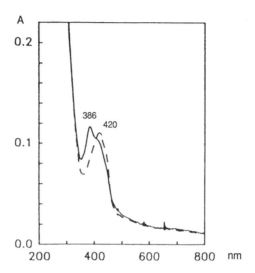

FIG. 3. UV-VIS spectrum of highly active methyl-CoM reductase. (Reproduced with permission from Ref. 35.)

In contrast to signal MCR_{red1}, which is consistent with an approximately square planar form of Ni(I)F430, similar to Ni(I)F430M as characterized in vitro, the nature of the species exhibiting the rhombic EPR signal MCR_{red2} is not understood at present. Both shape and g values of this spectrum point to a species with effective spin S = 1/2, which would be consistent with either Ni(I) or Ni(III).

5.   REDOX CHEMISTRY OF COENZYME F430 AND ITS DERIVATIVES

5.1.   Reduction of Ni(II)F430M to the Ni(I) Form

To our knowledge, the transformation catalyzed by methyl-CoM reductase (Scheme 4) has not been observed in nonenzymic chemistry until now.

In analogy to other enzymes with tetrapyrrolic metal complexes as cofactors, it is plausible to assume that the nickel center of

$$R_1\text{-S-CH}_3 + \text{HS-R}_2 \longrightarrow R_1\text{-S-S-R}_2 + CH_4$$

Scheme 4

coenzyme F430 is part of the active site of methyl-CoM reductase,
although at present there is no direct evidence for this hypothesis.
Several possible catalytic roles of F430 have been suggested: a
function similar to Raney nickel [73]; activation of the carbon-
sulfur bond in methyl-CoM through coordination of the sulfur to
Ni(II) [64]; coordination of the thiol HS-HTP by Ni(II)- or Ni(I)F430
[63,71]; the function of a hydride [64] or methyl [74] transfer agent;
and the role of a redox coenzyme.

The rather unreactive nature of thioethers and the observation
that Ni(II)F430M showed no tendency to coordinate to thioethers or
thiols prompted us to investigate the redox chemistry of F430. With
regard to the hypothesis that the nickel in F430 might need to change
its valence state in order to become reactive toward the carbon-sulfur
bond of methyl-CoM, it was of particular interest whether redox processes of F430 would lead to different valence states of the metal
or to ligand π radicals.

Since coenzyme F430 is only soluble in solvents with a restricted
electroneutrality window and because the protons of the carboxylic acid
functions are reduced at relatively positive potentials to give hydrogen,
most of the results described below have been obtained with the pentamethyl ester F430M (2) rather than with the pentaacid (1). In the
enzyme, most if not all of the carboxylic groups of F430 are likely to
be deprotonated. Multiple *tert*-ammonium salts of 1, which presumably
would be a better model than the pentaacid form, proved to be completely
insoluble in all solvents investigated.

Cyclic voltammetry in THF, DMF, and acetonitrile revealed that
F430M can be reduced reversibly at a potential of approximately -0.89 V
vs. SCE. At ambient temperature, the electron transfer process at Pt
and vitreous carbon electrodes was quasi-reversible. Chemically, however, the reduction was perfectly reversible in the absence of oxygen.
Kalousek techniques indicated a lifetime of the reduction product of
>10 sec. During bulk electrolysis of F430M in an electrochemical cell
with a large electrode, 1.05 F/mol was taken up, showing that a single
electron is transferred in the process.

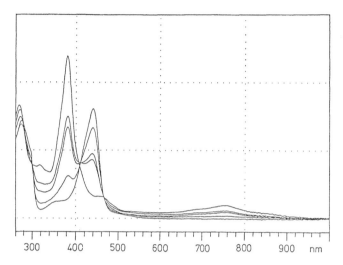

FIG. 4. Changes in the UV-VIS spectrum of F430M upon reduction with
NaHg in THF. (Reproduced with permission from Ref. 77.)

    F430M could also be reduced by chemical reducing agents. With
a slight excess of cobaltocene in THF, the Nernst equilibrium allowed
only ~5% of the F430M to be reduced. With liquid Zn/Hg (1.5% w/w) in
DMF, full and clean reduction was observed, whereas with liquid Na/Hg
(0.15% w/w) in THF, overreduction under attack of the ligand chromophore occurred after prolonged contact. Figure 4 shows the changes
observed in the UV-VIS spectrum during the reduction with Na/Hg.
Identical spectra were obtained if F430M was reduced in an optical
thin-layer electrolysis cell and, after reoxidation, F430M was quantitatively recovered.

    The EPR spectrum of the reduction product, measured in frozen
THF solution at 100 K, is shown in Fig. 5a. The signal is of nearly
axial type with $g_\perp < g_\parallel$. The partly resolved hyperfine splitting on
the $g_\perp$ line was assigned to the coupling with the four nitrogen nuclei
of the equatorial macrocycle. Figure 5b shows the best fit of a
computer-simulated powder spectrum obtained with the parameters
$g_{(xx,yy,zz)}$ = 2.065, 2.074, 2.250, and a single isotropic a_N value
of 0.95 mT.

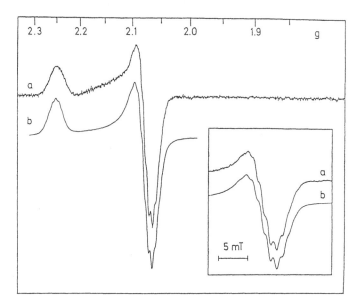

FIG. 5. EPR spectrum of Ni(I)F430M in THF at 100 K. (a) Experimental spectrum. (b) Computer simulation. (Reproduced with permission from Ref. 76.)

The strong anisotropy, the order of the g values, and the equatorial hyperfine coupling of this EPR spectrum are consistent with a species of effective spin S = 1/2 with its spin density residing predominantly in the $d_{x^2-y^2}$ type orbital of the central nickel ion. Similar spectra have been observed for the Ni(I) forms of synthetic N_4-macrocyclic nickel complexes [75]. Thus, the EPR spectrum demonstrates that reduction of F430M by one electron is metal-centered and leads to a species that is best described as square planar Ni(I)F430M [76]. The fact that further reduction of Ni(I)F430M is only observed at potentials cathodic of -2.0 V vs. SCE indicates that reduction of coenzyme F430 to the Ni(0) valence state is not possible under physiological conditions.

Hamilton et al. investigated the reduction of *sec*-amide derivatives of coenzyme F430 [59]. Both the pentaamides with native configuration and the 12,13-diepimers could be reduced to the Ni(I) form and

showed UV-VIS and EPR spectra that were practically identical to
those of Ni(I)F430M. The diepimer was found to be slightly more
difficult to reduce, which is in accordance with its lower tendency
to bind exogenous ligands in the axial positions.

5.2. Oxidation of Ni(II)F430M to the Ni(III) Form

The oxidative cyclic voltammogram of F430M in acetonitrile exhibits
a single, reversible, one-electron wave at +1.25 V vs. SCE. Coulom-
etry during bulk electrolysis confirmed that 1 F/mol is transferred
at +1.4 V vs. SCE. Isosbestic points in the overlay of UV-VIS spectra
measured at different degrees of oxidation demonstrate a clean transition into a single, stable product (Fig. 6). The EPR spectrum of the
oxidation product (Fig. 7) consists of a strongly anisotropic signal
of the axial type with a barely resolved fine structure on the $g_{\parallel}$ line

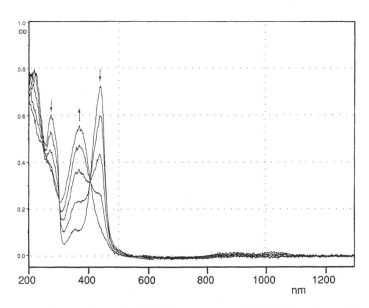

FIG. 6. Changes in the UV-VIS spectrum of F430M during oxidation
(OTLE cell, MeCN 0.1 M $TBABF_4$). (Reproduced with permission from
Ref. 77.)

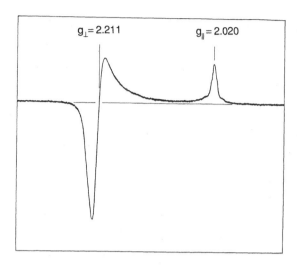

FIG. 7. EPR spectrum of Ni(III)F430M in frozen acetonitrile (100 K).
(Reproduced with permission from Ref. 77.)

and no resolved structure on the $g_\perp$ part. This type of spectrum is
typical for an S = 1/2 species with the unpaired electron in a nickel
d_{z^2} orbital, the configuration which is expected for a d^7 ion (Ni(III))
in an "elongated" tetragonal ligand field. With the exception of some
cases with very strong equatorial donors, which show spectra with $g_\perp$ <
$g_\parallel$ and have been attributed to square planar Ni(III), N_4-macrocyclic
Ni(III) complexes generally exhibit EPR spectra with $g_\parallel$ < $g_\perp$, similar
to that of Ni(III)F430M [75]. The fine structure of the $g_\parallel$ line is
presumably due to the coupling with the nitrogen nuclei of two acetonitrile solvent molecules coordinated in the axial positions. This
interpretation would be consistent with a six-coordinate complex with
a much weaker ligand field in the axial direction than in the equatorial plane [77].

     In contrast to other nickel hydroporphinates, e.g., isobacterio-
chlorinates, which are oxidized to the corresponding π-radical cations
[78], oxidation of F430M leads to the Ni(III) form without attacking
the ligand chromophore system. However, axially weakly coordinated
Ni(III)F430M is such a strong oxidant that it is not a plausible

intermediate in the enzymic process. Since Ni(II)F430 and Ni(II)F430M easily form five- and six-coordinate complexes with exogenous ligands, this tendency should be even more pronounced in the Ni(III) form. Preferential stabilization of the higher oxidation state by axial ligands is expected to reduce the oxidation potential toward more cathodic values. In the cyclic voltammograms, dramatic cathodic shifts of peak potentials of up to 0.5 V were indeed observed in the presence of potential ligands like Cl^-, pyridine, or imidazole. It was, however, not possible to deduce thermodynamic data from these experiments because the reversibility of the oxidation process was partly lost in the presence of these exogenous ligands.

5.3. Relationship Between the Structure of F430 and Its Redox Properties

The reduction potential of -0.89 V vs. SCE (approximately -600 mV vs. NHE) for the couple Ni(II)F430M/Ni(I)F430M is near the cathodic end of the range of potentials considered to be attainable under physiological conditions; it is comparable to the potential needed to reduce B_{12r} to B_{12s}. As discussed in Sec. 4, an EPR signal (MCR_{red1}) very similar to that of Ni(I)F430M has been observed with whole cells and, more recently, with highly active enzyme preparations [35,71]. This in vivo result demonstrates that in the enzyme the potential is sufficiently cathodic to reduce coenzyme F430 to the Ni(I) form. However, the standard potential of enzyme-bound coenzyme F430 is not known and it may differ from that of isolated F430. So far, reversible reduction or oxidation of the active enzyme in vitro has not been achieved. If obtainable, such potentiometric data would be particularly important with regard to the hypothesis that a thiolate is reducing coenzyme F430 to the Ni(I) form in the enzymic process (see Sec. 6).

In fact, coenzyme F430 and its pentaester or pentaamide derivatives are much more easy to reduce than the other square planar or quasi-octahedral nickel complexes of N_4-macrocyclic ligands that have

been shown to be reducible to the Ni(I) form. In view of the biophysical evidence, indicating that the Ni(I) form of the coenzyme is
an intermediate in the enzymic process, it is of particular interest
to ask which of the structural features of the hydrocorphinoid ligand
cause the nickel of factor 430 to be reducible at comparatively positive potentials.

In the Ni(I) form, the unpaired electron is located in the anti-
bonding $d_{x^2-y^2}$ orbital which lies in the equatorial plane. A weak
equatorial ligand field, lowering the energy of the $d_{x^2-y^2}$ orbital,
is expected to make reduction to the Ni(I) form easier. Since, upon
transition from low-spin square planar to high-spin five- or six-
coordinate forms, an electron is promoted from d_{z^2} to $d_{x^2-y^2}$, the same
argument holds with regard to the tendency of the Ni(II) form to add
ligands in the axial positions.

The strength of the equatorial ligand field is dependent on both
geometric and electronic factors. In a comprehensive study of a series
of saturated and partly unsaturated N_4-macrocyclic nickel complexes,
Busch and coworkers have shown that both the redox potentials and the
ligand field splittings depend on the size of the macrocycle: complexes
with 16-membered rings are easier to reduce to the Ni(I) form and more
difficult to oxidize to the Ni(III) form than those with 15-membered
macrocycles [75]. Based on a comparison of ligand field splittings
with those of monodentate and bidentate nickel amine complexes, Busch
concluded that 15-membered N_4-macrocycles are able to coordinate to
the small low-spin Ni(II) ion with nearly optimal Ni-N distances,
whereas larger macrocycles have to contract their coordination hole
in order to approach this distance. Accordingly, the equilibrium
Ni-N distance in a given complex is the result of a balance between
the tendency to attain optimal binding to Ni(II) and the strain introduced into the ligand by the contraction process. Therefore, the
equatorial ligand field should be weakest for those complexes with a
large coordination hole in the (hypothetical) strainless ligand and,
in the same time, a relatively rigid macrocycle.

For tetrapyrroles, increasing peripheral saturation leads to an
increase of both the size of the coordination hole and the flexibility

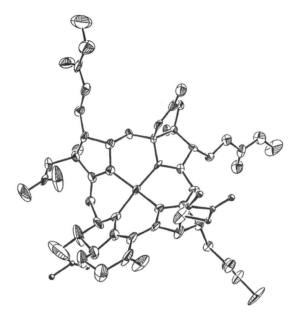

FIG. 8. Structure of 12,13-diepi-F430M (formula *4*, X = Br) as
determined by X-ray analysis. (Reproduced with permission from
Ref. 51.)

of the macrocycle. Through X-ray structure determination of an
extended series of low-spin Ni(II)hydroporphinates, Kratky and
Eschenmoser [79] showed that the equilibrium Ni-N distances decrease
with increasing saturation and that ligand hole contraction is
achieved through a saddle-shaped (S_4) deformation of the macrocycle.
The shortest Ni-N distances (1.86 Å) and the most extensive out-of-
plane deformation were found for 12,13-diepi-F430M (Fig. 8) [51].

While a large and relatively rigid macrocycle destabilizes the
small low-spin Ni(II) ion, thus increasing its reactivity toward both
reduction and axial ligation, it is equally important for these processes that the ligand allows strainless accommodation of the resulting Ni(I) or high-spin Ni(II) ions. As shown by the experimental
Ni-N distances compiled in Table 4 [80-82], the ionic radii of Ni(I)
and high-spin Ni(II) are comparable to each other and both considerably
larger than for low-spin Ni(II). Ni(II) porphyrins, for example, have

TABLE 4

Ni-N Distances in High-Spin Ni(II), Low-Spin Ni(II),
and Ni(I) Complexes

Compound	LS/HS,[a] axial ligands	r (Ni-N) (Å)	Method [Ref.]
Coenzyme F430	HS Ni(II) H_2O pH 7	2.10	EXAFS [62]
Ni(II)F430M (2)	LS THF	1.90	EXAFS [81]
Ni(I)F430M	THF (NaHg)	1.88 + 2.00[b]	EXAFS [81]
12,13-Diepi-F430	LS H_2O	1.89	EXAFS [63]
12,13-Diepi-F430	HS 1-Me-Imidazole	2.06	EXAFS [63]
12,13-Diepi-F430M (4)	LS bromide	1.86	X-Ray [51]
Ni tetrahydrocorphin	HS rhodanide	2.08	X-Ray [57]
Ni(II) isobacteriochlorin	LS Ni(II) THF	1.93	EXAFS [80]
Ni(I) isobacteriochlorin	THF (NaHg)	1.85 + 2.0[b]	EXAFS [80]
Ni(II) Me_6[14]4,11dieneN$_4$	LS	1.90	X-Ray [82]
Ni(II) Me_6[14]4,11dieneN$_4$	LS MeCN	1.93	EXAFS [82]
Ni(I) Me_6[14]4,11dieneN$_4$	perchlorate	1.98 + 2.07[b]	X-Ray [82]
Ni(I) Me_6[14]4,11dieneN$_4$	CH_3CN	1.97 + 2.06[b]	EXAFS [82]

[a]HS = high spin, LS = low spin.
[b]The best fit to EXAFS data was obtained with two sets of two N atoms
each and two different Ni-N distances.

quite large Ni-N distances and a very rigid macrocycle but they are
much less electrophilic than factor F430 [60]. In part, this may be
due to the fact that their rigid core is not able to expand to a size
that would be optimal for high-spin Ni(II).

The geometric requirements for a porphinoid nickel complex to
be reducible to the Ni(I) form at relatively positive potentials are
therefore not only a large coordination hole but also a critically
balanced macrocycle rigidity.

In addition to the geometric factors, there are important electronic contributions to be considered. Reduction to Ni(I) rather than
to a ligand π radical occurs only if the $d_{x^2-y^2}$ orbital is lower in

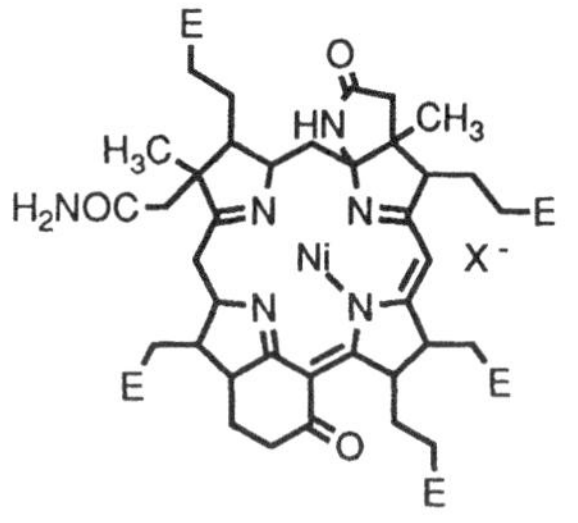

F430M (2) - 0.89 V

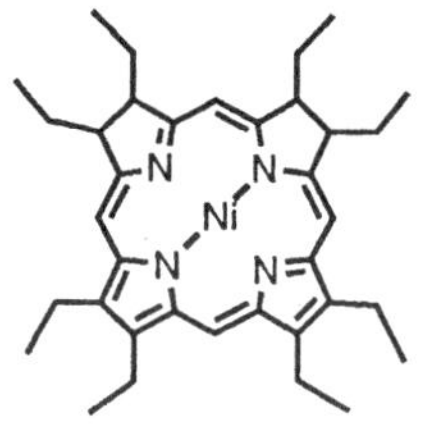

Ni-Octaethyl-iBchl [83]- 1.52 V

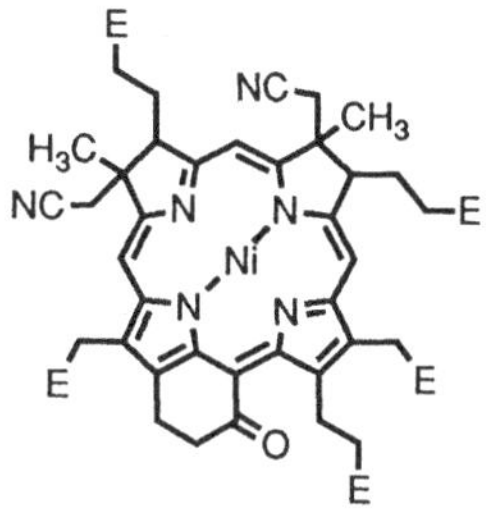

Ni-iBChl from F430 (5) [57] -1.15 V

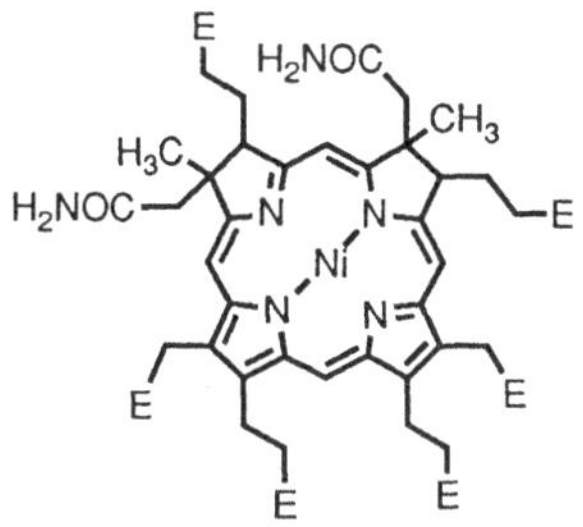

Ni-15, 17^3-seco-iBChl from F430 (6) [58] - 1.59 V

Scheme 5

energy than the lowest unoccupied π orbital. In F430 with its short
linear π chromophore, the gap between the highest occupied and the
lowest unoccupied π orbitals is expected to be larger than in porphi-
noids with cyclic π systems. Interestingly, molecular orbital calculations predict that in the isobacteriochlorins—the only other class
of hydroporphinoids for which reduction to the Ni(I) form was consistently observed [83]—the lowest unoccupied π orbital lies distinctly higher than in bacteriochlorins, chlorins, or porphyrins
[84]. However, the reduction potential of Ni(II) octaethylisobac-
teriochlorin is ~0.6 V more cathodic than that of F430M (Scheme 5).
In part, this difference is certainly due to the different formal
charges of the ligands: isobacteriochlorins are dianionic, while
F430M is monoanionic. Another very substantial contribution is
revealed if the reduction potentials of the isobacteriochlorins
5 and 6, both derived from F430, are compared: the carbonyl function

of the six-membered carbocycle causes an anodic shift of the potential
of more than 0.4 V. It follows that the presence of this structural
feature is crucial for the redox function of coenzyme F430 in the
enzyme.

6. METHANE FORMATION BY REACTION OF Ni(I)F430M WITH ELECTROPHILIC METHYL DONORS

6.1. Reactions of Ni(I)F430M with Electrophiles

With a method in hand to quantitatively prepare the Ni(I) form of
F430M, it was of interest to investigate the reactivity of Ni(I)F430M
toward electrophilic methyl donors, in particular toward methyl-sulfur
bonds. Of course, the fact that the nickel center of F430M does react
with a model substrate according to a certain mechanism does not necessarily mean that this mechanism is followed in the enzyme. To a large
extent, however, the enzymic mechanism must reflect the inherent reactivity of the free coenzyme. Understanding the nature of the reactions
of Ni(I)F430 or its derivatives should therefore contribute to the elucidation of the enzyme mechanism. As exemplified by the case of Ni(I)
F430, the spectroscopic characterization of possible intermediates in
vitro is often a prerequisite for their identification in the enzyme
or cell.

As of today, reaction of Ni(I)F430M with a thioether, including
methyl coenzyme M, could not be demonstrated unequivocally. With the
more reactive electrophiles shown in Scheme 6, however, Ni(I)F430M
reacted according to the indicated stoichiometry to give methane and
Ni(II)F430M [85].

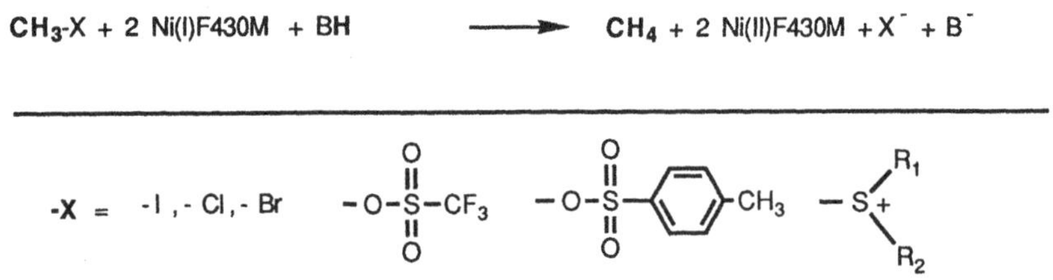

$$CH_3\text{-}X + 2\ Ni(I)F430M + BH \longrightarrow CH_4 + 2\ Ni(II)F430M + X^- + B^-$$

$$-X = -I, -Cl, -Br \quad -O\overset{O}{\underset{O}{\overset{\|}{\underset{\|}{S}}}}-CF_3 \quad -O\overset{O}{\underset{O}{\overset{\|}{\underset{\|}{S}}}}\!\!-\!\!\langle\rangle\!\!-CH_3 \quad -S\overset{R_1}{\underset{R_2}{\overset{+}{<}}}$$

Scheme 6

The order of reactivity observed was methyl iodide >> methyl triflate > methyl tosylate > methyl dialkyl sulfonium hexafluorophosphates. With protons, Ni(I)F430M slowly reacted to give hydrogen and Ni(II)F430M. In the presence of a 100-fold excess of a thiol or an ammonium salt (pK_a ~10), Ni(I)F430M had a half-life of more than 1 hr, whereas with the more acidic pyridinium tetrafluoroborate, it reacted within seconds. Since sulfonium ions do not react with zinc amalgam, it was possible to study their reaction with Ni(I)F430M using catalytic amounts of F430M in the presence of Zn/Hg. Under these conditions, the Ni(II)F430M formed in the reaction was immediately reduced back to the Ni(I) form.

6.2. Mechanistic Investigations

6.2.1. Reactions of Ni(I)F430M with Sulfonium Ions [86]

The first mechanistic question to be addressed was where the fourth hydrogen in the product methane is coming from. The answer followed from two experiments carried out in the presence of $(CH_3)_2CH-OD$ and $(CH_3)_2CD-OH$, respectively. Whereas with isopropanol-OD, the resulting methane was monodeuterated to >85%, no deuterium incorporation was observed in the presence of (2-D)-isopropanol.

Since alkyl radicals react with isopropanol by hydrogen abstraction from CH-2 whereas carbanions deprotonate the OH group, these results show that the hydrogen is introduced as a proton and that formation of free methyl radicals in this reaction can be excluded.

The reaction between sulfonium ions and Ni(I)F430M is not an outer sphere electron transfer followed by homolytic cleavage of the resulting sulfur-centered radical. This follows from the observation of a pronounced selectivity for cleavage of the methyl-sulfur over that of the alkyl-sulfur bonds in F430M-catalyzed reductions of mixed sulfonium ions (Scheme 7). In contrast, reduction by the homogeneous one-electron reagent sodium naphthalide gave product distributions that either are statistical (ethyl, n-propyl) or reflect the stability of the nascent alkyl radicals (i-propyl). The preferential methyl-

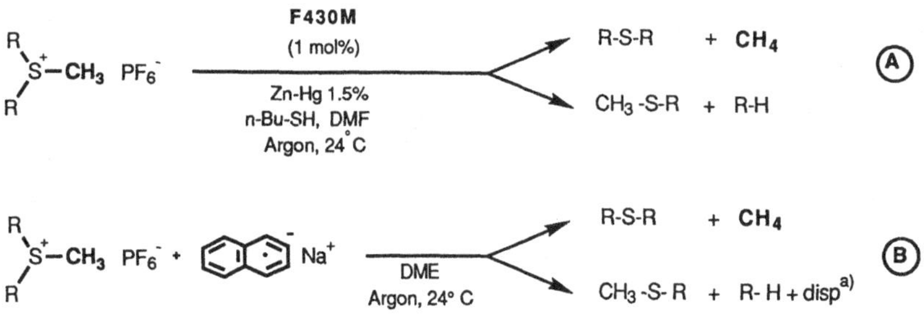

Scheme 7

Reaction →	A		B	
R-	CH$_4$ %	R-H %	CH$_4$ %	R-H + disp[a] %
CH$_3$CH$_2$ -	88	12	28	72
CH$_3$CH$_2$CH$_2$ -	98.7	1.3	38	62
(CH$_3$)$_2$CH -	99.5	0.5	8	92

a) products of radical disproportionation

Scheme 7

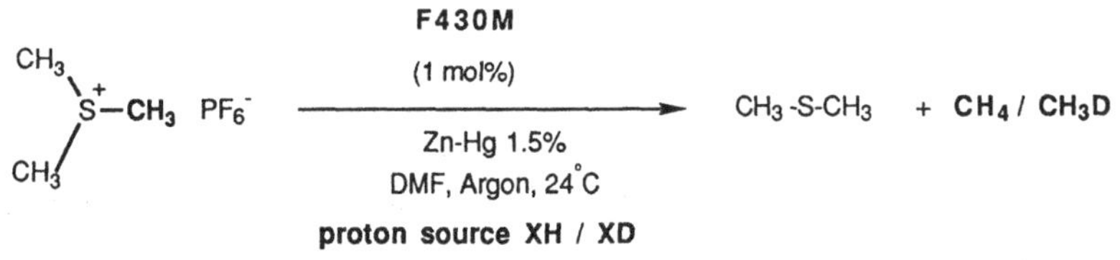

proton source XH / XD

Source of protons	XH / XD	(CH$_3$)$_2$CHOH (CH$_3$)$_2$CHOD	CH$_3$CH$_2$CH$_2$SH CH$_3$CH$_2$CH$_2$SD
Isotope effect on overall reaction rate	k$_H$ / k$_D$	2.8 ± 0.05	1.0 ± 0.05
Isotope effect on product distribution (competitive experiment)	CH$_4$ / CH$_3$D	11.0 ± 0.7	5.5 ± 0.3

Scheme 8

sulfur cleavage observed in the F430M catalyzed reduction indicates
that, in the product-determining step, there must be a close interaction between the Ni center and the sulfonium ion.

As with methyl iodide (see below), the reaction of Ni(I)F430M
with sulfonium ions proceeds via an intermediate from which methane
can be liberated through protonation. This follows from the observation of different isotope effects on the overall methane formation
rate, on the one hand, and on the product distribution in competitive
H/D experiments, on the other hand (Scheme 8).

6.2.2. Reaction of Ni(I)F430M with Methyl Iodide

If methyl iodide was allowed to react with Ni(I)F430M in a 2:1
stoichiometric ratio at low temperature, the color of the solution
changed within seconds from green (Ni(I)) to brown-orange (Ni(II)).
However, practically no methane could be detected at that time. If
acid was added after the color change, more than 80% of the theoretical amount of methane was immediately generated. Addition of deuter-
ated acid gave over 85% CH_3D (Scheme 9) [85].

This experiment proves that an intermediate is formed, which
can be dissociated to Ni(II)F430M and methane by protonation. Its
properties are consistent with those expected for a methyl-Ni(II)
derivative with an axial nickel-carbon bond. Several authors have
postulated the intermediate formation of methyl-Ni(II) species in
the reactions of synthetic macrocyclic Ni(I) complexes with organic
halides. Gosden et al. [87] reported that electrolysis of n-alkyl
halides in acetonitrile, if catalyzed by a Ni-hexamethyl cyclam,

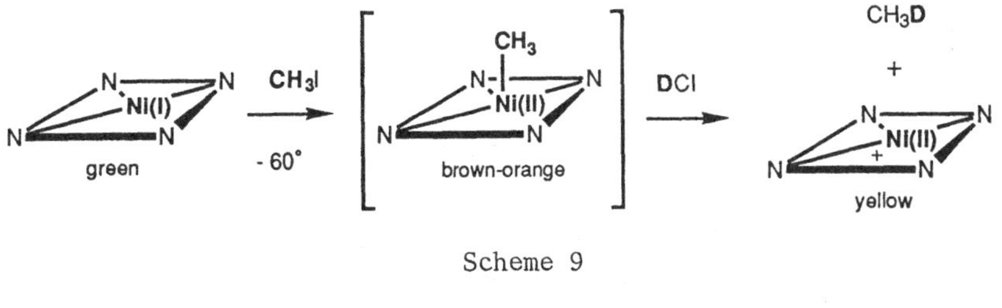

Scheme 9

$$Ni(I)L^+ + R\text{-}I \xrightarrow{-I^-} R\text{-}Ni(III)L^{2+} \xrightarrow{e^-} R\text{-}Ni(II)L^+ \xrightarrow{H^+} RH + Ni(II)L^{2+} \quad (a)$$

$$Ni(I)L^+ + R\text{-}I \xrightarrow{-I^-} R^\bullet + Ni(II)L^{2+}$$

$$Ni(I)L^+ + R^\bullet \longrightarrow R\text{-}Ni(II)L^+ \xrightarrow{H^+} RH + Ni(II)L^{2+} \qquad (b)$$

Scheme 10

gave a product distribution that is consistent with an anionic but not with a free radical mechanism. Stolzenberg et al. [83] investigated the reaction of Ni(I) isobacteriochlorinates with different halides in acetonitrile and concluded from the relative rates observed for primary, secondary, and tertiary halides that the reaction did not proceed via free radicals. On the other hand, Espenson and coworkers, who observed a metastable intermediate attributed to methyl-Ni(II)tmc in the reaction of (R,R,S,S)-Ni(I)tetramethylcyclam with n-alkyl iodides in alkaline aqueous solution, concluded from kinetic analysis that formation of the intermediate occurs via a free radical mechanism [74]. Whereas in all three studies the formation of alkyl-Ni(II) intermediates was proposed, two different mechanisms according to Scheme 10 were postulated for their formation.

One possible explanation might be that two different solvent systems were used in these studies. While a mechanism as shown in Scheme 10b, where free methyl radicals have to survive a diffusion step to react with a second molecule of the Ni(I) complex, may be acceptable for a solvent with a high bond dissociation energy like water, it would certainly be unreasonable for acetonitrile or THF, where hydrogen abstraction from the solvent would be expected to be very fast.

6.3. Methyl-Ni(II) Derivatives of Model Compounds and of F430M

The intermediate suggested by the mechanistic results has properties that are consistent with an organometallic methyl-Ni(II)F430M species

7

containing a methyl group which is directly bound to the nickel. In
contrast to the alkyl derivatives of other transition metals, in particular the well-studied alkyl-Co(III) corrins, practically nothing is
known about alkyl-nickel(II) complexes with equatorial N_4-macrocyclic
ligands. The only isolated compound for which such a structure was
proposed is the paramagentic complex 7 obtained by D'Aniello and
Barefield upon methylation of (R,R,S,S)-Ni(II)tetramethylcyclam with
dimethylmagnesium [88].

In order to identify CH_3-Ni(II)F430M as an intermediate in the
reactions of Ni(I)F430M with electrophiles, a spectroscopic technique
providing direct proof for a carbon-nickel bond in this hypothetical
compound had to be found. Because a paramagnetic high-spin d^8 species
was expected, neither high-resolution NMR nor EPR was a suitable technique. However, high-spin Ni(II) complexes often show quite well-
dispersed, isotropically shifted NMR spectra. To test the method,
the synthesis of 7 as described by Barefield was repeated, using
$(CD_3)_2Mg$ and $(CH_3)_2Mg$, respectively, and the ^2H- and ^1H-NMR spectra
were measured. The very broad and strongly temperature-dependent
signal of the nickel-bound methyl group was found at -320 ppm (r.t.)
in both the ^2H- and the ^1H spectrum (Fig. 9). Although less sensitive, ^2H-NMR proved to be preferable over ^1H-NMR because the lines
were up to 20 times more narrow (in Hz) and the assignment of the
CD_3 signal was straightforward.

Using the same technique, Ni(II)F430M was reacted in a sealed
glass system at -78°C with 0.95 eq of $(CD_3)_2Mg$. As with the model

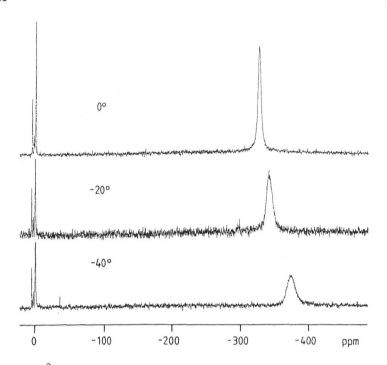

FIG. 9. ²H-NMR spectrum (46 MHz) of CD₃-Ni(II)tmc (formula 7) at different temperatures. (Reproduced with permission from Ref. 89.)

8

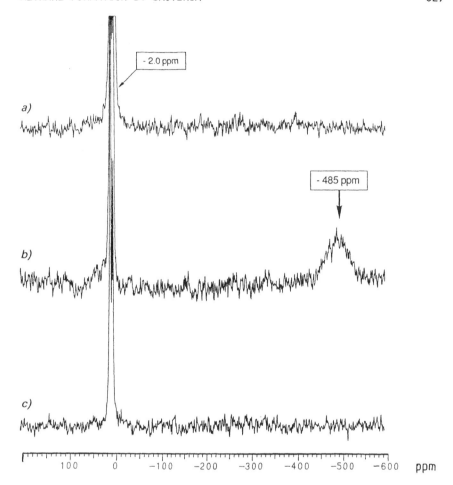

FIG. 10. ^{2}H-NMR spectra demonstrating formation of CD_3-Ni(II)F430M (formula *8*); (46 MHz, -40°C, THF). (a) Control of reagent; $(CD_3)_2$Mg solution in THF. (b) Spectrum of CD_3-Ni(II)F430M. (c)Same solution after addition of acid. (Reproduced with permission from Ref. 89.)

compound *7*, the ^{2}H-NMR spectrum of the reaction mixture at -40°C exhibited a very broad, temperature-dependent D signal at high field (-490 ppm, Fig. 10). After addition of acid, this signal disappeared and more than 95% of the original F430M was recovered unchanged. This result demonstrates that CD_3-Ni(II)F430M (*8*) exists and is stable below -20°C. At higher temperatures, the methyl-nickel derivative slowly decayed by proton transfer from the hydrocorphinoid ligand to the methyl group, giving methane and a deprotonated form of F430M. At present, it is not known whether the methyl group in CD_3-Ni(II)F430M (8) occupies the axial position on the α or on the β side of the macrocyle. In view of the very large linewidth of the NMR signal, the possibility cannot be excluded that a mixture of the two stereoisomers was obtained [89].

7. CONCLUSIONS: POSSIBLE CATALYTIC MECHANISMS FOR METHYL-COENZYME M REDUCTASE

The spectroscopic characterization of a methylnickel derivative of F430M, as well as the mechanistic evidence for its formation in the reaction of Ni(I)F430M with electrophilic methyl donors, indicates that such an organometallic compound may be the last intermediate in the enzymic process. This postulate is consistent with the finding of Floss and coworkers that the conversion of ethyl-CoM to ethane by methyl-CoM reductase occurs under retention of configuration at the carbon originally bound to sulfur [90]. Based on the properties of methyl-Ni(II)F430M, the nickel-carbon bond in such an intermediate is expected to be very electron-rich. Because even a weak acid would be able to irreversibly dissociate methyl-Ni(II)F430 into methane and Ni(II)F430, it is doubtful whether it will be possible to trap the enzyme in this state for direct spectroscopic characterization.

The identification of the species giving rise to EPR signal MCR$_{red1}$ as square planar Ni(I)F430 together with the fact that this signal is present in the highly active enzyme indicates that this valence state of the coenzyme is active in the enzyme process.

Therefore, in our view, the major current question regarding the enzyme mechanism is how the Ni(I) form of coenzyme F430 reacts with the thioether methyl-CoM to give a methyl-nickel intermediate and the heterodisulfide of CoM and HS-HTP. In this context, the nature of the species giving rise to the EPR signal MCR$_{red2}$ is of central importance. The fact that upon addition of methyl-CoM to the highly active enzyme the intensity of MCR_{red2} decreased while MCR_{red1} increased [35] points to the possibility that it is the intermediate exhibiting signal MCR_{red2} that attacks Me-CoM. In this case, MCR_{red2}, which corresponds to a species with $S = 1/2$, must be attributed to a second form of Ni(I)F430. The presently available data do not allow, however, the definitive exclusion of the possibility that MCR_{red2} is the spectrum of the first intermediate after the attack on methyl-CoM. In that case, a Ni(III) species like methyl-Ni(III)F430 would have to be considered also.

If the rhombic signal MCR_{red2} is due to a second form of Ni(I)F430, this species must have a coordination geometry that is distinctly different from that of square planar Ni(I)F430. Ligand field analysis shows that for d^9 metal ions, rhombicity is only expressed in the EPR spectrum if the in-plane anisotropy is comparable to the main axial ligand field splitting [91]. Rhombic EPR spectra have been observed for pentacoordinate square pyramidal Ni(I) complexes with axial π-acceptor ligands [92]. However, experiments performed in our laboratory showed that the EPR spectrum of Ni(I)F430M does not change in the presence of π acceptors like CO, imidazole, or triethylphosphite. Therefore, it remains to be demonstrated whether ligands exist that can coordinate to Ni(I)F430 in such a way that the EPR spectrum becomes rhombic. Based on force field calculations, Zimmer and Crabtree suggested that a trigonal bipyramidal structure of Ni(I)F430 might be close in energy to the square planar conformation [93]. Although such a strong out-of-plane distortion of the equatorial macrocycle would be consistent with a rhombic EPR spectrum, no experimental evidence pointing in this direction is available.

If a non-square planar form of Ni(I)F430 (MCR_{red2}) is reacting with methyl-CoM in the enzyme, this could explain why, in homogeneous solution, Ni(I)F430M did not react with methyl-CoM and other thioethers. On the other hand, it is conceivable that in the enzyme coenzyme M needs to be activated before reacting with Ni(I)F430. One possible mechanism of activation, proposed by us [77] and independently by Berkessel [94], is shown in Scheme 11.

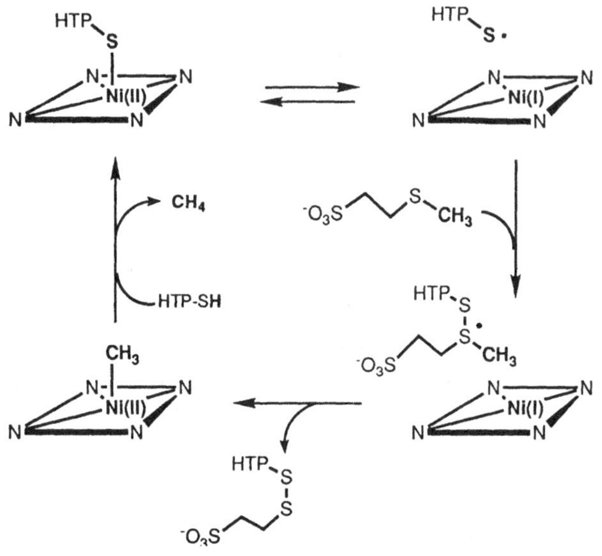

Scheme 11 (Reproduced with permission from Ref. 77.)

Formation of sulfuranyl radicals by addition of a thiyl
radical to a thioether as well as radical substitution on sulfur
has been observed in gas phase photochemical experiments [95].
Further work both with the isolated coenzyme and with the enzyme
is needed to determine whether one of the mechanisms discussed
above corresponds to the catalytic action of coenzyme F430 in
methyl-CoM reductase.

ABBREVIATIONS

ATP	adenosine 5'-triphosphate
CD	circular dichroism
CoA	coenzyme A
CoM	coenzyme M (2-mercaptoethanesulfonate)
CoM-SS-HTP	mixed disulfide of the thiols CoM and HS-HTP
DME	dimethoxyethane
DMF	dimethylformamide

EPR	electron paramagnetic resonance (electron spin resonance)
2D-NOS	two-dimensional nuclear Overhauser spectroscopy
EXAFS	extended X-ray absorption fine structure
F430M	coenzyme F430 pentamethyl ester
F560	12,13-didehydro-F430
FAB	fast atom bombardment
HS-HTP	7-mercaptoheptanoylthreonine phosphate (see Scheme 2)
H_4MPT	tetrahydromethanopterin (see Scheme 2)
MCD	magnetic circular dichroism
MCR	methyl-coenzyme M reductase
MCR_{ox}	disactivated, aerobically purified methyl-coenzyme M reductase
MCR_{red1}	axial EPR-signal from whole cells and highly active MCR
MCR_{red2}	rhombic EPR-signal from whole cells and highly active MCR
MFR	methanofuran (see Scheme 2)
Me-CoM	S-methyl-coenzyme M
methyl tosylate	p-toluenesulfonic acid methyl ester
methyl triflate	trifluoromethonesulfonic acid methyl ester
NADPH	reduced form of nicotinamide adenine dinucleotide phosphate
NHE	normal hydrogen reference electrode
NMR	nuclear magnetic resonance
NOE	nuclear Overhauser effect
OTLE	optical thin-layer electrolysis
RNA	ribonucleic acid
RPG effect	kinetic coupling of the first step of methanogenesis to MCR activity
SCE	saturated calomel reference electrode
$TBABF_4$	tetrabutylammonium tetrafluoroborate
THF	tetrahydrofuran
tmc	(R,R,S,S)-N,N',N'',N'''-tetramethyl-[14]-ane-N_4
UV-VIS	electronic absorption spectroscopy in the ultraviolet and visible
XAS	X-ray absorption spectroscopy

REFERENCES

1. A. J. B. Zehnder, K. Ingvorsen, and T. Marti, in *Anaerobic Digestion*, Proc. of the 2nd Int. Symposium on Anaerobic Digestion (D. E. Hughes, ed.), Elsevier, Amsterdam, 1982, pp. 45-66.

2. K. Wuhrmann, *Experientia, 38,* 193-198 (1982).

3. Ch. T. P. G. Schnellen, Ph.D. thesis, Techn. Univ. Delft, NL, 1947.

4. P. H. Smith and R. E. Hungate, *J. Bacteriol., 75,* 713-718 (1958).

5. R. Huber, M. Kurr, H. W. Jannasch, and K. O. Stetter, *Nature, 342,* 833-834 (1989).

6. O. Kandler, *Zbl. Bakt. Hyg. I. Abt. Orig. C3,* 149-160 (1982).

7. T. A. Langworthy, T. G. Tornabene, and G. Holzer, *Zbl. Bakt. Hyg. I. Abt. Orig. C3,* 228-244 (1982).

8. C. R. Woese, *Microbiol. Rev., 51,* 221-271 (1987).

9. C. R. Woese and G. E. Fox, *Proc. Natl. Acad. Sci. USA, 74,* 5088-5090 (1977).

10. R. S. Wolfe, in *The Molecular Basis of Bacterial Metabolism* (G. Hauska and R. Thauer, eds.), Springer-Verlag, Berlin, 1990, pp. 1-12.

11. R. K. Thauer, A. Brandis-Heep, G. Diekert, H. H. Gilles, E.-G. Graf, R. Jaenchen, and P. Schönheit, in *Environmental Regulation of Microbial Metabolism* (I. S. Kulaev, E. A. Dawes, and D. W. Tempest, eds.), Academic Press, New York, 1985, pp. 231-239.

12. J. A. Leigh, K. L. Rinehart, and R. S. Wolfe, *Biochemistry, 24,* 995-999 (1985).

13. M. Karrasch, G. Börner, M. Enssle, and R. K. Thauer, *Eur. J. Biochem., 194,* 367-372 (1990); M. Karrasch, G. Börner, and R. K. Thauer, *FEBS Lett., 274,* 48-52 (1990).

14. P. van Beelen, J. W. van Neck, R. M. de Cook, G. D. Vogels, W. Guijt, and A. G. Hasnoot, *Biochemistry, 23,* 4448-4454 (1984).

15. J. C. Escalante-Semerena, K. L. Rinehart, and R. S. Wolfe, *J. Biol. Chem., 259,* 9447-9455 (1984).

16. L. D. Eirich, G. D. Vogels, and R. S. Wolfe, *Biochemistry, 17,* 4583-4593 (1978).

17. W. M. H. van de Wijngaard, R. L. Lugtigheid, and C. van der Drift, *Antonie van Leeuwenhoek, 60,* 1-6 (1991); and refs. cited therein.

18. K. Laufer, B. Eikmanns, U. Frimmer, and R. K. Thauer, *Z. Naturforsch., 42c,* 360-372 (1987); R. K. Thauer, D. Möller-Zinkhan, and A. M. Spormann, *Ann. Rev. Microbiol., 43,* 43-67 (1989).

19. S. W. Ragsdale, H. G. Wood, T. A. Morton, L. G. Ljungdahl, and D. V. DerVartanian, in *The Bioinorganic Chemistry of Nickel* (J. R. Lancaster, ed.), VCH, New York, 1988, pp. 311-332.

20. C. D. Taylor and R. S. Wolfe, *J. Biol. Chem., 249*, 4879-4885 (1974).

21. K. M. Noll, K. L. Rinehart, Jr., R. S. Tanner, and R. S. Wolfe, *Proc. Natl. Acad. Sci. USA, 83*, 4238-4242 (1986); J. Ellermann, A. Kobelt, A. Pfaltz, and R. K. Thauer, *FEBS Lett., 220*, 358-362 (1987).

22. R. Hedderich and R. K. Thauer, *FEBS Lett., 234*, 223-227 (1988).

23. R. Hedderich, S. P. J. Albracht, D. Linder, J. Koch, and R. K. Thauer, *FEBS Lett., 298*, 65-68 (1992).

24. R. P. Gunsalus and R. S. Wolfe, *Biochem. Biophys. Res. Commun., 76*, 790-795 (1977).

25. T. A. Bobik and R. S. Wolfe, *Proc. Natl. Acad. Sci. USA, 85*, 60-63 (1988).

26. R. K. Thauer, K. Jungermann, and K. Decker, *Bacteriol. Rev., 41*, 100-180 (1977); L. Daniels, *Trends Biotechnol., 2*, 91-97 (1984); L. Daniels, R. Sparling, and G. D. Sprott, *Biochim. Biophys. Acta, 768*, 113-163 (1984).

27. M. Blaut and G. Gottschalk, *Trends Biochem. Sci., 10*, 486-489 (1985); B. Kaesler and P. Schönheit, *Eur. J. Biochem., 174*, 189-197 (1988).

28. W. L. Ellefson and R. S. Wolfe, *J. Biol. Chem., 256*, 4259-4262 (1981).

29. A. Klein, R. Allmansberger, M. Bokranz, S. Knaub, B. Müller, and E. Muth, *Mol. Gen. Genet., 213*, 409-420 (1988).

30. P. L. Hartzell and R. S. Wolfe, *Proc. Natl. Acad. Sci. USA, 83*, 6726-6730 (1986).

31. R. P. Gunsalus and R. S. Wolfe, *J. Bacteriol., 225*, 1891-1895 (1980); D. P. Nagle, Jr., and R. S. Wolfe, *Proc. Natl. Acad. Sci. USA, 80*, 2151-2155 (1983).

32. W. L. Ellefson and R. S. Wolfe, *J. Biol. Chem., 255*, 8388-8389 (1980).

33. D. Ankel-Fuchs and R. K. Thauer, *Eur. J. Biochem., 156*, 171-177 (1986).

34. J. Ellermann, R. Hedderich, R. Böcher, and R. K. Thauer, *Eur. J. Biochem., 172*, 669-677 (1988).

35. S. Rospert, R. Böcher, S. P. J. Albracht, and R. K. Thauer, *FEBS Lett., 291*, 371-375 (1991).

36. R. P. Gunsalus and R. S. Wolfe, *J. Bacteriol., 135*, 851-857 (1978).

37. L. P. Wackett, J. F. Honek, T. P. Begley, V. Wallace, W. H. Orme-Johnson, and C. T. Walsh, *Biochemistry, 26,* 6012-6018 (1987).

38. J. Ellermann, S. Rospert, R. K. Thauer, M. Bokranz, A. Klein, M. Voges, and A. Berkessel, *Eur. J. Biochem., 184,* 63-68 (1989).

39. A. Kobelt, A. Pfaltz, D. Ankel-Fuchs, and R. K. Thauer, *FEBS Lett., 214,* 265-268 (1987).

40. R. Ossmer, T. Mund, P. L. Hartzell, U. Konheiser, G. W. Kohring, A. Klein, and R. S. Wolfe, *Proc. Natl. Acad. Sci. USA, 83,* 5789-5792 (1986).

41. F. Mayer, M. Rohde, M. Salzmann, A. Jussofie, and G. Gottschalk, *J. Bacteriol., 170,* 1438-1444 (1988).

42. Unpublished result referred to by R. S. Wolfe, *Trends Biochem. Sci., 10,* 396-399 (1985).

43. W. L. Ellefson, W. B. Whitman, and R. S. Wolfe, *Proc. Natl. Acad. Sci. USA, 79,* 3707-3710 (1982).

44. P. Schönheit, J. Moll, and R. K. Thauer, *Arch. Microbiol., 123,* 105-107 (1979).

45. G. Diekert, B. Klee, and R. K. Thauer, *Arch. Microbiol., 124,* 103-106 (1980).

46. G. Diekert, R. Jaenchen, and R. K. Thauer, *FEBS Lett., 119,* 118-120 (1980); R. Jaenchen, G. Diekert, and R. K. Thauer, *FEBS Lett., 130,* 133-135 (1981).

47. A. Pfaltz, B. Jaun, A. Fässler, A. Eschenmoser, R. Jeanchen, H. H. Gilles, G. Diekert, and R. K. Thauer, *Helv. Chim. Acta, 65,* 828-865 (1982).

48. J. T. Keltjens, C. G. Caerteling, A. M. van Kooten, H. F. van Dijk, and G. D. Vogels, *Biochim. Biophys. Acta, 743,* 351-358 (1983); J. T. Keltjens, W. B. Whitman, C. G. Caerteling, A. M. van Kooten, R. S. Wolfe, and G. D. Vogels, *Biochem. Biophys. Res. Commun., 108,* 495-503 (1982).

49. D. A. Livingston, A. Pfaltz, J. Schreiber, A. Eschenmoser, D. Ankel-Fuchs, J. Moll, R. Jaenchen, and R. K. Thauer, *Helv. Chim. Acta, 67,* 334-351 (1984); see also R. P. Hausinger, W. H. Orme-Johnson, and C. T. Walsh, *Biochemistry, 23,* 801-804 (1984).

50. A. Fässler, A. Kobelt, A. Pfaltz, A. Eschenmoser, C. Bladen, A. R. Battersby, and R. K. Thauer, *Helv. Chim. Acta, 68,* 2287-2298 (1985).

51. G. Färber, W. Keller, C. Kratky, B. Jaun, A. Pfaltz, C. Spinner, A. Kobelt, and A. Eschenmoser, *Helv. Chim. Acta, 74,* 697-716 (1991).

52. K. D. Olson, H. Won, R. S. Wolfe, D. R. Hare, and M. F. Summers, *J. Am. Chem. Soc., 112,* 5884-5886 (1990).

53. A. Eschenmoser, *Ann. N.Y. Acad. Sci.*, *471*, 108-129 (1986).

54. A. Pfaltz, A. Kobelt, R. Hüster, and R. K. Thauer, *Eur. J. Biochem.*, *170*, 459-467 (1987).

55. G. Diekert, U. Konheiser, K. Piechulla, and R. K. Thauer, *J. Bacteriol.*, *148*, 459-464 (1981).

56. A. Pfaltz, D. A. Livingston, B. Jaun, G. Diekert, R. K. Thauer, and A. Eschenmoser, *Helv. Chim. Acta*, *68*, 1338-1358 (1985).

57. A. Fässler, Ph.D. thesis, ETH Nr. 7799, Zürich, 1985.

58. A. Kobelt, Ph.D. thesis, ETH Nr. 8509, Zürich, 1988.

59. C. L. Hamilton, M. W. Renner, and R. A. Scott, *Biochim. Biophys. Acta*, *1074*, 312-319 (1991).

60. F. A. Walker, E. Hui, and J. M. Walker, *J. Am. Chem. Soc.*, *97*, 2390-2397 (1975); B. D. McLees and W. S. Coughey, *Biochemistry*, *7*, 642-652 (1968).

61. C. L. Hamilton, R. A. Scott, and M. K. Johnson, *J. Biol. Chem.*, *264*, 11605-11613 (1989); G. P. Diakun, B. Piggott, H. J. Tinton, D. Ankel-Fuchs, and R. K. Thauer, *Biochem. J.*, *232*, 281-284 (1985).

62. A. K. Shiemke, J. A. Shelnutt, and R. A. Scott, *J. Biol. Chem.*, *264*, 11236-11245 (1989).

63. A. K. Shiemke, W. A. Kaplan, C. L. Hamilton, J. A. Shelnutt, and R. A. Scott, *J. Biol. Chem.*, *264*, 7276-7284 (1989).

64. L. P. Wackett, J. F. Honek, T. P. Begley, S. L. Shames, E. C. Niederhoffer, R. P. Hausinger, W. H. Orme-Johnson, and C. T. Walsh, in *The Bioinorganic Chemistry of Nickel* (J. R. Lancaster, ed.), VCH, New York, 1988, pp. 249-274.

65. M. K. Eidsness, R. J. Sullivan, J. R. Schwartz, P. L. Hartzell, R. S. Wolfe, A.-M. Flank, S. P. Cramer, and R. A. Scott, *J. Am. Chem. Soc.*, *108*, 3120-3121 (1986).

66. J. A. Shelnutt, *J. Phys. Chem.*, *93*, 6283-6290 (1989).

67. A. K. Shiemke, R. A. Scott, and J. A. Shelnutt, *J. Am. Chem. Soc.*, *110*, 1645-1646 (1988).

68. A. K. Shiemke, C. L. Hamilton, and R. A. Scott, *J. Biol. Chem.*, *263*, 5611-5616 (1988).

69. G. J. Colpas, M. J. Maroney, C. Bagyinka, M. Kumar, W. S. Willis, S. L. Suib, N. Baidya, and P. K. Mascharak, *Inorg. Chem.*, *30*, 920-928 (1991).

70. S. P. J. Albracht, D. Ankel-Fuchs, J. W. van der Zwaan, R. D. Fontijn, and R. K. Thauer, *Biochem. Biophys. Acta*, *870*, 50-57 (1986).

71. S. P. J. Albracht, D. Ankel-Fuchs, R. Böcher, J. Ellermann, J. Moll, J. W. van der Zwaan, and R. K. Thauer, *Biochim. Biophys. Acta*, *955*, 86-102 (1988).

72. J. A. Krzycki and R. C. Prince, *Biochim. Biophys. Acta, 1015*, 53-60 (1990).

73. P. E. Rouvière and R. S. Wolfe, *J. Biol. Chem., 263*, 7913-7916 (1988).

74. M. S. Ram, A. Bakac, and J. H. Espenson, *Inorg. Chem., 25*, 3267-3272 (1986).

75. F. V. Lovecchio, E. S. Gore, and D. H. Busch, *J. Am. Chem. Soc., 96*, 3109-3118 (1974).

76. B. Jaun and A. Pfaltz, *J. Chem. Soc., Chem. Commun.*, 1327-1329 (1986).

77. B. Jaun, *Helv. Chim. Acta, 73*, 2209-2217 (1990).

78. A. M. Stolzenberg and M. T. Stershic, *Inorg. Chem., 27*, 1614-1620 (1988).

79. C. Kratky, R. Waditschatka, C. Angst, J. E. Johansen, J. C. Plaquevent, J. Schreiber, and A. Eschenmoser, *Helv. Chim. Acta, 68*, 1312-1337 (1985).

80. M. W. Renner, L. R. Furenlid, K. M. Barkigia, A. Foreman, H.-K. Shim, D. J. Simpson, and K. M. Smith, *J. Am. Chem. Soc., 113*, 6891-6898 (1991).

81. L. R. Furenlid, M. W. Renner, and J. Fajer, *J. Am. Chem. Soc., 112*, 8987-8989 (1990).

82. L. R. Furenlid, M. W. Renner, D. J. Szalda, and E. Fujita, *J. Am. Chem. Soc., 113*, 883-892 (1991).

83. A. Stolzenberg and M. T. Stershic, *Inorg. Chem., 26*, 3082-3083 (1987).

84. C. K. Chang, L. K. Hanson, P. F. Richardson, R. Young, and J. Fajer, *Proc. Natl. Acad. Sci. USA, 78*, 2652-2656 (1981).

85. B. Jaun and A. Pfaltz, *J. Chem. Soc., Chem. Commun.*, 293-294 (1988).

86. S.-K. Lin and B. Jaun, *Helv. Chim. Acta* (1992), in press.

87. C. Gosden, K. P. Healy, and D. Pletcher, *J. Chem. Soc., Dalton*, 972-976 (1978).

88. M. J. D'Aniello, Jr., and E. K. Barefield, *J. Am. Chem. Soc., 98*, 1610-1611 (1976).

89. S.-K. Lin and B. Jaun, *Helv. Chim. Acta, 74*, 1725-1738 (1991).

90. Y. Ahn, J. A. Krzycki, and H. G. Floss, *J. Am. Chem. Soc., 113*, 4700-4701 (1991).

91. J. C. Salerno, in *The Bioinorganic Chemistry of Nickel* (J. R. Lancaster, ed.), VCH, New York, 1988, pp. 53-71.

92. R. R. Gagne and D. M. Ingle, *J. Am. Chem. Soc.*, *102*, 1444-1446 (1980); P. Chmielewski, M. Grzeszczuk, L. Latos-Grazynski, and J. Lisowski, *Inorg. Chem.*, *28*, 3546-3552 (1989).

93. M. Zimmer and R. H. Crabtree, *J. Am. Chem. Soc.*, *112*, 1062-1066 (1990).

94. A. Berkessel, *Bioorg. Chem.*, *19*, 101-115 (1991).

95. E. Anklam and S. Steenken, *J. Photochem. A*, *43*, 233-235 (1988); T. Yokota and O. P. Strausz, *J. Phys. Chem.*, *83*, 3196-3202 (1979).

10

Synthesis and Degradation of Organomercurials by Bacteria

A Comment by The Editors

Helmut Sigel and Astrid Sigel

Institute of Inorganic Chemistry
University of Basel
Spitalstrasse 51
CH-4056 Basel, Switzerland

The author, who had undergone the commitment to write the above-titled Chapter 10 was unable for various reasons to deliver his manuscript until it was too late to engage another competent author for this contribution. As *organomercurials* are important compounds, we feel that they should not be completely missing in a volume focusing on the "Biological Properties of Metal Alkyl Derivatives". For this reason we decided to provide the following short information.

The main justification of the scheduled author for not keeping his commitment, we believe, is of general interest to the readership, as it reflects the present scientific controversy about the formation of organomercurials in nature. Therefore, the decisive part of his FAX message (which reached us several months after the deadline) is given below (*in italics*) in an edited version:

There were [several] *reasons for my failure to get the expected chapter on methylmercury in on time. . . . First and most important . . . was my lack of comfort with the subject matter that I wished to understand and present. It meant coming to grips in a much*

deeper fashion with a range of materials that had been . . . accepted
based on work some fifteen [or more] *years ago, whereas people truly*
involved in the subject . . . today . . . believe that the old pic-
ture [may be] *incorrect.* [The question is:] *how is methylmercury*
. . . made and cycled in the environment? . . . Newer [studies,]
emphasizing sulfate-reducing bacteria tend to ignore . . . [the older]
. . . picture of biological synthesis of methyl Co-B$_{12}$ followed by
abiotic transfer of the methyl group to mercury. . . . Yet, how
methylmercury is synthesized is still unclear. As of today, we still
do not understand the evidence in favor of a primarily biotic origin
of methylmercury versus arguments for a substantially or totally
abiotic pathway. That is obviously crucial to what we want to con-
sider. . . . The situation is unresolved . . . [and there are]
new data supporting a direct synthesis of dimethylmercury, with mono-
methylmercury coming only as a secondary breakdown product. . . .
This clearly requires understanding that takes time and energy on a
large scale—and we have not completed that process yet.

This appears as an honest description of the scheduled author's
dilemma that we had to respect. Yet, to help the interested reader
in his efforts to find access to the pertinent literature, we are providing below a selection of references (covering both aspects and
views indicated above). We are doing this also in the hope of promoting somewhat the resolution of the indicated problems regarding
the origin of organomercurials in nature. Some further information
may also be obtained via the Subject Index of this volume where the
appearance of *Mercury* in other chapters of this book is summarized;
in fact, *Mercury* appears in the Index of practically every volume of
this series, demonstrating its various effects on life processes.

From a literature search we selected the following references
[1-45], which are listed in a historical order beginning with the
year 1975 and an emphasis on more recent publications. To increase
the information content, and hence the usefulness, of the list, the
complete title of a citation is also given and the length of each
article is indicated.

REFERENCES

1. J. M. Wood, "Biological Cycles for Elements in the Environment", *Naturwissenschaften, 62,* 357-364 (1975).

2. K. Reisinger, M. Stoeppler, and H. W. Nuernberg, "Biological Methylation of Inorganic Mercury by *Saccharomyces cerevisiae*— A Possible Environmental Process?", *Fresenius' Z. Anal. Chem., 316,* 612-615 (1983).

3. G. C. Compeau and R. Bartha, "Effects of Sea Salt Anions on the Formation and Stability of Methylmercury", *Bull. Environ. Contam. Toxicol., 31,* 486-493 (1983).

4. *"Methods Involving Metal Ions and Complexes in Clinical Chemistry",* Vol. 16 of *Metal Ions in Biological Systems* (H. Sigel and A. Sigel, eds.), Marcel Dekker, New York, 1983.

5. J. B. Robinson and O. H. Tuovinen, "Mechanisms of Microbial Resistance and Detoxification of Mercury and Organomercury Compounds: Physiological, Biochemical, and Genetic Analyses", *Microbiol. Rev., 48,* 95-124 (1984).

6. J. M. Wood, "Microbiological Strategies in Resistance to Metal Ion Toxicity", *Met. Ions Biol. Syst., 18,* 333-351 (1984); see also Ref. 7.

7. *"Circulation of Metals in the Environment",* Vol. 18 of *Metal Ions in Biological Systems* (H. Sigel and A. Sigel, eds.), Marcel Dekker, New York, 1984.

8. G. C. Compeau and R. Bartha, "Sulfate-reducing Bacteria: Principal Methylators of Mercury in Anoxic Estuarine Sediment", *Appl. Environ. Microbiol., 50,* 498-502 (1985).

9. L. Magos, A. W. Brown, S. Sparrow, E. Bailey, R. T. Snowden, and W. R. Skipp, "The Comparative Toxicology of Ethyl- and Methylmercury", *Arch. Toxicol., 57,* 260-267 (1985).

10. T. P. Begley, A. E. Walts, and C. T. Walsh, "Mechanistic Studies of a Protonolytic Organomercurial Cleaving Enzyme: Bacterial Organomercurial Lyase", *Biochemistry, 25,* 7192-7200 (1986).

11. S. M. Callister and M. R. Winfrey, "Microbial Methylation of Mercury in Upper Wisconsin River Sediments", *Water, Air, Soil Pollut., 29,* 453-465 (1986).

12. *"Concepts on Metal Ion Toxicity",* Vol. 20 of *Metal Ions in Biological Systems* (H. Sigel and A. Sigel, eds.), Marcel Dekker, New York, 1986.

13. G. C. Compeau and R. Bartha, "Effect of Salinity on Mercury-methylating Activity of Sulfate-reducing Bacteria in Estuarine Sediments", *Appl. Environ. Microbiol., 53,* 261-265 (1987).

14. E. T. Korthals and M. R. Winfrey, "Seasonal and Spatial Varia-
    tions in Mercury Methylation and Demethylation in an Oligotrophic
    Lake", *Appl. Environ. Microbiol.*, *53*, 2397-2404 (1987).

15. L. Xun, N. E. R. Campbell, and J. W. M. Rudd, "Measurements of
    Specific Rates of Net Methylmercury Production in the Water
    Column and Surface Sediments of Acidified and Circumneutral
    Lake", *Can. J. Fish. Aquat. Sci.*, *44*, 750-757 (1987).

16. K. Miura and N. Imura, "Mechanism of Methylmercury Cytotoxicity",
    CRC Crit. Rev. Toxicol., *18*, 161-188 (1987).

17. J. M. Wood, "Mechanisms for B_{12}-dependent Methyl Transfer to
    Heavy Elements", in *The Biological Alkylation of Heavy Elements*
    (P. J. Craig and F. Glockling, eds.), Royal Society of Chemistry,
    London, Spec. Publ. No. 66, 1988, pp. 62-76.

18. S. Silver and T. K. Misra, "Plasmid-Mediated Heavy Metal
    Resistances" with a section on "Mercury and Organomercurial
    Resistances", *Ann. Rev. Microbiol.*, *42*, 717-743 (1988).

19. F. Baldi, E. Cozzani, and M. Filippelli, "Gas Chromatography-
    Fourier Transform Infrared Spectroscopy for Determining Traces
    of Methane from Biodegradation of Methylmercury", *Environ. Sci.
    Technol.*, *22*, 836-839 (1988).

20. M. Horvat, K. May, M. Stoeppler, and A. R. Byrne, "Comparative
    Studies of Methylmercury Determination in Biological and Environ-
    mental Samples", *Appl. Organometal. Chem.*, *2*, 515-524 (1988).

21. R. J. Steffan, E. T. Korthals, and M. R. Winfrey, "Effects of
    Acidification on Mercury Methylation, Demethylation, and Vola-
    tilization in Sediments from an Acid-susceptible Lake", *Appl.
    Environ. Microbiol.*, *54*, 2003-2009 (1988).

22. C. T. Walsh, M. D. Distefano, M. J. Moore, L. M. Shewchuk, and
    G. L. Verdine, "Molecular Basis of Bacterial Resistance to
    Organomercurial and Inorganic Mercuric Salts", *FASEB J.*, *2*,
    124-130 (1988).

23. A. E. Walts and C. T. Walsh, "Bacterial Organomercurial Lyase:
    Novel Enzymatic Protonolysis of Organostannanes", *J. Am. Chem.
    Soc.*, *110*, 1950-1953 (1988).

24. F. E. Brinckman and G. J. Olson, "Global Biomethylation of the
    Elements: Its Role in the Biosphere Translated to New Organo-
    metallic Chemistry and Biotechnology", in *The Biological Alkyla-
    tion of Heavy Elements* (P. J. Craig and F. Glockling, eds.),
    Royal Society of Chemistry, London, Spec. Publ. No. 66, 1988,
    pp. 168-196.

25. L. Magos, "Mercury", in *Handbook on Toxicity of Inorganic Com-
    pounds* (H. G. Seiler, H. Sigel, and A. Sigel, eds.), Marcel
    Dekker, New York, 1988, pp. 419-436.

26. N. S. Bloom, "Determination of Picogram Levels of Methylmercury
    by Aqueous Phase Ethylation, Followed by Cryogenic Gas Chroma-
    tography with Cold Vapor Atomic Fluorescence Detection", *Can.
    J. Fish. Aquat. Sci.*, *46*, 1131-1140 (1989).

27. U. Harms, "Determination of Methylmercury in Biological Materials by Gas Chromatography-Atomic Absorption Spectrometry", in *5th Colloq. Atomspectrom. Spurenanal.* (B. Welz, ed.), Bodenseewerk Perkin-Elmer, Überlingen (FRG), 1989, pp. 737-746.

28. N. W. Revis, T. R. Osborne, G. Holdsworth, and C. Hadden, "Distribution of Mercury Species in Soil from a Mercury-contaminated Site", *Water, Air, Soil Pollut., 45,* 105-113 (1989).

29. M. Berman, T. Chase, Jr., and R. Bartha, "Carbon Flow in Mercury Biomethylation by *Desulfovibrio desulfuricans*", *Appl. Environ. Microbiol., 56,* 298-300 (1990).

30. M. R. Winfrey and J. W. M. Rudd, "Environmental Factors Affecting the Formation of Methylmercury in Low pH Lakes", *Environ. Toxicol. Chem., 9,* 853-869 (1990).

31. Y. L. Tsai and B. H. Olson, "Effects of Mercury, Methyl Mercury, and Temperature on the Expression of Mercury Resistance Genes in Environmental Bacteria", *Appl. Environ. Microbiol., 56,* 3266-3272 (1990).

32. R. D. Wilken, H. Hintelmann, and R. Ebinghaus, "Biologically Induced Mercury Interconversions in the Elbe River", *Vom Wasser, 74,* 383-392 (1990).

33. F. Baldi, "Microbial Transformation of Mercury", in *The Accumulation and Transformation of Chemical Contaminants by Biotic and Abiotic Processes in the Marine Environment* (G. Gabrielides, ed.), FAO, UNEP, IAEA Proceed. MAP Technical Reports No. 59, 1991, pp. 57-74.

34. F. M. D'Itri, "Mercury Contamination. What We Have Learned Since Minamata", *Environ. Monitoring Assessment, 19,* 165-182 (1991).

35. Y.-H. Lee and Å. Iverfeldt, "Measurement of Methylmercury and Mercury in Run-off, Lake and Rain Waters", *Water, Air, Soil Pollut., 56,* 309-321 (1991).

36. R. P. Mason and W. F. Fitzgerald, "Mercury Speciation in Open Ocean Waters", *Water, Air, Soil Pollut., 56,* 779-789 (1991).

37. C. Henriette, E. Petitdemange, G. Raval, and R. Gay, "Mercuric Reductase Activity in the Adaptation to Cationic Mercury, Phenyl Mercuric Acetate and Multiple Antibiotics of a Gram-negative Population Isolated from an Aerobic Fixed-bed Reactor", *J. Appl. Bacteriol., 71,* 439-444 (1991).

38. F. Baldi and M. Filippelli, "New Method for Detecting Methylmercury by Its Enzymic Conversion to Methane", *Environ. Sci. Technol., 25,* 302-305 (1991).

39. R. S. Oremland, C. W. Culbertson, and M. R. Winfrey, "Methylmercury Decomposition in Sediments and Bacterial Cultures: Involvement of Methanogens and Sulfate Reducers in Oxidative Demethylation", *Appl. Environ. Microbiol., 57,* 130-137 (1991).

40. R. Cela, R. A. Lorenzo, E. Rubi, A. Botana, M. Valino, C. Casais,
     M. S. Garcia, M. C. Mejuto, and M. H. Bollain, "Mercury Specia-
     tion in Raw Sediments of the Pontevedra Estuary (Galicia, Spain)",
     Environ. Technol., *13*, 11-22 (1992).

41. S. Silver and M. Walderhaug, "Gene Regulation of Plasmid- and
     Chromosome-Determined Inorganic Ion Transport in Bacteria" with
     a section on "Mercury Resistance", *Microbiol. Rev.*, *56*, 195-228
     (1992).

42. S. Krishnamurthy, "Biomethylation and Environmental Transport
     of Metals", *J. Chem. Educ.*, *69*, 347-350 (1992).

43. S. Padberg, Å. Iverfeldt, Y.-H. Lee, F. Baldi, and M. Filippelli,
     "Determination of Low-Level Methylmercury Concentrations in Water
     and Complex Matrices by Different Analytical Methods. A Methodo-
     logical Intercomparison", conference presentation during May 31-
     June 4, 1992 in Monterey, California; cf. also in Ref. 44.

44. F. Baldi and M. Filippelli, "Importance of New Specific Analytical
     Procedures in Determining Organic Mercury Species Produced by
     Microorganism Cultures", conference presentation during May 31-
     June 4, 1992 in Monterey, California; cf. also in Ref. 44.

45. "*Mercury as a Global Pollutant—Toward Integration and Synthesis*"
     (J. Huckabee and C. Watras, eds.), Lewis Publishers, 1993, in
     press.

11

Biogenesis and Metabolic Role of Halomethanes in Fungi and Plants

David B. Harper

Department of Food and Agricultural Chemistry
The Queen's University of Belfast
Newforge Lane
Belfast, BT9 5PX, Northern Ireland

1. INTRODUCTION

The realization that halogenated gases such as the man-made chlorofluorocarbons have increased the rate of ozone destruction in the
stratosphere has stimulated a growing interest in identifying and
quantifying the natural sources of volatile halogenated compounds.
The most important of these compounds in terms of atmospheric abundance are undoubtedly the gaseous monohalomethanes (CH$_3$Cl, CH$_3$Br,
and CH$_3$I), but less volatile polyhalogenated compounds, in particular
CHBr$_3$, are also believed to be formed in nature in substantial quantities. The mean concentrations of these compounds in the atmosphere
and in seawater are given in Table 1 together with estimates for each
of the annual global inputs from various sources both natural and
anthropogenic. An approximate residence time for each compound in
the atmosphere is also included.

The predominant volatile halohydrocarbon in the atmosphere is
CH$_3$Cl. Between 2.5 and 5 million tonnes/year must originate from
natural sources according to estimates based on environmental concentrations [1,2]; the contribution from industrial sources of 30 thousand tonnes/year is insignificant in comparison [3]. When vegetation
is burned a small proportion of the Cl$^-$ present is converted to CH$_3$Cl
and the compound is therefore released as a product of forest fires
and slash-and-burn agriculture in the tropics [4]. However, calculations by Crutzen et al. [5] indicate that release of CH$_3$Cl by biomass
burning is at least an order of magnitude less than that required to
account for the observed atmospheric concentration of CH$_3$Cl. It has
also been suggested that volcanic emissions may contain significant
quantities of CH$_3$Cl but again thermodynamic calculations by Symonds
et al. [6] demonstrate that the annual flux from this source is

TABLE 1

Environmental Concentrations and Annual Global Inputs to the
Atmosphere of the Most Abundant Naturally
Occurring Halomethanes

Halomethane	Air conc. (pptv)	Seawater conc. (ng/liter)	Atmos. residence time (years)	Global input to atmosphere (10^6 ton/year)
CH_3Cl	630	11.5	1.5	Natural 2.5-5.0 oceanic and terrestial 0.45 veg. burning Anthropogenic 0.03
CH_3Br	23	1.2	1.5	Natural 0.1-0.3 oceanic Anthropogenic 0.08
CH_3I	2	1.6	0.02	Natural 0.3-1.3 oceanic
$CHBr_3$	2	10	0.04	Natural 1.0-2.0 oceanic

Source: Data compiled from Refs. 1-3, 5, 12, 14, 15, 17, 22, 23, 32.

negligible and that elevated CH_3Cl in the plume arising from a volcanic eruption is likely to be due to the burning of vegetation associated with lava flows. The oceans appear to be the dominant source
of CH_3Cl and indeed the other monohalomethanes [1,2]. Zafiriou [7]
proposed that CH_3Cl is formed in seawater by reaction of biologically
formed CH_3I with Cl^-, a reaction which has a half-life of ~20 days at
19°C, but Singh et al. [2] could find no relationship between oceanic
CH_3I and CH_3Cl concentrations. Furthermore the coexistence of high
levels of CH_3I and relatively low levels of CH_3Cl and vice versa
suggested independent oceanic sources of CH_3Cl and CH_3I. However,
Zika et al. [8] argued that the rate of Cl^- reaction with CH_3I
will vary considerably with temperature and hence latitude from a
half-life of 5-6 days in tropical waters to one of 150 days in

polar regions, so that a direct correlation between CH_3I and CH_3Cl
or CH_3Br may be masked by other processes. Regardless of the extent
of this reaction CH_3Cl is known to originate in the oceans by direct
biological synthesis by marine macroalgae and possibly marine microorganisms [9-11]. Several terrestrial biological sources of CH_3Cl
have also been identified. Thus CH_3Cl production has been shown to
occur in at least 34 species of wood-rotting fungi [12] and CH_3Cl
release from certain higher plants has also been observed [11,13].
However, the relative contributions of the different biological
sources to the total global flux of the compound is still far from
clear and there are insufficient data on CH_3Cl release rates by producer organisms under natural conditions to be certain that all major
biological sources have been identified. As there is some evidence
of enhanced CH_3Cl concentrations over equatorial regions [1] it would
be logical to concentrate the search for novel biological CH_3Cl producers on such areas.

For CH_3Br there would appear to be a global oceanic source of
between 90 and 300,000 tonnes/year [2,14]. Interestingly, a high
correlation was found between CH_3Cl and CH_3Br concentrations in seawater, suggesting a common origin [2]. This finding is consistent
with the observation that several macroalgae capable of producing
CH_3Cl also release CH_3Br though in somewhat smaller quantities [10].
The estimated anthropogenic emissions of CH_3Br (mainly derived from
the use of CH_3Br as a fumigant) constitute a relatively large proportion of the total global flux of the compound. The significantly
higher CH_3Br levels in the northern vs. the southern hemisphere
(26-30 pptv as opposed to 19 pptv) also imply a major man-made
northern hemispheric source [2,14]. No major biological source of
CH_3Br on land has yet been identified.

CH_3I also has an oceanic origin. The first estimate of the
annual production rate by Lovelock et al. [16] from measurements of
air and seawater concentrations was 40 million tonnes/year, using a
value for atmospheric residence time of 50 hr. However, more recent
estimates of annual production, based on a residence time of between
5 and 10 days, range from 0.3 to 1.3 million tonnes/year [2,17].

Substantial variations in CH_3I concentration in seawater provide strong evidence of a biological origin. Lovelock [9] noted a 1000-fold elevation in CH_3I concentrations near kelp beds while Rasmussen et al. [17] in a global survey showed that atmospheric CH_3I levels were markedly higher (10-20 pptv) in oceanic regions of high biomass productivity such as the areas off the Peruvian, South African, and Icelandic coasts. The latter workers calculated that these areas, although comprising only 10% of the ocean surface, might produce as much as 80% of the total CH_3I flux. Macroalgae are now well established as CH_3I producers [10,18] but it is nevertheless debatable whether all CH_3I produced in the marine environment arises by direct biosynthesis by marine organisms.

White [19] has suggested that CH_3I and other halomethanes can arise by reaction of dimethylsulfonium compounds present in many marine algae and phytoplankton with the appropriate halide ion in seawater. Brinkman et al. [20] have confirmed in the laboratory that one of the most commonly occurring dimethylsulfonium compounds, dimethylpropiothetin, does indeed react with I^- quite readily, so that it is possible that extracellular reactions of this type may be as important in marine production of CH_3I as direct biosynthesis of the compound.

It is only comparatively recently that the scale of the global flux of $CHBr_3$ has been appreciated. Using air and seawater measurements of $CHBr_3$ collected by Penkett et al. [14] and Dyrssen and Fogelqvist [21,22], Liss [23] computed an annual sea-to-air flux of 2 million tonnes/year. Atmospheric concentrations of the compound show no significant N-S interhemispheric differences suggesting a widespread natural source dominating input into the atmosphere. Biological formation is clearly a major source, biosynthesis of $CHBr_3$ having been observed in numerous marine algae [18,24-26]. Interestingly, atmospheric levels of $CHBr_3$ in the Arctic during March and April were found to average 38 pptv, approximately 10-fold higher than normal [27], and $CHBr_3$ concentrations of up to 300 ng/liter in surface seawater off Greenland have been recorded [21]. Cicerone et al. [28] also observed marked seasonal fluctuations in $CHBr_3$

concentrations in the atmosphere of the Canadian Arctic with a winter maximum and a summer minimum. Berg et al. [27] considered that emissions from marine organisms such as red benthic algae, molluscs, and sponges were the most likely explanation for these elevated $CHBr_3$ concentrations, while Sturges and Barrie [29] speculated that an oceanic bloom after the Arctic dawn in March-April is responsible for the pulse of $CHBr_3$ observed in these months. An algal source is clearly implicated by studies of the variation of $CHBr_3$ concentration in the Arctic Ocean with depth which show a strong correlation between $CHBr_3$ concentration and algal biomass [21]. Wever et al. [30] postulated that $CHBr_3$ may also arise abiotically in the marine environment by reaction of HOBr released to seawater by algae with dissolved organic matter giving rise by the haloform reaction to unstable brominated compounds which decompose to form $CHBr_3$.

In addition to those listed in Table 1, other halogenated methanes with possible natural sources are widespread in the environment but data on most of these are sparse. $CHCl_3$ is present at a background concentration of about 20 pptv [31] in air and while a significant proportion of this atmospheric burden is believed to be anthropogenic in origin, several authors have concluded that much of the CH_3Cl in the atmosphere is of natural origin [31]. Brominated species such as CH_2BrCl (3 pptv), $CHBr_2Cl$ (0.9 pptv), and CH_2Br_2 (3-60 pptv) are also widely present in the atmosphere, the latter occurring at particularly high levels in the Arctic atmosphere during the spring period when $CHBr_3$ concentrations are markedly enhanced [27,32].

2. BIOLOGICAL HALOMETHANE PRODUCTION

2.1. Fungi

Hutchinson [34] first reported the presence of CH_3Cl in the headspace above cultures of several species of the classical genus *Fomes*, a widespread group of bracket fungi. In a later survey Cowan et al. [35] using gas chromatography (GC) identified significant amounts of CH_3Cl

above malt extract-grown cultures of 6 of 32 *Fomes* species examined.
Trace quantities of CH_3Cl have also been reported in compost after
commercial cultivation of *Agaricus bisporus* [36]. Harper [12]
adapted headspace techniques devised for determination of volatile
bacterial metabolites to quantitative measurement of CH_3Cl production
by the fungus *Phellinus pomaceus* syn *Fomes pomaceus*, a white-rot
fungus typically found on trees of the Rosaceae, especially *Prunus*.
Fungal cultures were grown on a variety of substrates in stoppered
flasks each fitted with an outlet comprising a glass tube packed
with cotton wool permitting a limited exchange of gases between the
internal and external atmospheres. The concentration of gaseous
CH_3Cl in the flask headspace was measured by GC at regular intervals.
From these values and a knowledge of the rate of diffusion of CH_3Cl
from the flask (obtained in a calibration procedure prior to fungal
culture), the total CH_3Cl generated at any given point during the
growth cycle was calculated using a computer integrating program.
All rubber fittings employed were coated with a coherent layer of
polytetrafluoroethylene (PTFE) as CH_3Cl is strongly absorbed by
untreated rubber. It was found that failure to observe this pre-
caution not only invalidated measurements but also that the swelling
of the rubber stopper associated with CH_3Cl uptake was such that
fracture of the neck of the flask frequently occurred.

With glucose as fungal growth substrate, CH_3Cl biosynthesis
was clearly confined to the period after exponential growth and
before autolysis, a pattern characteristic of secondary metabolite
production (Fig. 1). Cl^- uptake from the medium closely paralleled
CH_3Cl release. CH_3Cl yield, i.e., CH_3Cl release expressed as a percentage of Cl^- originally present in the medium, varied from 90 to
100% at Cl^- concentrations <4 mM to 20% at 50 mM. However, with more
natural growth substrates, such as wood and cellulosic material, CH_3Cl
yields remained substantial even at higher Cl^- concentrations although
production continued over a longer period. Thus CH_3Cl release typically extended over 4 weeks with cotton wool, 8 weeks with filter
paper, and 6 months with *Prunus domestica* sawdust, and CH_3Cl yields
could exceed 90% at 50 mM Cl^-. This contrast in yield with glucose-

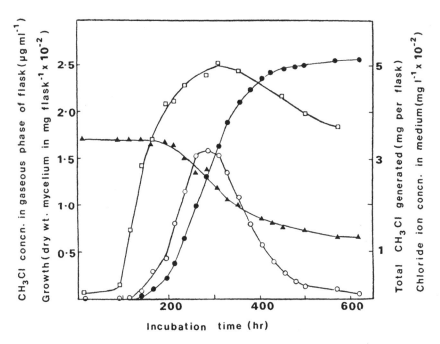

FIG. 1. Growth of *P. pomaceus* on glucose-based medium containing
9.6 mM NaCl in relation to CH₃Cl release and Cl⁻ loss from the medium.
Growth, □; CH₃Cl conc. in gaseous phase of culture flask, o; total
CH₃Cl generated per culture, ●; Cl⁻ conc. in culture medium, ▲.
Reproduced by permission from Ref. 12.)

based medium was ascribed to the suppression of secondary metabolism
by glucose.

In addition to Cl⁻, both Br⁻ and I⁻ but not F⁻ acted as
acceptors for the methylation system. The effect of halide ion concentration in the medium on the yield of different halomethanes with
filter paper as the fungal growth substrate is shown in Fig. 2 [37].
Br⁻ was methylated almost as efficiently as Cl⁻ but a sharp decline
in I⁻ methylation was noted at concentrations greater than 1 mM, a
finding attributed to the toxicity of the ion. When equimolar concentrations of the three halide ions were present in the medium, I⁻
was initially the preferred substrate. However, as the incubation
proceeded, Br⁻ and Cl⁻ were successively methylated as the concen-

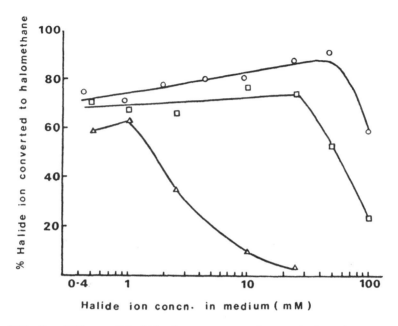

FIG. 2. Effect of halide ion concentration on halomethane release
by *P. pomaceus*. Fungus was grown on cellulose-based medium supplemented with Cl⁻, o; Br⁻, □; or I⁻, △. (Reproduced by permission
from Ref. 37.)

tration of first I⁻ and then Br⁻ decreased on conversion to
halomethane.

2.1.1. Distribution of the CH_3Cl Biosynthesis Trait

Harper and Kennedy [37] using gas chromatography/mass spectrometry
(GC/MS) could confirm release of CH_3Cl by only four of the six
species reported by Cowan et al. [35] as producing the compound.
Nevertheless a comprehensive survey of the distribution of the trait
among 90 species of polypore by Harper et al. [38] indicated that
the ability to convert Cl⁻ to CH_3Cl is widespread in the Hymenochaetaceae. Of 63 species from this family screened on three
different culture media, 34 (54%) were capable of CH_3Cl biosynthesis
including representatives of the genera *Phellinus, Inonotus, Hymeno-
chaete, Onnia,* and *Fomitoporia.* Quantitative differences in CH_3Cl

TABLE 2

Chloromethane Production on Three Different Media
by Selected *Phellinus* and *Inonotus* Species

Species	Genus of host tree from which culture isolated	Percentage Cl⁻ converted to CH₃Cl		
		Glucose/ mycol. peptone/ agarose 9.5 mM Cl⁻	Malt extract/ agarose 9.5 mM Cl⁻	Filter paper mycol. peptone 10 mM Cl⁻
P. ignarius	*Pyrus*	16	39	63
P. lundelli	*Betula*	55	54	27
P. occidentalis	*Crataegus*	43	29	79
P. pachyphloes	*Mangifera*	42	4	18
P. pini	*Tsuga*	5	70	47
P. pomaceus	*Prunus*	57	46	82
P. populicola	*Populus*	6	43	22
P. ribis	*Crataegus*	21	50	82
P. trivialis	*Salix*	20	19	47
I. andersonii	*Quercus*	18	18	8
I. hispidus	*Populus*	3	43	11

Source: Data compiled from Ref. 38.

production between different isolates of the same species were noted
in several instances.

Biosynthesis of CH_3Cl was found to be particularly common in
the widely distributed genera *Phellinus* and *Inonotus*, white-rot
fungi characterized by bracket-like perennial fruiting bodies. In
these two genera, 61% of species examined released CH_3Cl during culture. The magnitude of CH_3Cl production varied with the species and
was dependent on the nature of the medium, biosynthesis being particularly favored by cellulose-based media. Over 58% of CH_3Cl-
producing species were found capable of converting more than 10% of
Cl^- in the medium to CH_3Cl and in many instances CH_3Cl yields were
much higher. Table 2 shows the percentage of Cl^- in the medium

converted to CH_3Cl on three different culture media by several of the more prolific CH_3Cl producing *Phellinus* and *Inonotus* species.

Of 27 non-Hymenochaetaceous species of polypore screened from the other major families of the suborder Polyporineae, namely, the Ganodermataceae and the Polyporaceae, only one—*Fomitopsis cytisina* syn *Fomes fraxineus*—was confirmed as releasing CH_3Cl during culture. This species gave a 10% conversion of Cl^- to CH_3Cl when grown on filter paper/mycological peptone medium. The significance of this sole non-Hymenochaetaceous producer is difficult to evaluate. Another species of the genus examined, *F. pinicola*, did not display CH_3Cl production nor did members of other genera belonging to this subfamily of the Polyporaceae, the Fomitoideae. It is possible that CH_3Cl biosynthesis by *F. cytisina* is indicative of a closer taxonomic relationship between the species and the Hymenochaetaceae than has hitherto been appreciated. Indeed in the wider context the CH_3Cl biosynthesis trait could prove a useful diagnostic character in the classification of the Hymenochaetaceae.

2.1.2. *Other Related Volatiles*

Harper and Kennedy [37] identified 13 compounds in addition to CH_3Cl in headspace collected from above *P. pomaceus*, including among the major components methyl esters of benzoic, salicylic, and 2-furoic acids. Both methyl benzoate and methyl salicylate had previously been reported as important odorous constituents of other *Phellinus* species [39]. Harper and Kennedy [37] speculated that the presence of such esters in species of this genus might be biochemically associated with the production of halomethanes, a proposal given some support by the results of experiments involving supplementation of culture medium with the pseudohalide ion, thiocyanate (SCN^-).

As the structural similarity of SCN^- to halide ion often results in similar chemical behavior it was considered possible that biologically the ion might act as an alternative substrate or a competitive inhibitor of the halomethane-generating system. Although Harper and Kennedy [37] could find no evidence for the methylation

TABLE 3

Effect of Thiocyanate on Halomethane and Methyl Benzoate Production
by *P. pomaceus* Growing on Glucose-Based Medium

Halide present	Conc. (mM)	Sodium thiocyanate (6 mM)	Total halomethane produced [μg/g mycelium]	Methyl benzoate [ng/g mycelium] after 40 days incubation
Cl^-	1	-	645	275
Cl^-	1	+	185	26
Cl^-	18	-	3998	151
Cl^-	18	+	418	36
Br^-	10	-	3135	73
Br^-	10	+	1887	38

Source: Data from Ref. 37.

of SCN^-, it was clear that the ion was a powerful inhibitor of halomethane biosynthesis. The competitive nature of this inhibition was
demonstrated by the observation that the production of halomethane
from halide ions with a higher affinity for the methylating system
than chloride was not curtailed as drastically by the presence of
SCN^- (Table 3). Significantly, SCN^- supplementation of culture
medium also dramatically reduced the biosynthesis of methyl benzoate,
suggesting that the methylation of halide ions and aromatic acids
might be mediated by the same biochemical system.

A comparison of methyl ester and CH_3Cl production in *Phellinus*
species revealed that, of the 23 CH_3Cl-releasing species discovered
in the genus, 12 were capable of producing methyl benzoate and also,
in most instances, methyl esters of salicylic and 2-furoic acids [38].
Ester production was not observed in any of the 14 *Phellinus* species
examined which did not produce CH_3Cl, a finding regarded as providing
further evidence of a biochemical link between CH_3Cl and ester biosynthesis.

2.2. Algae

2.2.1. Monohalomethanes

Lovelock [9] in 1975 noted that CH_3I was present in seawater in the vicinity of *Laminaria digitata* kelp beds off southwestern Ireland at a level approximately 1000-fold higher than that in the open ocean. Subsequently, CH_3I was identified as a minor component of fresh *Asparagopsis armata*, a red alga collected on the Spanish Mediterranean coast [24], while Gschwend et al. [18] reported but did not quantify the production of CH_3I by several brown, green, and red macroalgae from the coast of northeastern United States (see also Sec. 2.2.2). On the basis of measurements of CH_3I concentration in both the atmosphere and in seawater, Rasmussen et al. [17] concluded that regions of the ocean with high biomass productivity were a major source of CH_3I.

A detailed study of monohalomethane production by a macroalga did not appear until 1987 when Manley and Dastoor [10] published an investigation of the giant kelp *Macrocystis pyrifera*, which is an important and often dominant primary producer in Californian coastal waters. Mature blades of the kelp were placed in Plexiglas aquaria filled with seawater and illuminated for 6 hr/day. The aquaria were sealed with commercial plastic wrapping and a Plexiglas lid. Halomethanes were extracted from the seawater into a known volume of N_2 and subsequently cryogenically concentrated before analysis by GC using an electron capture detector.

During the first 3 days of culture CH_3Cl and CH_3I concentrations increased dramatically with initial production rates of 20 ng/g wet tissue per day and 4 ng/g wet tissue per day, respectively. In further experiments with tissue slices incubated in bottles filled with seawater and closed with rubber stoppers, rates of CH_3Cl, CH_3Br, and CH_3I production of 14, 4.3, and 14 ng/g wet tissue per day, respectively, were observed in the light. No significant difference in rates was exhibited by tissue incubated in the dark. Halomethane production did not appear to be a response to tissue wounding or

desiccation. The similarity in CH_3Cl and CH_3I release rates does
not support the proposition of Zafiriou [7] that CH_3Cl in the marine
environment is a product of the reaction of Cl^- in seawater and CH_3I
as the half-life of the latter reaction is reportedly ~20 days (see
also Sec. 1).

Field measurements of halomethane concentrations in surface
seawater in the kelp canopy indicated levels of CH_3Cl, CH_3Br, and
CH_3I of 34.0, 2.1, and 3.5 ng/liter, respectively, compared with
control values of 28.3, 1.6, and 2.8 ng/liter from a site 18 km
distant from the kelp bed. The failure to observe the large concentration differences reported by Lovelock [9] was tentatively ascribed
to differences in coastal typography, the relatively exposed site in
California perhaps allowing greater water exchange than the Irish
site. Making several assumptions and approximations, Manley and
Dastoor [10] estimated halomethane production by the kelp on the
basis of the field measurements and found that, while the tissue-
normalized CH_3Br and CH_3I production rates so calculated were similar
to those observed experimentally in tissue incubations in the laboratory, the estimate of CH_3Cl production was an order of magnitude
larger. The authors surmised that this discrepancy might imply an
additional source of CH_3Cl. However, this anomaly also raises the
possibility that CH_3Cl production in the laboratory experiments was
underestimated due to sorption of CH_3Cl by plastic materials used in
the construction of aquaria, a risk to which the workers allude
briefly but which they do not attempt to quantify. Additionally,
the use of uncoated rubber stoppers can lead to major errors in
CH_3Cl determination as such fitments are capable of not only absorbing CH_3Cl most efficiently (see Sec. 2.1) but also under appropriate
circumstances releasing gas previously absorbed. Although the latter
possibility was minimized in the experiments described by Manley and
Dastoor by baking the rubber stoppers at 80°C under vacuum for 16 hr,
no attempt was apparently made to prevent absorption of CH_3Cl.

More recently, Nightingale [40] reported CH_3I levels in *Laminaria digitata* kelp beds off the western coast of Scotland of about

4-5 ng/liter—about twice that measured away from the bed. This
finding in conjunction with the observations of Manley and Dastoor
[10] suggests that the exceptional enhancement of CH_3I concentrations
in the vicinity of the kelp beds studied by Lovelock [9] may not be
typical of kelp beds generally. In laboratory experiments Nightingale
[40] noted CH_3I release from a wide variety of macroalgae, brown algae
being among the most prolific producers with release rates ranging
from 30 ng/g dry tissue per day for *Laminaria saccharina* to 4 ng/g dry
tissue per day for *Fucus serratus*. In experiments with *Ascophyllum
nodosum* there was some evidence that CH_3I production was associated
with reimmersion of the plant after partial desiccation on exposure
by the retreating tide and also that CH_3I release was enhanced in
the dark and in older tissue.

Wuosma and Hager [11] mentioned that a survey of marine algae
collected randomly off California had shown that 22 out of 44 species
were CH_3Cl producers. The red alga *Endocladia muricata* was reported
to release 1.5 ng/g/day when incubated in 100 mM KCl. However, the
methods employed in this investigation are open to criticism (see
Sec. 2.3) and these results must await independent confirmation.

2.2.2. *Polyhalogenated Methanes*

The Hawaiian seaweed *Asparagopsis taxiformis* is highly prized for
its aroma and flavor. The essential oil which composes about 0.4%
of this red alga has been shown to consist predominantly of bromine-
and iodine-containing haloforms, principally $CHBr_3$ (80%) and $CHBr_2I$
(5%) [41]. The related species *A. armata* contained in addition small
quantities of polychlorinated compounds such as $CHCl_3$ and CCl_4 [24].

The release of a number of halogenated organic compounds to
seawater by a range of temperate marine monoalgae was quantitatively
determined by Gschwend et al. [18]. In these experiments algae were
incubated for 24 hr in seawater in Teflon-lined screw-capped jars in
situ at the collection site. Volatiles accumulating in the water
were concentrated by cryogenic trapping and transferred by closed-loop
stripping to Tenax-GC traps from which the volatiles were thermally

TABLE 4

Release Rates of Polyhalogenated Methanes
to Seawater by Temperate Macroalgae

Algal species	No. of samples	Release rates (ng/g dry wt per day)		
		$CHBr_3$	$CHBr_2Cl$	CH_2Br_2
Brown				
Ascophyllum nodosum	8	150-12,500	ND-3,000	ND-2,100
Fucus vesiculosus	7	140-4,700	ND-820	ND-590
Green				
Enteromorpha linza	2	ND-850	ND	ND-300
Ulva lacta	2	1,700-14,000	590-4,300	ND-250
Red				
Gigartina stellata	3	ND-21,000	ND-3,000	ND

Note: ND = not detected.
Source: Data taken with permission from Ref. 18 (Copyright 1985 by the AAAS).

desorbed and analyzed by GC/MS. Unfortunately, the method precluded the measurement of compounds with boiling points below 60°C. Hence the monohalomethanes were not quantified although CH_3I was detected in many samples (see Sec. 2.2.1). The release of polyhalogenated methanes—principally $CHBr_3$, $CHBr_2Cl$, and CH_2Br_2—was observed in five out of six species examined which included brown, green, and red algae (see Table 4). The halogenated volatile content of the algal tissue, quick frozen by dousing in liquid N_2, was also determined after vacuum distillation of the frozen tissue and cryogenic trapping. In addition to the compounds noted in Table 4, the presence of several other polyhalogenated methanes such as $CHBrCl_2$ and CH_2I_2 was noted. Some longer chain haloalkanes were also identified mainly bromo and iodo derivatives of propane and butane. Recoveries of poly-halogenated compounds were found to be very much higher (500%) when vacuum distillation was performed on fresh tissue. This loss of

volatiles on rapid freezing was considered by the authors as evidence
that the compounds were associated with the algal surface and readily
driven off by evaporation of liquid N_2 [18,42].

In a later study Gschwend and Macfarland [43] investigated the
seasonal variation in release rates in the brown algae A. *nodosum* and
F. *vesiculosus*. Release continued throughout most of the year although
there was some evidence of higher rates of emission in autumn but no
pronounced trend. These workers were surprised by the absence of
marked seasonality in polybromomethane release in view of the major
changes in algal growth, reproductive physiology, and peroxidase
enzyme status which occur during the year. It was proposed that epiphytic fungi such as *Mycosphaerella ascophylli* growing symbiotically
on the algae might be responsible for biosynthesis of polybromo-
methanes. Inoculation of seawater with surface material scraped from
A. *nodosum* or F. *vesiculosus* led to significant increases in $CHBr_3$
concentration lending some support for this hypothesis. Investigations by Nightingale [40] on algal species from the western coast of
Scotland showed rates of polyhalogenated methane release similar to
those noted by Gschwend et al. [18]. The observation by this worker
that older tissue had the highest output of halocarbons is consistent
with an epiphytic origin for these compounds as such communities might
reasonably be expected to be more strongly established on such tissue.

2.3. Higher Plants

In a careful and well-designed investigation Varns [13] showed that
freshly harvested tubers of the potato (*Solanum tuberosum*) released
CH_3Cl during the suberization process. Volatiles from the air above
the tubers were concentrated by adsorption on Tenax-GC at -40°C and
the adsorbent transferred to vials fitted with PTFE septa which were
heated at 140°C to allow desorption. Headspace samples were removed
from the vials and analyzed by GC using a flame ionization detector.
Identities of unknown compounds were confirmed by MS. The rate of
release of CH_3Cl attained a maximum of 100-700 ng/kg tuber per hour

within 2 or 3 days following harvest. Thereafter the release rate declined sharply falling to between 2 and 10% of the maximum by the seventh day after harvest. This period of rapid decrease coincided with the interval required for development of resistance to water loss by surfaces cut or damaged during harvesting. Indeed the high rate of CH_3Cl release could be reestablished by recutting or wounding of tubers. As it is known that increase in resistance of potato tissue to water loss is directly correlated with the formation of periderm and the deposition of suberin, Varns concluded that CH_3Cl release was associated with the suberization process. Somewhat surprisingly he interpreted his results as signifying that developing tuber tissue possesses a mechanism for concentrating natural CH_3Cl from the atmosphere and that the compound so sequestered is released on damage to the tuber. However, it is difficult to provide a rationale for such a mechanism. Also, as Varns admits, exchange by passive diffusion of accumulated CH_3Cl from tissue to the atmosphere should take only hours not the days reported. A more plausible explanation of his findings in the opinion of the reviewer is that CH_3Cl is actually synthesized de novo by potato tissue during the suberization process. The compound may be a metabolic intermediate itself or a byproduct of the biosynthesis of some other metabolite. A detailed study of the biochemical mechanism underlying release of CH_3Cl by potato tubers is clearly required and could yield some unexpected insights into plant metabolism.

Wuosma and Hager [11] reported in 1990 that whole cells of *Mesembryanthemum crystallinum*, the California ice plant, released CH_3Cl when incubated with 100 mM KCl and that cell-free extracts of the plant could convert *S*-adenosylmethionine (SAM) to CH_3Cl under similar conditions. Although the full experimental details of this work have yet to be published, the preliminary report indicated a rate of CH_3Cl release for whole cells of 19 pmol/g/day and for cell-free extracts of 3 fmol/mg/min. As *M. crystallinum* is a succulent species growing abundantly in the saline soils of the Californian coast the authors suggested that a survey of CH_3Cl release in other succulents growing in saline-rich environments might be rewarding.

Unfortunately, the validity of the measurement of such low rates of CH_3Cl release within the experimental context described by Wuosma and Hager is questionable. Calculations based on the reported rates of CH_3Cl release indicate that, even after making the most favorable assumptions regarding the experimental parameters, the concentration of CH_3Cl measured by the headspace technique used would be in the region of 1 ppb (v/v). It would seem doubtful that such concentrations could be determined with any degree of accuracy by direct GC on packed columns even if the unspecified detector was of the electron capture type. In these circumstances a preliminary concentration stage is normally essential prior to GC if adequate sensitivity is to be achieved. Considerable uncertainty is introduced merely by the fact that ambient atmospheric concentrations of CH_3Cl are also in the region of 1 ppb. As previously mentioned in Sec. 2.1 the use of vials with rubber septa untreated with PTFE will also lead to gross error. Wuosma and Hager also do not appear to take account of the fact that methylsulfonium salts such as SAM will decompose chemically to CH_3Cl under the conditions described in their experiments [44]. In this laboratory rates of abiotic release of CH_3Cl from 0.5 mM SAM in phosphate buffer at pH 6.8 containing 250 mM KCl, of as much as 1.5 nmol/mmol SAM per hr were measured (unpublished observations). Such rates are more than adequate to account for all CH_3Cl release recorded by Wuosma and Hager.

3. MECHANISMS OF HALOMETHANE BIOSYNTHESIS

Biochemically the enzymic incorporation of halide ion into a one-carbon compound poses some interesting questions. Although a large number of organohalogen compounds have been isolated from natural sources, until very recently the sole biological route identified for insertion of halogen has been the haloperoxidase-catalyzed introduction of halide ion in the presence of hydrogen peroxide [45] (1).

$$\text{Substrate + halide ion} + H^+ + H_2O_2 \longrightarrow \text{halometabolite} + 2H_2O$$

$$(1)$$

The halide ion may be Cl$^-$, Br$^-$, I$^-$, or even the pseudohalide ion SCN$^-$. The carbon substrate can be an alkyne, cyclopropane, phenol, aniline, or β-diketone. Whether the haloperoxidase generates free hypohalous acid which then reacts relatively unspecifically with a broad range of nucleophilic acceptors or whether an enzyme-bound halogenating intermediate is involved is still open to debate. Active incorporation of halogen to form a halonium ion is followed by passive incorporation of a nucleophile which in dilute aqueous solution is usually OH$^-$ leading to the formation of the halohydrin (2).

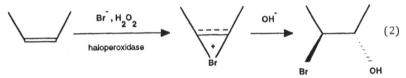

Obviously such a mechanism is not tenable if incorporation is directly into a one-carbon compound. The formation of a halomethane if haloperoxidase were to be responsible must necessarily involve a two- or more carbon intermediate which either spontaneously decomposes or is enzymically converted to halomethane. Such a mechanism has indeed been invoked to explain the algal production of polyhalogenated methanes and this biosynthetic route is discussed in Sec. 3.2. However there is no evidence that this pathway operates in the formation of monohalomethanes. Attention in this area has been directed to identifying novel enzyme systems catalyzing the direct methylation of halide ion by methyl donors such as SAM and other methylsulfonium compounds or N^5-methyl FH$_4$ and methylcobalamin, mechanisms reviewed in Sec. 3.1.

3.1. Direct Methylation of Halide Ion

Investigations by White [46] using cultures of *P. pomaceus* grown in the presence of L-[methyl-^{2}H$_3$]methionine demonstrated that the amino acid could act as a source of the methyl group in CH$_3$Cl, all of the deuterium atoms of the methyl group of the labeled compound being incorporated as a unit into the halomethane. When the culture medium was supplemented with DL-$[3,3$-$^2H_2]$serine, only two deuterium atoms were incorporated into the halomethane formed. Again this is consistent with biosynthesis of CH_3Cl via methionine as the β carbon of serine is utilized in the synthesis of N^5-methyl FH_4 which is responsible for methylation of homocysteine to methionine (Fig. 3).

Harper and Hamilton [47] compared the incorporation of deuterium from labeled D- and L-methionine into CH_3Cl at various stages during growth of $P. pomaceus$ (see Sec. 4.1). Very high incorporation (~80%) from the L compound was recorded in the early stages of growth but, as incubation proceeded, labeling fell probably due to dilution by unlabeled compound formed from endogenous L-methionine. Incorporation of label from the D compound was also noted, initially at a lower level (~20%) than with the L isomer but increasing to over 50% later in growth. This unusual incorporation pattern was explained in terms of the induction of either a racemase or an uptake system for D-methionine. Harper and Hamilton [47] also observed that the presence of L-methionine in the culture medium of the fungus did not stimulate CH_3Cl production, suggesting that biosynthesis of the compound was not restricted by the endogenous concentration of the amino acid. In contrast D-methionine gave significant increases in CH_3Cl release as did the racemic compound. Whether this promotion of CH_3Cl production indicates that the D isomer can act per se as a precursor of CH_3Cl or is affecting metabolism so as to enhance the availability of L-methionine for CH_3Cl biosynthesis is not clear.

On the basis of his results White [46] concluded that the actual methyl donor was most probably the general biological methylating agent S-adenosylmethionine (SAM), but he did not exclude the possibility that SAM could transfer a methyl group to FH_4 with the formation of N^5-methyl FH_4 which then methylated Cl^- with the formation of CH_3Cl. In support of this latter hypothesis he cited an apparent stimulation of CH_3Cl biosynthesis in fungal cultures by folic acid. However, the concentration of the compound used (0.23 mM) was far in excess of normal physiological requirements. Harper and Hamilton [47]

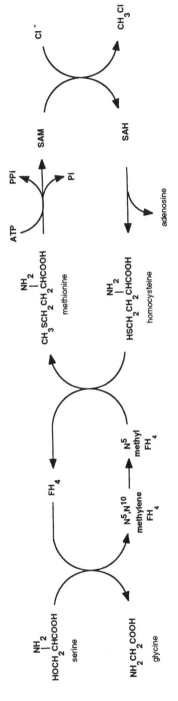

FIG. 3. Possible route for formation of CH₃Cl in *P. pomaceus.*

using a more accurate assay system were unable to demonstrate any
stimulation by folic acid at concentrations ranging from 0.023 to
230 μM. There is therefore little evidence that N^5-methyl FH_4 is
directly involved in methylation of Cl^-.

After a detailed in vitro study of the nucleophilic methylation
of iodine by methylcobalamin, Wood et al. [48] proposed that in the
marine environment vitamin B_{12} could be implicated in CH_3I formation.
However, neither White [46] nor Harper and Hamilton [47] could show
any effect of vitamin B_{12} or cobalt on CH_3Cl release by cultures of
P. pomaceus rendering it unlikely that methylcobalamin participates
in fungal CH_3Cl biosynthesis. White [19] also suggested that dimethylsulfonium compounds such as dimethylpropiothetin in marine algae may
react with I^- to form CH_3I (3). Many algae and marine phytoplankton
contain dimethylsulfonium compounds at levels of up to 3% [49] and I^-
is concentrated by numerous algae [50]. Although this reaction has
been observed in vitro (see Sec. 1), no firm evidence has been
adduced for its occurrence in vivo.

$$\begin{array}{c} CH_3 \\ \diagdown \overset{+}{} \\ SCH_2CH_2COOH + I^- \longrightarrow CH_3I + CH_3S(CH_2)_2COOH \quad (3) \\ \diagup \\ CH_3 \end{array}$$

Dimethylpropiothetin

Recently, Wuosma and Hager [11] in a preliminary report claimed
enzymatic synthesis of halomethanes from SAM by crude cell-free systems
from several organisms. Rates of CH_3Cl production of 25, 670, and 3
fmol/min/mg, respectively, were measured when cell-free extracts of
the fungus P. pomaceus, the alga E. muricata, and the succulent plant
M. crystallinum were incubated in phosphate buffer pH 6.8 containing
250 mM KCl and 0.5 mM SAM. For reasons explained in Sec. 2.3 there
must be some doubt regarding the soundness of the methodology employed
in this investigation and the conclusions of the authors must therefore
be treated with caution. Nevertheless these workers were apparently
successful in purifying a SAM/halide ion methyltransferase 800-fold
from crude extracts of E. muricata. The molecular weight of the

enzyme determined by gel filtration was 20,000-25,000. The enzyme
displayed maximal activity at pH 7.5 and could utilize Cl^-, Br^-, and
I^- as substrates though I^- was by far the preferred acceptor, the
rate of reaction with I^- exceeding the rate with Br^- by over two
orders of magnitude. The K_m values of the enzyme with Cl^-, Br^-, and
SAM were 5 mM, 40 mM, and 16 µM, respectively. In view of the preliminary nature of this report and the apparent deficiencies in
methodology, it is not possible at present to assess the significance of these findings.

3.2. The Haloperoxidase Route

Bonnemaisonia hamifera is a strongly smelling tropical red alga which
produces a variety of halogenated ketones including di-, tri-, and
tetrabromo-2-heptanone [51]. Theiler et al. [25] demonstrated that a
bromoperoxidase isolated from the species was capable of incorporating
bromine into a number of organic compounds, β-keto acids being particularly good acceptors. When a partially purified preparation of
bromoperoxidase was incubated at pH 5.8 with H_2O_2, Br^-, and 3-
oxooctanoic acid, 1-bromo-, 1,1-dibromo-, and 1,1,1-tribromoheptanone
were identified as products. However, when the reaction was conducted
at pH 7.3, CH_2Br_2, $CHBr_3$, and 1-bromopentane were formed in significant quantities. This observation was rationalized in terms of multiple halogenation on the α-carbon atoms of the keto acid yielding the
brominated ketones listed above, which at pH 7.3 then underwent
hydrolysis either enzyme catalyzed or spontaneous to give the bromo-
alkanes (see Fig. 4). The authors suggested that bromo- and iodo-
methanes found in natural waters and in algae are products of the
enzyme-catalyzed bromination of ketones.

Subsequently Beissner et al. [26] reported that extracts of the
green marine alga *Penicillus capitatus* incorporated Br^- into 3-
oxooctanoic acid producing $CHBr_3$ and several brominated ketones
similar to those biosynthesized by *B. hamifera*. In contrast to the

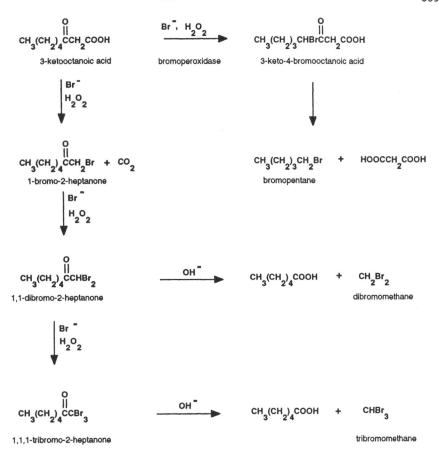

FIG. 4. Postulated metabolic pathway for formation of brominated alkanes by extracts of *B. hamifera*.

latter species there was no evidence of formation of CH_2Br_2 or 1-bromopentane. These workers concluded that in the formation of $CHBr_3$, decarboxylation does not occur until after 3-oxooctanoic acid has undergone mono- and dibromination, i.e., the precursor of 1,1-dibromo-2-heptanone in Fig. 4 is 2,2-dibromo-3-oxooctanoic acid rather than 1-bromo-2-heptanone.

Marine algae in general are known to be excellent sources of bromoperoxidases [52] whether they are of the typical hemoprotein type of *P. capitatus* [53] or contain vanadium at the active site as

with recently described enzymes from the brown algae A. *nodosum,*
L. *saccharina,* and F. *vesiculosus* [54,55]. In A. *nodosum* at least
two different vanadium bromoperoxidases are present, one of which is
located within the thallus especially around the conceptacles and
displays seasonal variations in activity [56,57]. The other enzyme
is situated on the thallus surface [58], which may explain the observations of Gschwend et al. [18] suggesting that biosynthesis of brominated compounds occurs on the surface of the algae. Indeed Wever
et al. [30] postulated that HOBr could be released by this peroxidase
directly into seawater where it reacts with dissolved organic matter
forming unstable brominated compounds which decompose by the haloform
reaction to give $CHBr_3$. It would therefore seem likely that the haloperoxidase pathway can readily account for the widespread occurrence
of polyhalogenated methane biosynthesis in the marine environment.
However, there is no indication that monohalomethanes can be formed
by this route.

4. METABOLIC ROLE OF CH_3Cl IN FUNGI

4.1. Precursor Studies

As indicated in Sec. 2.1.2 methyl esters are important natural products
in *Phellinus* species and their biosynthesis like that of halomethanes
is inhibited by thiocyanate. In the light of this observation and the
finding that ester biosynthesis does not occur in *Phellinus* species
lacking the ability to biosynthesize CH_3Cl, Harper et al. [38] proposed
that methylation of halide ion and aromatic acids was closely linked.
This metabolic relationship was further explored by measuring incorporation of C^2H_3 from L- and D-[methyl-$^2H_3$]methionine into both CH_3Cl and
methyl esters of benzoic, salicylic, and 2-furoic acids at various
stages of growth of P. *pomaceus*. The fungus was grown in sealed vials
on a cellulose-based medium containing labeled methionine and NaCl.
The isotopic compositions of methyl benzoate and CH_3Cl produced by the
fungus were determined by GC/MS from the point at which CH_3Cl was first

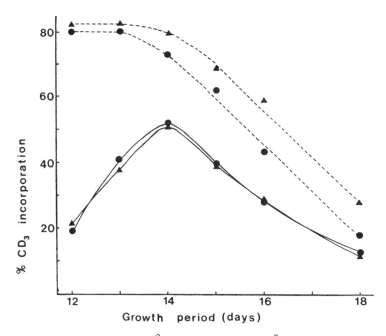

FIG. 5. Incorporation of C^2H_3 from L-[methyl-$^2H_3$]methionine (-----)
and D-[methyl-$^2H_3$]methionine (————) into CH_3Cl (●) and methyl benzoate
(▲) by $P. pomaceus$. (Reproduced by permission from Ref. 47.)

detectable in the headspace. The incorporation of label from L- and
D-methionine into CH_3Cl under these conditions was considered in Sec.
3.1. Intriguingly, when labeling of methyl benzoate was monitored in
these experiments a striking similarity emerged between the pattern
of incorporation into the ester and that into CH_3Cl (Fig. 5). Indeed
the levels of incorporation of C^2H_3 from labeled D-methionine into
CH_3Cl and methyl benzoate did not differ significantly at any stage
during the experiment. Harper and Hamilton [47] offered two possible
explanations for this remarkable correspondence:

1. The same enzyme system requiring an identical methyl donor
   was being utilized for methylation of both Cl^- and benzoate.
2. Rapid transmethylation occurred between CH_3Cl and benzoic
   acid or between methyl benzoate and Cl^- [reactions 1 and 2
   in Eq. (4)].

$$CH_3Cl \; + \; \underset{\text{COOH}}{\bigcirc} \; \underset{2}{\overset{1}{\rightleftharpoons}} \; \underset{\text{COOCH}_3}{\bigcirc} \; + \; HCl \qquad (4)$$

An activated derivative of benzoic acid such as benzoyl-CoA rather than the acid per se might conceivably be the substrate for the enzyme in this reaction.

The pattern of incorporation of label from L-[methyl-$^2H_3$]methionine into methyl-2-furoate was practically identical to that found with methyl benzoate suggesting a similar origin, but the pattern observed with methyl salicylate was quite different. Initial labeling of the ester was low (~40%) but later rose to a maximum (~60%) in midincubation before falling again. This divergent behavior was interpreted as signifying that methyl salicylate is almost certainly not biosynthesized by direct methylation of the corresponding acid. Instead the response is consistent with formation of this ester by o-hydroxylation of methyl benzoate, a pathway which would lead to the observed lag in labeling.

4.2. Methylation of Carboxylic Acids

In order to distinguish between the various possibilities outlined in 1 and 2 above, Harper et al. [59] incubated washed mycelia of P. pomaceus with either [C$^2H_3$]-methyl benzoate and NaCl or C$^2H_3$Cl and benzoic acid. No incorporation of C$^2H_3$ from the labeled ester into CH$_3$Cl was found but high levels of C$^2H_3$ incorporation from C$^2H_3$Cl into methyl benzoate were recorded demonstrating that the transmethylation 2 was occurring in the direction of reaction 1 in Eq. (4). transmethylation system exhibited broad substrate specificity methylating not only benzoic and 2-furoic acids but also a wide range of substituted aromatic acids. Some aliphatic acids also acted as good substrates; thus butyric acid was methylated at a rate threefold greater than benzoic acid under the assay conditions given in Table 5.

TABLE 5

Utilization of C^2H_3Cl for Methylation of Carboxylic Acids
by Washed Mycelia of *P. pomaceus*

Substrate acid	Methyl ester formed (nmol/g mycelium/hr $\pm$ SD)	% Ester labeled with $C^2H_3 \pm$ SD
Benzoic	82 $\pm$ 20	50 $\pm$ 8
o-Toluic	30 $\pm$ 2	83 $\pm$ 1
m-Toluic	15 $\pm$ 3	42 $\pm$ 2
o-Chlorobenzoic	22 $\pm$ 12	58 $\pm$ 9
m-Chlorobenzoic	13 $\pm$ 7	38 $\pm$ 5
p-Chlorobenzoic	33 $\pm$ 7	62 $\pm$ 5
n-Butyric	244 $\pm$ 48	61 $\pm$ 4
n-Hexanoic	57 $\pm$ 4	74 $\pm$ 5
n-Heptanoic	30 $\pm$ 8	81 $\pm$ 2
2-Furoic	94 $\pm$ 20	68 $\pm$ 9

Note on assay conditions: 1.23 mM C^2H_3Cl, 2 mM carboxylic acid in
20 mM citrate buffer pH 4.0 for 12 hr at 25°C.
Source: Data taken with permission from Ref. 59.

This observation allowed these workers to develop a sensitive and
accurate assay for the activity of the system based on headspace
analysis of the methyl butyrate formed when butyric acid was substrate. Significantly, salicylic acid was not a substrate, a finding
in accord with the conclusion of Harper and Hamilton [47] that methyl
salicylate was of secondary origin probably formed by *o*-hydroxylation
of methyl benzoate.

Surprisingly there was considerable variation in the level of
incorporation of C^2H_3 into esters of different aromatic acids. One
possible explanation of these differences proffered by Harper et al.
[59] was the inhibition or stimulation of endogenous CH_3Cl biosynthesis by some substrate acids and later study of the effect of substrate
concentration on CH_3Cl release by the mycelia corroborated this interpretation (see Sec. 4.2.1).

Incorporation of label from C^2H_3Cl into ester by washed mycelia was found by McNally et al. [60] to be very rapid. Thus 17% incorporation of C^2H_3 into methyl butyrate was recorded after 5 min incubation with butyric acid and a plateau value of 40% was achieved within 1 hr. The speed with which label was incorporated and the relatively short period before attainment of maximum labeling of ester was considered to render it unlikely that CH_3Cl was converted to methanol or other diffusible intermediate prior to use as a methyl donor. The absence of an initial hydrolytic dehalogenation was confirmed by the failure of unlabeled methanol to affect either the rate of biosynthesis or the labeling of methyl butyrate formed. Butyryl-CoA was not a substrate suggesting that the intermediate formation of CoA derivatives is also improbable.

In addition to CH_3Cl both CH_3Br and CH_3I acted as methyl donors though neither was as readily utilized as CH_3Cl. Under the conditions given in Table 5 with butyric acid as substrate, rates of methylation with CH_3Br and CH_3I relative to that with CH_3Cl were 77% and 38%, respectively.

4.2.1. Effect of Substrate Concentration

McNally et al. [60] investigated the effect of benzoic acid concentration on the labeling and rate of synthesis of methyl benzoate in the presence and absence of C^2H_3Cl (Fig. 6). A small but significant increase in the rate of ester biosynthesis was obtained in the presence of exogenous C^2H_3Cl suggesting some limitation of methylation by the endogenous C^2H_3Cl biosynthesis. The optimum concentration of benzoic acid was 0.5 mM and the rate of methylation fell rapidly at higher concentrations, indicating powerful inhibition of the methylating system probably of an allosteric nature. Incorporation of C^2H_3 into methyl benzoate remained relatively constant at approximately 30% up to 1 mM benzoic acid but above this concentration it rose sharply reaching 60% at 1.5 mM. McNally et al. suggested that this increase was due to inhibition of CH_3Cl biosynthesis at concentrations above 1 mM, the active site of the methylating enzyme thereby

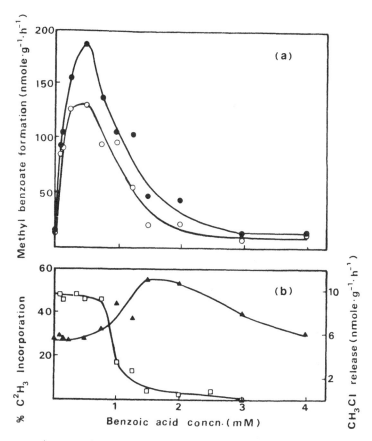

FIG. 6. Effect of benzoic acid concentration on (a) production of methyl benzoate in the presence and absence of C^2H_3Cl (b) incorporation of C^2H_3 label into methyl benzoate and release of CH_3Cl by washed mycelia of *P. pomaceus*. Methyl benzoate production in absence of exogenous C^2H_3Cl, o; methyl benzoate production with 1.23 mM C^2H_3Cl, •; percentage of methyl benzoate labeled with C^2H_3, ▲; CH_3Cl released by mycelia in absence of exogenous C^2H_3Cl, □. (Reproduced by permission from Ref. 60.)

becoming more accessible to exogenous C^2H_3Cl. To test this hypothesis the rate of release of CH_3Cl from mycelium incubated in the absence of CH_3Cl at various benzoic acid concentrations was measured (Fig. 6b). A very marked decline in gaseous CH_3Cl release was evident between 1 and 1.5 mM exactly coincident with the concentration range over which

incorporation increased, almost certainly betokening a real reduction in the overall rate of biosynthesis sufficient to result in the
greater C^2H_3 incorporation observed. Although these workers were
unable to offer a comprehensive explanation for this subtle and complex regulation of both methylation and CH_3Cl biosynthesis, they tentatively suggested that the inhibition of methylation at higher concentrations of benzoic acid might represent a means of diverting
benzoic acid into alternative metabolic pathways at certain stages
of growth.

In studies of the effect of exogenous C^2H_3Cl concentrations on
the system it was shown that whereas methyl benzoate biosynthesis was
stimulated by increase in C^2H_3Cl concentration up to 10 μM, methyl
butyrate biosynthesis was not affected by the concentration of exogenous C^2H_3Cl. Nevertheless in experiments with saturating concentrations of C^2H_3Cl, a linear logarithmic relationship ($p < 0.001$) was
found between percent C^2H_3-incorporation and exogenous C^2H_3Cl concentration with both benzoic and butyric acid as substrate (5).

$$\log (\% \ C^2H_3\text{-incorp.}) = \log 0.56 + x \log (C^2H_3Cl \text{ conc.}) \qquad (5)$$

where $x = 0.47$ for benzoic acid and 0.58 for butyric acid.

McNally et al. [60] concluded that the nature of the relationship, which is identical to the Freundlich adsorption isotherm,
favored location of the CH_3Cl-biosynthesizing system and the CH_3Cl-
utilizing system on either side of, and probably physically associated with, a membrane within the cell through which CH_3Cl diffuses.

4.2.2. *Changes in Methylating Activity During Growth*

Harper et al. [59] monitored methyl benzoate concentrations during
growth of *P. pomaceus* cultures and demonstrated that, in contrast to
CH_3Cl release, initial methyl benzoate biosynthesis was growth-
related. Initial accumulation of methyl benzoate was paralleled by
an increase in the activity of the CH_3Cl-utilizing methylating system
in the mycelium. A temporary fall in ester concentration in the later
stages of growth was attributed by the authors to diversion of the

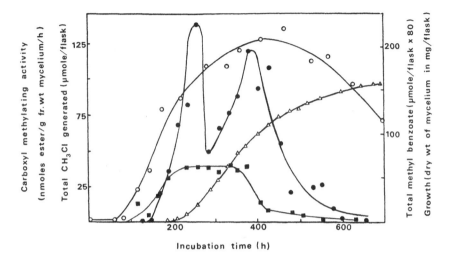

FIG. 7. Growth of *P. pomaceus* in relation to methyl benzoate accumulation, CH_3Cl release, and benzoic acid methylating activity in the
mycelium: o, growth; △, total CH_3Cl generated; ■, benzoic acid methylating activity; ●, total methyl benzoate. (Reproduced with modification by permission from Ref. 59.)

compound to the biosynthesis of other secondary metabolites (Fig. 7).
Changes in the incorporation of exogenous C^2H_3Cl into ester by mycelia
harvested at various stages of growth were also measured in a separate
experiment (Fig. 8). C^2H_3 incorporation remained relatively low (~30%)
during early exponential growth suggesting that endogenous biosynthesis
and utilization of CH_3Cl by the fungus were closely coupled during this
period, an interpretation supported by the fact that no CH_3Cl release
occurred at this stage of growth. As growth proceeded incorporation
of C^2H_3 into ester increased rising to approximately 75% as maximum
growth was achieved, the methylating system apparently becoming more
accessible to exogenous C^2H_3Cl. Actual CH_3Cl release by the fungus
began as incorporation started to rise and attained a maximum rate
as incorporation reached a peak value.

Harper et al. speculated that the initial tight channeling of
CH_3Cl utilization could either take the form of direct transfer of
CH_3Cl from the biosynthesizing enzyme to the methylating system or

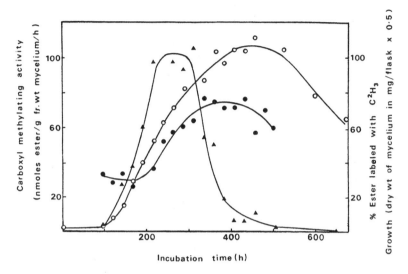

FIG. 8. Growth of *P. pomaceus* in relation to carboxyl methylating activity and % C^{2}H$_3$ incorporation into ester from exogenous C^{2}H$_3$Cl: ○, growth; ▲, carboxyl methylating activity with butyric acid as substrate; ●, C^{2}H$_3$ incorporation into ester. (Reproduced by permission from Ref. 59.)

alternatively might involve confinement of endogenously synthesized CH$_3$Cl to an unmixed layer not readily accessible to exogenous C^{2}H$_3$Cl. Both mechanisms presuppose the existence of a multienzyme complex associated with a membrane as postulated by McNally and Harper [60]. It was further postulated that later in growth, as a consequence of the dissociation of the putative multienzyme complex, CH$_3$Cl biosynthesis becomes progressively uncoupled from CH$_3$Cl utilization leading to the breakdown in channeling. This would result not only in CH$_3$Cl release but in greater accessibility of the methylating system to exogenous C^{2}H$_3$Cl. Thus although CH$_3$Cl ostensibly appears a typical secondary metabolite, its release in the idiophase would seem merely to reflect the breakdown of the strict coordination of biosynthesis and utilization which exists in the primary metabolic phase.

4.3. Methylation of Phenols

A CH_3Cl-utilizing system capable of methylating a variety of substituted phenols and thiophenol was identified in *P. pomaceus*
(Table 6) [59]. It was clearly biochemically distinct from the
carboxylic acid methylating system as maximum activity was attained
in the idiophase. Although the natural substrate of the system has
yet to be identified, the latter finding implies that the enzyme is
involved in secondary metabolism. Levels of C^2H_3 incorporation from
exogenous C^2H_3Cl into anisole were generally lower than those into
methyl ester, suggesting an even tighter channeling of CH_3Cl during
phenol methylation than during carboxylic acid methylation. Nevertheless as with carboxylic acid methylation, incorporation of C^2H_3
into methylated product increased during growth suggesting some
uncoupling of CH_3Cl biosynthesis from CH_3Cl utilization. Detailed
study of the kinetics of the system in washed mycelia by McNally

TABLE 6

Utilization of C^2H_3Cl for Methylation of Phenols
by Washed Mycelia of *P. pomaceus*

Substrate phenol or thiophenol	Anisole or thioanisole formed (nmol/g mycelia/hr) ± SD	% Product labeled with C^2H_3 ± SD
Phenol	9 ± 1	30 ± 3
Thiophenol	37 ± 15	70 ± 4
2-Chlorophenol	16 ± 2	31 ± 6
4-Chlorophenol	3 ± 0.7	25 ± 6
2,6-Dichlorophenol	7 ± 2	35 ± 8
2-Iodophenol	1 ± 0.3	16 ± 5
Methyl 3-hydroxybenzoate	16 ± 6	38 ± 9

Note on assay conditions: 1.23 mM C^2H_3Cl, 2 mM phenol in 20 mM citrate
buffer pH 4.0 for 12 hr at 25°C.
Source: Data taken with permission from Ref. 59.

and Harper [61] demonstrated that phenol does not exercise the tight
regulation of the activity of the system and the rate of CH_3Cl biosynthesis that is shown by benzoic acid in carboxylic acid methylation. However, the effect of exogenous C^2H_3Cl concentrations on C^2H_3
incorporation was broadly similar leading these workers to conclude
that, as with the carboxylic acid methylating system, the CH_3Cl-
biosynthesizing and CH_3Cl-utilizing elements are associated in a
membrane-bound multienzyme complex.

4.4. Veratryl Alcohol Biosynthesis

Harper et al [59] noted that the non-Hymenochaetaceous fungus *Fomitopsis pinicola* could utilize exogenous C^2H_3Cl in methylation of
benzoic acid despite the fact that at no stage of growth did the
species release CH_3Cl. These workers reasoned that the presence of
a CH_3Cl-utilizing methylating system in such a species raised the
possibility that CH_3Cl may be a metabolic intermediate in fungi and
even higher plants. The compound might have escaped detection hitherto
because (1) it is a chemically unreactive gas at normal temperatures
which is readily overlooked by normal analytical techniques and (2) it
is generated and utilized in tightly channeled systems in which
release of free CH_3Cl is seldom or never observed. Metabolism of
CH_3Cl in the Hymenochaetaceae may be unusual only in that uncoupling
of CH_3Cl biosynthesis from utilization occurs in the final stages of
growth leading to release of the compound as a natural product.

Harper and coworkers [62] therefore examined the possibility
that CH_3Cl might have a role in the biosynthesis of the key secondary
metabolite veratryl alcohol (3,4-dimethoxybenzyl alcohol) by the non-
Hymenochaetaceous species *Phanerochaete chrysosporium*, a white-rot
fungus frequently employed as a model organism in studies of lignin
degradation. Veratryl alcohol induces elements of the lignolytic
system and not only acts as a substrate for lignin peroxidase but
prevents its inactivation by hydrogen peroxide. It also may function
as a one-electron redox mediator in lignin depolymerization. Shimada

TABLE 7

Incorporation of C^2H_3 into Veratryl Alcohol After Addition
of C^2H_3Cl or L-[methyl-$^2H_3$]methionine to
72-hr Cultures of $P.$ $chrysosporium$

Time after addition of precursor (hr)	% C^2H_3 substitution in 3 and 4 positions of veratryl alcohol					
	C^2H_3Cl (0.77 mM)			L-[Methyl-$^2H_3$]methionine (1 mM)		
	$DiCH_3$	$CH_3C^2H_3$	DiC^2H_3	$DiCH_3$	$CH_3C^2H_3$	DiC^2H_3
0	100	0	0	100	0	0
2	67	16	17	97	1	1
4	47	20	33	96	2	2
8	26	20	54	93	3	4
12	19	20	61	89	3	7

$Source$: Data taken with permission from Ref. 62.

et al. [63] showed that the 3- and 4-o-methyl groups of this important
metabolite are derived from methionine.

Labeling of veratryl alcohol isolated from $P.$ $chrysosporium$
cultures in which the growth media had been supplemented with L-
[methyl-$^2H_3$]methionine was compared with that from cultures supplemented with C^2H_3Cl [62]. The results indicated quite clearly that
C^2H_3Cl was as effective a precursor as L-methionine for 3- and 4-o-
methylation. Indeed in some circumstances CH_3Cl was much more readily
utilized than the amino acid. Thus when C^2H_3Cl was added in mid-growth
cycle to 72-hr cultures actively synthesizing veratryl alcohol, C^2H_3
incorporation was very rapid with 81% of veratryl alcohol labeled
after 12 hr. By contrast incorporation of C^2H_3 from labeled methionine
was comparatively slow, attaining only 10% after a similar period of
time (Table 7). Incorporation of C^2H_3 from C^2H_3Cl into veratryl
alcohol was also noted in two other white-rot fungi widely used in
studies of lignin degradation, $Phlebia$ $radiata$ and $Coriolus$ $versicolor$.

Despite this unambiguous evidence for CH_3Cl as a metabolic intermediate the release of CH_3Cl as a natural product could not be detected at any stage of growth in any of the species, a finding which inevitably leads to the conclusion that biosynthesis of CH_3Cl in these species, as with *F. pinicola*, is closely coupled to its utilization by a high-affinity methylation system throughout the growth cycle. Thus the compound is at all times enzyme-bound and never achieves the status of a freely diffusible intermediate. Recently it was reported [64] that supplementation of the culture medium of *P. chrysosporium* with 0.5 mM CH_3Cl advances initiation of veratryl alcohol biosynthesis by up to 36 hr and also causes the earlier appearance of lignin peroxidase activity. The rate of methylation of veratryl alcohol precursors such as caffeic and ferulic acids may therefore be restricted by the metabolic availability of CH_3Cl in the early stages of growth so limiting veratryl alcohol biosynthesis.

Participation of CH_3Cl in the biosynthesis of a compound of such central importance in lignin degradation affords strong support for the proposal by Harper et al. [59] that CH_3Cl may have a widespread if previously unsuspected role in methylation processes in many fungi and possibly higher plants. The nature of the relationship between CH_3Cl and SAM, the activated metabolic intermediate hitherto believed to be responsible for almost all biological methylation, is not yet clear. Coulter and Harper [65] recently reported preliminary findings of an investigation to determine whether CH_3Cl acts as a methyl donor per se in *P. chrysosporium* or is converted to SAM before utilization in methylation processes. The 4-*O*-methylation of the synthetic substrate acetovanillone (4-hydroxy-3-methoxyacetophenone) by washed mycelia harvested at various stages of growth was monitored using as methyl donor either C^2H_3Cl or C^2H_3-labeled SAM in the presence and absence of *S*-adenosylhomocysteine (SAH) which acts as a competitive inhibitor of most methylations involving SAM. A parallel experiment was performed using cell-free extracts prepared from these mycelia. The results of this study suggest that two independent methylating systems exist in *P. chrysosporium*: (1) an enzyme utilizing SAM as methyl donor which is present throughout the fungal

growth cycle and can be inhibited by SAH, and (2) a CH_3Cl-utilizing
methylation system of broader substrate specificity not susceptible
to inhibition by SAH and absent in early growth but attaining peak
activity in mid-growth phase about 24 hr later than the SAM-utilizing
enzyme.

Cell-free extracts could only use SAM as methyl donor implying
that the CH_3Cl-utilizing system is highly labile or requires an as
yet unidentified cofactor. On the basis of these findings Coulter
and Harper [65] consider it unlikely that CH_3Cl is converted to
SAM prior to utilization in methylation and so it would appear that
separate methylating systems using CH_3Cl and SAM as methyl donor can
coexist in lignin-degrading fungi. The reason for the existence of
a separate CH_3Cl-requiring system in such species is not immediately
obvious but undoubtedly it has a wider significance in lignin degradation than the mere methylation of veratryl alcohol precursors.

5. OTHER POSSIBLE METABOLIC ROLES FOR CH_3Cl

The production of certain halomethanes by algae may prevent microbial
attack or provide protection against herbivore feeding. Thus CH_2I_2
found in some algal species acts as a feeding deterrent for the herb-
ivorous snail *Littorina littorea* (periwinkle) [18]. The antibacterial
activity of extracts of the red algae *A. taxiformis* and *A. armata* was
attributed *inter alia* to halomethanes such as $CHBr_3$ produced by the
species [24]. The abundance of *Asparagopsis* species in areas of
extremely high algal predation was believed by McConnell and Fenical
[24] to be not unrelated to the general biocidal nature of these compounds. In contrast Wever et al. [30] argued that halomethanes are
generated merely as a byproduct of the action of HOBr formed by algal
peroxidases on dissolved organic matter in seawater, the strongly
biocidal HOBr being released as part of a host defense system by which
algae prevent bacterial and fungal infection or herbivore feeding.
The production of CH_3Cl by *M. crystallinum*, a succulent growing in
saline rich environments, could represent part of a halotolerance

mechanism for the species [11], though the elimination or segrega-
tion of cations would also have to be accomplished by the plant.

6. CONCLUDING REMARKS

Atmospheric concentrations of CH_3Cl have latterly attained particular
significance because of the importance attached to chlorine derived
from halocarbons in hastening the rate of destruction of ozone in the
stratosphere. Consequent on the relatively short residence time of
CH_3Cl in the lower atmosphere only a small proportion ($\sim$6%) of natural
CH_3Cl reaches the stratosphere [3]. Nevertheless because of the large
quantities of CH_3Cl produced in nature a substantial proportion of
stratospheric chlorine is derived from this source. Currently CH_3Cl
accounts for 20% of the chlorine in the stratosphere, the remainder
being provided mainly by chlorofluorocarbons $11(CFCl_3)$ and $12(CF_2Cl_2)$
which have much lower global fluxes (360 and 450 thousand tonnes/year,
respectively), but much longer residence times (60 and 120 years)
[33]. A predominantly biological origin for natural CH_3Cl has substantial implications in this area. Computer models of the upper
atmosphere assume that natural CH_3Cl production has remained constant
at current levels throughout previous millenia. It is possible that
biological production from both oceanic and terrestial sources has
varied considerably over geological time through the effects of cli-
matic change on ocean temperature and forest cover. In the more
recent past the activities of humans have obviously also had a major
impact on the latter. If conditions were more favorable in the past
to biological production of CH_3Cl and the input of chlorine into the
stratosphere from this source much higher, present perceptions of the
threat posed by man-made halocarbons might be markedly altered.

    The biological production of CH_3I may be important in another
environmental context. Brinkman et al. [20] proposed that CH_3I generated in the oceans may have a major role in the solubilization and
transport of refractory minerals containing metals such as tin, lead,
and mercury. Thus SnS, the main tin mineral in anoxic sulfidic

sediment, can be readily oxidized and methylated by CH_3I with the formation of $(CH_3)_3SnI$ [66]. Aqueous CH_3I also readily solubilizes the common oxide of tin, cassiterite, SnO_2 (see Chap. 1).

CH_3I synthesized in the oceans also appears to play a key role in geochemical cycling of iodine. It has been estimated that a sea-to-air flux of iodine of about 0.5 million tonnes is required for geochemical balance [67]. The flux of CH_3I from the oceans is clearly of this order and so it appears that oceanic emissions of the compound are probably dominant in the geochemical cycling of iodine, an element of vital importance to higher terrestial life forms.

At the biochemical level the enzymes involved in utilizing CH_3Cl in methylation reactions in fungi may have biotechnological application. Broad-specificity enzyme systems capable of methylating phenols or carboxylic acids in aqueous solution at room temperature using gaseous CH_3Cl as methyl donor may have some commercial potential, particularly in the form of immobilized bacterial preparations containing the appropriate cloned fungal gene or genes.

ABBREVIATIONS

FH_4	tetrahydrofolic acid
GC	gas chromatography
MS	mass spectrometry
pptv	parts per 10^{12} by volume
PTFE	polytetrafluoroethylene
SAH	*S*-adenosylhomocysteine
SAM	*S*-adenosylmethionine
SD	standard deviation

REFERENCES

1. R. A. Rasmussen, L. E. Rasmussen, M. A. K. Khalil, and R. W. Dalluge, *J. Geophys. Res.*, *85*, 7350 (1980).

2. H. B. Singh, L. J. Salas, and R. E. Stiles, *J. Geophys. Res.*, *88*, 3684 (1983).

3. P. R. Edwards, I. Campbell, and G. S. Milne, *Chem. Ind. (London)*, *1982*, 619 (1982).

4. T. Y. Palmer, *Nature, 263*, 44 (1976).

5. P. J. Crutzen, L. E. Heidt, J. P. Krasnec, and W. H. Pollock, *Nature, 282*, 253 (1979).

6. R. B. Symonds, W. I. Rose, and M. H. Reed, *Nature, 334*, 415 (1988).

7. O. C. Zafiriou, *J. Mar. Res., 33*, 75 (1975).

8. R. G. Zika, L. T. Gidel, and D. D. Davis, *Geophys. Res. Lett., 11*, 353 (1984).

9. J. E. Lovelock, *Nature, 256*, 193 (1975).

10. S. L. Manley and M. N. Dastoor, *Limnol. Oceanogr., 32*, 709 (1987).

11. A. M. Wuosma and L. P. Hager, *Science, 249*, 160 (1990).

12. D. B. Harper, *Nature, 315*, 55 (1985).

13. J. L. Varns, *Am. Potato J., 59*, 593 (1982).

14. S. A. Penkett, B. M. R. Jones, M. J. Rycroft, and D. A. Simmons, *Nature, 318*, 550 (1985).

15. S. C. Wofsy, M. B. McElroy, and Y. L. Yung, *Geophys. Res, Lett., 2*, 215 (1975).

16. J. E. Lovelock, R. J. Maggs, and R. J. Wade, *Nature, 241*, 194 (1973).

17. R. A. Rasmussen, M. A. K. Khalil, R. Gunawardena, and S. D. Hoyt, *J. Geophys. Res., 87*, 3086 (1982).

18. P. M. Gschwend, J. K. Macfarlane, and K. A. Newman, *Science, 215*, 923 (1985).

19. R. H. White, *J. Mar. Res., 40*, 529 (1982).

20. F. E. Brinckman, G. J. Olson, and J. S. Thayer, in *Marine and Estuarine Geochemistry* (A. Sigleo and A. Hattori, eds.), Lewis, Chelsea, Michigan, 1985, p. 227.

21. D. Dyrssen and E. Fogelqvist, *Oceanol. Acta, 4*, 313 (1981).

22. E. Fogelqvist, *J. Geophys. Res., 90*, 9181 (1985).

23. P. S. Liss, in *The Role of Air-Sea Exchange in Geochemical Cycling* (P. Buat-Ménard, ed.), Reidel, Boston, 1986, p. 283.

24. O. McConnell and W. Fenical, *Phytochemistry, 16*, 367 (1977).

25. R. Theiler, J. C. Cook, L. P. Hager, and J. F. Siuda, *Science, 202*, 1094 (1978).

26. R. S. Beissner, W. J. Guilford, R. M. Coates, and L. P. Hager, *Biochemistry, 20,* 3724 (1981).

27. W. W. Berg, L. E. Heidt, W. Pollock, P. Sperry, and R. J. Cicerone, *Geophys. Res. Lett., 11,* 429 (1984).

28. R. J. Cicerone, L. E. Heidt, and W. H. Pollock, *J. Geophys. Res., 93,* 3745 (1988).

29. W. T. Sturges and L. A. Barrie, *Atmos. Environ., 22,* 1179 (1988).

30. R. Wever, M. G. M. Tromp, B. E. Krenn, A. Marjani, and M. Van Tol, *Environ. Sci. Technol., 25,* 446 (1991).

31. M. A. K. Khalil, R. A. Rasmussen, and S. D. Hoyt, *Tellus, 35B,* 266 (1983).

32. World Meterological Organization, *Atmospheric Ozone 1985, Global Ozone Research and Monitoring Report No. 16,* WMO, Geneva, 1986, p. 70.

33. M. J. Prather and R. T. Watson, *Nature, 344,* 729 (1990).

34. S. A. Hutchinson, *Trans. Br. Mycol. Soc., 57,* 185 (1971).

35. M. I. Cowan, T. A. Glen, S. A. Hutchinson, M. E. MacCartney, J. M. Mackintosh, and A. M. Moss, *Trans. Br. Mycol. Soc., 60,* 347 (1973).

36. E. M. Turner, M. Wright, T. Ward, D. J. Osborne, and R. J. Self, *J. Gen. Microbiol., 91,* 167 (1975).

37. D. B. Harper and J. T. Kennedy, *J. Gen. Microbiol., 132,* 1231 (1986).

38. D. B. Harper, J. T. Kennedy, and J. T. G. Hamilton, *Phytochem., 27,* 3147 (1988).

39. R. P. Collins and A. F. Halim, *Can. J. Microbiol., 18,* 65 (1972).

40. P. D. Nightingale, Low molecular weight halocarbons in seawater, Ph.D. thesis, University of East Anglia, 1991.

41. B. J. Burreson, R. E. Moore, and P. P. Roller, *J. Agric. Food Chem., 24,* 856 (1976).

42. K. A. Newman and P. M. Gschwend, *Limnol. Oceanogr., 32,* 702 (1987).

43. P. M. Gschwend and J. K. Macfarlane, in *Organic Marine Geochemistry,* Am. Chem. Soc. Symp. Ser., Washington, 1986, p. 314.

44. C. J. M. Stirling, in *Organic Chemistry of Sulfur* (S. Oae, ed.), Plenum Press, New York, 1977, p. 492.

45. S. L. Neidleman and J. Geigert, *Biohalogenation: Principles, Basic Roles and Applications,* Ellis Horwood Ltd., Chichester, 1986, p. 46.

46. R. H. White, *Arch. Microbiol., 132,* 100 (1982).

47. D. B. Harper and J. T. G. Hamilton, *J. Gen. Microbiol.*, *134*, 2831 (1988).

48. J. M. Wood, H. J. Segall, W. P. Ridley, A. Cheh, W. Chudyk, and J. S. Thayer, *Proc. Int. Conf. on Heavy Metals in the Environment* (T. C. Hutchinson, ed.), Plenum Press, New York, 1975, p. 49.

49. R. G. Ackman, C. S. Tocher, and J. McLachlan, *J. Fish. Res. Bd. Canada*, *23*, 357 (1966).

50. T. I. Shaw, in *Physiology and Biochemistry of Algae* (R. A. Lewin, ed.), Academic Press, New York, 1962, p. 247.

51. J. F. Siuda, G. R. Van Blamcom, P. D. Shaw, R. D. Johnson, R. H. White, L. P. Hager, and R. L. Rinehart, Jr., *J. Am. Chem. Soc.*, *97*, 937 (1975).

52. W. D. Hewson and L. P. Hager, *J. Phycol.*, *16*, 340 (1980).

53. J. A. Manthey and L. P. Hager, *J. Biol. Chem.*, *256*, 11232 (1981).

54. R. Wever, E. de Boer, H. Plat, and B. E. Krenn, *FEBS Lett.*, *216*, 1 (1987).

55. E. de Boer and R. Wever, *J. Biol. Chem.*, *263*, 12326 (1988).

56. H. Vilter, K.-W. Glombitza, and A. Grawe, *Bot. Mar.*, *26*, 331 (1983).

57. R. Wever, *Nature*, *335*, 501 (1988).

58. B. E. Krenn, M. G. M. Tromp, and R. Wever, *J. Biol. Chem.*, *264*, 19287 (1989).

59. D. B. Harper, J. T. G. Hamilton, J. T. Kennedy, and K. J. McNally, *Appl. Environ. Microbiol.*, *55*, 1981 (1989).

60. K. J. McNally, J. T. G. Hamilton, and D. B. Harper, *J. Gen. Microbiol.*, *136*, 1509 (1990).

61. K. J. McNally and D. B. Harper, *J. Gen. Microbiol.*, *137*, 1029 (1991).

62. D. B. Harper, J. A. Buswell, J. T. Kennedy, and J. T. G. Hamilton, *Appl. Environ. Microbiol.*, *56*, 3450 (1990).

63. M. Shimada, F. Nakatsubo, T. K. Kirk, and T. Higuchi, *Arch. Microbiol.*, *129*, 321 (1981).

64. D. B. Harper, J. A. Buswell, and J. T. Kennedy, *J. Gen. Microbiol.*, *137*, 2867 (1991).

65. C. Coulter and D. B. Harper, *Abst. 1st U.K. Biotechnology Congress*, Leeds, 1991, B1.

66. W. F. Manders, G. T. Olson, F. E. Brinckman, and T. M. Bellama, *J. Chem. Soc. Chem. Commun.*, *1984*, 538 (1984).

67. Y. Miyake and S. Tsunogai, *J. Geophys. Res.*, *68*, 3989 (1963).

Author Index

Numbers in parentheses are reference numbers and indicate that an author's work is referred to although his name may not be cited in the text. Underlined numbers give the page on which the complete reference is listed.

Woods, D. D., 262(90), 284
Woodward, H. E., 174(105), 183
Woolson, E. A., 163(7,14), 179
Wrench, J. J., 194(58), 221
Wright, M., 351(36), 387
Wu, L., 192(43), 221
Wu, R., 266(114,115), 267(114,
  115), 268(115), 270(115), 285
Wuertz, S., 43(15), 73
Wuhrmann, K., 288(2), 332
Wuosma, A. M., 348(11), 359(11),
  362(11), 367(11), 384(11), 386

                    X

Xavier, A. V., 234(12), 263(95),
  280, 284
Xun, L., 340(15), 342

                    Y

Yakoyama, T., 45(34), 73
Yamada, H., 206(114), 224; 235
(16), 280
Yamaga, S., 149(53), 150(53),
  159
Yamamoto, H., 45(34), 73; 265
  (107), 266(107), 284
Yamamura, Y., 6(10), 9(25), 30,
  31; 140(11), 141(11,25), 142
  (11,28,34), 144(28,36,41), 145
  (43), 146(11,28,34), 148(28,
  48), 149(56), 152(56), 153(41),
  155(56), 157, 158, 159; 162(3),
  165(3), 167(3), 169(87,92), 170
  (3,92), 171(87), 173(3,92),
  178, 182, 183
Yamanaka, H., 164(31), 180
Yamanaka, K., 167(63-66), 170
  (64), 181
Yamato, N., 169(87), 171(87,99),
  182, 183
Yamauchi, H., 9(25), 31; 140(11),
  141(11,25), 142(11), 144(41),
  145(43), 146(11), 149(56), 152
  (56), 153(41), 155(56), 157,
  158, 159; 162(3), 165(3), 167
  (3,57), 169(87,92), 170(3,92),
  171(87), 173(3,92), 178, 181,
  182, 183

Yamazaki, N., 103(5-8), 106(5-8),
  107(8), 122(80), 124(80), 133,
  136
Yancey, P. H., 163(21), 179
Yang, G., 199(95), 223
Yannai, S., 15(66), 33
Yaromich, J., 48(51), 55(75), 56
  (75), 70(51), 74, 75; 140(20),
  158
Ylaranta, T., 202(105), 205(105),
  224
Yokota, T., 330(95), 337
Yokoyama, T., 265(107), 266(107),
  284
Yonemoto, J., 13(48), 32; 195
  (74,75), 222
Yong, C., 14(59), 32
Yoshida, M., 141(25), 142(28),
  144(28,36), 145(43), 146(28),
  147(28), 149(56), 152(56), 155
  (56), 158, 159
Young, R., 319(84), 336
Young, T. W. K., 191(35), 196
  (35), 220
Young, V. R., 195(80), 223
Yu, T. H., 103(1,5-8), 106(5-8),
  107(8,40), 110(40,45,46), 114
  (59), 122(80,83), 124(80,83),
  125(83), 128(83), 133, 135, 136
Yung, Y. L., 347(15), 386

                    Z

Zafiriou, O. C., 347(7), 358(7),
  386
Zangrando, E., 249(151), 286
Zapf, P., 16(69), 33
Zappia, V., 8(24), 9(24), 20(24),
  31
Zehnder, A. J. B., 288(1), 332
Zehr, J. P., 212(128), 225
Zerner, B., 257(83), 283
Zhang, L., 206(113), 207(113),
  224
Zhang, X., 199(95), 223
Zhao, L., 13(46), 32; 206(113),
  207(113), 213(142), 224, 226
Zickler, J., 241(54), 282
Zieve, R., 191(33,35), 192(33,
  48), 193(48), 196(35), 198(48),
  199(33), 201(33), 204(33), 210

Printed and bound by CPI Group (UK) Ltd, Croydon, CR0 4YY

17/10/2024

01775700-0017